The Song of the Cell

THE SONG OF THE CELL

AN EXPLORATION OF MEDICINE AND THE NEW HUMAN

SIDDHARTHA MUKHERJEE

THORNDIKE PRESS
A part of Gale, a Cengage Company

GALE
A Cengage Company

**LIBRARY OF CONGRESS CIP DATA ON FILE.
CATALOGUING IN PUBLICATION FOR THIS BOOK
IS AVAILABLE FROM THE LIBRARY OF CONGRESS.**

ISBN-13: 979-8-8857-8619-5 (hardcover alk. paper)

Published in 2023 by arrangement with Scribner, a Division of Simon & Schuster, Inc.

Printed in Mexico
Print Number: 3   Print Year: 2023

To W.K. and E.W. — among the
first to cross

In the sum of the parts, there are only the parts.
The world must be measured by eye.
— Wallace Stevens

[Life] is a continuing rhythmic movement,
of the pulse, of the gait, even of the cells.
— Friedrich Nietzsche

CONTENTS

9

PRELUDE
"THE ELEMENTARY PARTICLES OF ORGANISMS"

"Elementary," he said. "It is one of those instances where the reasoner can produce an effect which seems remarkable to his neighbor, because the latter has missed the one little point which is the basis of the deduction."
— Sherlock Holmes to Dr. Watson, in Sir Arthur Conan Doyle, "The Crooked Man"

The conversation took place over dinner in October 1837. Dusk had likely fallen, and the city's gas lamps had lit up the central streets of Berlin. Only scattered memories of the evening survive. No notes were taken, and no scientific correspondence ensued. What remains is the story of two friends — lab mates — discussing experiments over a casual meal, and the exchange of one crucial idea.* One

*Footnotes for the Prelude have been moved to the endnotes of the book.

13

of the two diners, Matthias Schleiden, was a botanist. He had a prominent, disfiguring scar across his forehead, the remnant blemish of a prior suicide attempt. The other, Theodor Schwann, a zoologist, had sideburns that descended to his jowls. Both worked under Johannes Müller, the eminent physiologist at the University of Berlin.

Schleiden, a lawyer turned botanist, had been studying the structure and development of plant tissues. He had been "hay gathering" *("Heusammelei")*, as he called it, and collected hundreds of specimens from the plant kingdom: tulips, dog hobble, spruce, grasses, orchids, sage, linanthus, peas, and dozens of kinds of lilies. His collection was prized among botanists.

That evening, Schwann and Schleiden were discussing phytogenesis — the origin and development of plants. And what Schleiden told Schwann was this: in looking through all his plant specimens, he had found a "unity" in their construction and organization. During the development of plant tissues — leaves, roots, cotyledons — a subcellular structure, called the nucleus, became prominently visible. (Schleiden did not know the function of the nucleus but recognized its distinctive form.)

But perhaps more surprisingly, there was a deep uniformity in the construction of the tissues. Each part of the plant was built, bricolage-like, out of autonomous, independent units — *cells*. "Each cell leads a double life," Schleiden would write a year later, "an entirely independent one, belonging to its own development alone; and an incidental one, in so far as it has become part of a plant."

A life within a life. An independent living being — a unit — that forms a part of the whole. A living building block contained within the larger living being.

Schwann's ears pricked up. He, too, had noted the prominence of the nucleus, but in the cells of a developing *animal,* a tadpole. And he, too, had noted the uniformity in the microscopic construction of animal tissues. The "unity" that Schleiden had observed in plant cells was, perhaps, a deeper unity that ran through life.

An inchoate but radical thought — one that would swerve the history of biology and medicine — began to form in his mind. Perhaps that very evening, or soon after, he invited Schleiden (or dragged him, possibly) to the lab at the anatomical theater, where Schwann kept his specimens. Schleiden looked through the scope. The developing animal's microscopic structure, including the prominently visible nucleus, Schleiden confirmed, looked almost identical to that of the plant's.

Animals and plants — as seemingly different as living organisms could be. Yet, as both Schwann and Schleiden had noticed,

15

the similarity of their tissues under the microscope was uncanny. Schwann's hunch had been right. That evening in Berlin, he would later recall, the two friends had converged on a universal and essential scientific truth: both animals and plants had a "common means of formation through cells."

In 1838, Schleiden collected his observations in an expansive paper entitled *Contributions to Our Knowledge of Phytogenesis.* A year later, Schwann followed Schleiden's work on plants with his tome on animal cells: *Microscopical Researches into the Accordance in the Structure and Growth of Animals and Plants.* Both plants and animals, Schwann posited, were similarly organized — each an "aggregate of fully individualized independent beings."

In two seminal works, published about twelve months apart, the living world converged to a single, sharp point. Schleiden and Schwann weren't the first to see cells, or to realize that cells were the fundamental units of living organisms. The acuity of their insight was in the proposition that a deep unity of organization and function ran through living beings. "A bond of union" connects the different branches of life, Schwann wrote.

Schleiden left Berlin for a position at the University of Jena in late 1838. And in 1839, Schwann left, too, for a position at the Catholic University in Leuven, Belgium. Despite their dispersal out of Müller's lab, they kept up a lively correspondence and friendship. Their seminal work on the foundations of cell theory is indubitably traced back to Berlin, where they had been intimate colleagues, collaborators, and friends. They had found, in Schwann's words, the "elementary particles of organisms."

This book is the story of the cell. It is a chronicle of the discovery that all organisms, including humans, are made of these "elementary particles." It's a story of how cooperative, organized accumulations of these autonomous living units — tissues, organs, and organ systems — enable profound forms of physiology: immunity, reproduction, sentience, cognition, repair, and rejuvenation. Conversely, it is the story of what happens when cells become dysfunctional, tipping our bodies from cellular physiology into cellular pathology — the malfunctioning of cells precipitating the malfunction of the body. And finally, it is a story about how our deepening understanding of cellular physiology and pathology has sparked a revolution in biology and medicine, leading to the birth of transformational medicines, and of human beings transformed by these medicines.

Between 2017 and 2021, I wrote three articles for the *New Yorker* magazine. The first was about cellular medicine and its future — in particular, about the invention of T cells reengineered to attack cancers. The second concerned a new vision of cancer centered on the idea of the *ecology* of cells — not cancer cells in isolation, but cancer in situ, and why specific locations in the body seem so much more hospitable to malignant growth than other organs do. The third, written in the early days of the Covid-19 pandemic, was about how viruses behave in our cells and bodies, and how that behavior might help us understand the physiological devastation caused by some viruses in humans.

I wondered about the thematic links among these three pieces. At the center of all of them, it seemed, was the story of cells and cellular reengineering. There was a revolution in the making, and a history (and future) that had been unwritten: of cells, of our capacity to manipulate cells, and of the transformation of medicine that is unfolding as this revolution unfolds.

From the seed of those three pieces, this book grew stalks, roots, and tendrils on its own. This chronicle begins in the 1660s and 1670s, when a reclusive Dutch cloth seller and an unorthodox English polymath, working independently, and about two hundred miles apart, looked down their handcrafted microscopes and discovered the first evidence of cells. It moves to the present — a time when human stem cells are being manipulated by scientists and infused into patients with chronic, potentially life-threatening diseases such as diabetes and sickle cell anemia, and electrodes are being inserted into cellular circuits of the brains of men and women with recalcitrant neurological illnesses. And it brings us to the precipice of an uncertain future, in which "maverick" scientists (one of whom was jailed for three years and has been permanently disbarred from performing experiments) are designing gene-edited embryos, and using cell transplantation to blur the boundaries between the natural and the augmented.

I draw from an array of sources: interviews; patient encounters; itinerant walks with scientists (and their dogs); visits to labs; visions through a microscope; conversations with nurses, patients, and doctors; historical sources; scientific papers; and personal letters. My purpose is not to write a comprehensive history of medicine or of the birth of cell biology. Roy Porter's *The Greatest Benefit to Mankind: A Medical History of Humanity,* Henry Harris's *The Birth of the Cell,* and Laura Otis's *Müller's Lab* are exemplary accounts. This, rather, is the story of how the concept of the cell, and our comprehension of cellular physiology, altered medicine, science, biology, social structures, and culture. It culminates in the vision of a future in which we learn to manipulate these units into new forms, or perhaps even create synthetic versions of cells, and parts of humans.

There are, inevitably, gaps and lacunae in this version of the story of the cell. Cell biology is inextricably linked with genetics, pathology, epidemiology, epistemology, taxonomy, and anthropology. Aficionados of particular niches in medicine or cell biology, legitimately partial to a particular cell type, might have viewed this history through a very different eyepiece; botanists, bacteriologists, and mycologists will doubtless miss adequate focus on the plants, bacteria, and fungi. To enter each of these fields in a non-desultory manner would be to enter labyrinths that fork into further labyrinths. I have moved many aspects of the story to footnotes and endnotes. I urge readers to read them seriously.

Throughout this journey, we will meet many patients, including some of my own. Some are named; others chose to be anonymous, with their names and identifying details removed. I feel an immeasurable gratitude to these men and women who have ventured into uncharted territories, entrusting their bodies and minds to an evolving and uncertain realm of science. And I feel an exhilaration, just as immeasurable, as I witness cell biology come to life in a new kind of medicine.

INTRODUCTION
"WE SHALL ALWAYS RETURN TO THE CELL"

*No matter how we twist and turn,
we shall eventually come back to the cell.*
— Rudolf Virchow, 1858

In November 2017, I watched my friend Sam P. die because his cells had rebelled against his body.

Sam had been diagnosed with a malignant melanoma in the spring of 2016. The cancer had first appeared as a coin-shaped mole, purple-black with a halolike aureole, near his cheek. His mother, Clara, a painter, had first noticed it during a late-summer vacation on Block Island. She had cajoled — and then begged and threatened — him to have it examined by a dermatologist, but Sam was a busy, active sportswriter for a big newspaper, with little time to worry about a pesky spot on his cheek. By the time I saw and examined him in March 2017 — I was not his oncologist, but a friend had asked me to look at his case — the tumor had grown into

21

a thumb-sized, oblong mass, and there was evidence of a metastasis in his skin. When I touched the growth, he winced in pain.

It is one thing to encounter a cancer, it is quite another to bear witness to its mobility. The melanoma had begun to travel across Sam's face toward his ear. If you looked closely, it had marked its progression like a ferry moving across the water, leaving a wake of stippled, purple dots behind it.

Even Sam, the sportswriter who had spent his life learning about speed, motility, and agility, was astonished by the pace of the melanoma's progression. How, he asked me insistently — *how, how, how* — had a cell that had sat perfectly still in his skin for decades suddenly acquired the properties of a cell capable of careening along his face while also dividing furiously?

But cancer cells don't "invent" any of these properties. They don't build anew, they hijack — or, more accurately, the cells that are fittest for survival, growth, and metasisis are naturally selected. The genes and proteins that cells use to generate the building blocks required for growth are appropriated from the genes and cells that a developing embryo uses to fuel its fierce burst of expansion during the first days of life. The pathways used by the cancer cell to move across vast bodily spaces are commandeered from those that allow inherently mobile cells in the body to move.

The genes that enable unfettered cell division are distorted, mutated versions of genes that allow cell division in normal cells. Cancer, in short, is cell biology visualized in a pathological mirror. And as an oncologist, I am, first, a cell biologist — except one who perceives the normal world of cells reflected and inverted in a looking glass.

In late spring 2016, Sam was prescribed a medicine to turn his own T cells into an army to fight the rebel army that was growing in his body. Consider this thought: for years, perhaps decades, Sam's melanoma and his T cells had coexisted, essentially ignoring each other. His malignancy was invisible to his immune system. Millions of his T cells had brushed past his melanoma every day and just moved on, bystanders that had turned their faces away from a cellular catastrophe.

The drug that Sam had been prescribed would hopefully uncloak the tumor's invisibility and make his T cells recognize the melanoma as a "foreign" invader and reject it, much as T cells reject microbe-infected cells. The passive bystanders would become active effectors. We were engineering the cells in his body to make visible what had previously been invisible.

The discovery of this "uncloaking" medicine was the culmination of radical advances in cellular biology that date back to the 1950s:

an understanding of the mechanisms used by T cells to discriminate the self from the nonself, the identification of the proteins that these immune cells use to detect foreign invaders, the uncovering of pathways by which our normal cells resist being attacked by this detection system, the way cancer cells co-opt it to make themselves invisible, and the invention of a molecule that would strip the malignant cells of their cloak of invisibility — each insight, built atop an earlier insight and each dug by cell biologists out of hard, cold earth.

Almost immediately after Sam began his treatment, a civil war unfolded in his body. His T cells, shaken awake to the presence of the cancer, were pitched against his malignant cells, their vengeance provoking further cycles of vengeance. The crimson boil on his cheek turned hot one morning because the immune cells had infiltrated the tumor and unleashed a cycle of inflammation; then the malignant cells folded camp and left, leaving smoldering, dying campfires. When I saw him again a few weeks later, the oblong mass and the stipples behind it had vanished. Instead, there was just the dying remnant of a tumor, shriveled like a large raisin. He was in a remission.

We shared a coffee to celebrate. The remission had not just changed Sam physically; it had charged him psychologically. For the first time in weeks, I saw the creases of worry in his face relax. He laughed.

But then things turned: April 2016 was a cruel month. The T cells that attacked his tumor turned on his own liver, provoking an autoimmune hepatitis, an inflammation of the liver that could barely be controlled with immune-suppressive drugs. In November, we discovered that the cancer — in remission just weeks prior — had battened to his skin, muscles, and lungs, hidey-holing in new organs and finding new niches to survive the attack of his immune cells.

Sam maintained a steely dignity through these victories and setbacks. At times, his withering humor seemed like its own form of counterattack: *he would desiccate the cancer to death*. When I visited him at his desk in the newsroom one day, I asked if he'd like a private space — the men's bathroom, perhaps — where he could show me where the new tumors had arisen. He laughed breezily. "By the time we get to the bathroom, it will have moved to a new site. Better look at it while it's still here."

The doctors blunted the immune assault to control the autoimmune hepatitis, but then the cancer grew back. They restarted the immunotherapy to attack the cancer, and the fulminant hepatitis returned. It was like watching some kind of sport of bestial warfare: put the immune cells on a leash, and the animals would strain against their chains to attack and kill. Unleash them, and they would indiscriminately attack both the cancer and the liver. Sam died on a spring morning, about six months after I had first felt his tumor. In the end, the melanoma won.

On a blustery afternoon in 2019, I attended a conference at the University of Pennsylvania, in Philadelphia. Nearly a thousand scientists, doctors, and biotech researchers converged on a brick-and-stone auditorium on Spruce Street. They were there to discuss advances in a bold frontier in medicine: the use of cells, genetically modified and transplanted into humans, to cure diseases. There were talks on T cell modifications, on new viruses that could deliver genes into cells, and on the next major steps in cellular transplantation. The language, on and off stage, felt as if biology, robotics, science fiction, and alchemy had gotten together on an ecstatic evening and produced a precocious child. *"Reboot the immune system." "Therapeutic cellular reengineering." "Long-term persistence of grafted cells."* It was a conference about the future.

But the present was also present. Sitting just a few rows ahead of me was Emily Whitehead, then fourteen, a year older than my elder daughter. She had tousled brown hair, wore a yellow-and-black shirt and dark pants, and was in her seventh year of remission

26

from leukemia. "She was happy to miss a day of school," her father, Tom, told me. Emily smiled at the thought.

Emily was Patient No. 7, treated at the Children's Hospital of Philadelphia (CHOP). Nearly everyone in the audience knew her or knew of her: she had altered the history of cellular therapy. In May 2010, Emily had been diagnosed with acute lymphoblastic leukemia (ALL). Among the most rapidly progressive forms of cancer, this leukemia tends to afflict young children.

The treatment for ALL ranks among the most intensive chemo regimens ever devised: seven or eight drugs given in combination, some injected directly into the spinal fluid to kill any cancer cells hiding in the brain and spine. Although the collateral damage of the treatment — permanent numbness in the fingers and toes, brain damage, stunted growth and life-threatening infections, to name just a few — can be daunting, the treatment cures about 90 percent of pediatric patients. Unfortunately, Emily's cancer fell in the remnant 10 percent, proving unresponsive to standard therapy. She relapsed sixteen months into treatment. She was listed for a bone marrow transplant — the only option for a cure — but her condition worsened while she awaited a suitable donor.

"The doctors told me not to Google" her chances of survival, Emily's mother, Kari, told me. "So, of course, I did that right away."

What Kari found on the web was chilling: of the children who relapse early, or relapse twice, almost none survive. When Emily arrived at Children's Hospital in early March 2012, nearly every one of her organs was packed with malignant cells. She was seen by a pediatric oncologist, Stephan Grupp, a gentle, burly man with an expressive, ever-moving mustache, and then enrolled in a clinical trial.

Emily's trial involved infusing her body with her own T cells. But these T cells had to be weaponized, via gene therapy, to recognize and kill her cancer. Unlike Sam, who had received drugs to activate immunity *inside* his body, Emily's T cells had been extracted and grown *outside* her body. This form of treatment had been pioneered by the immunologist Michel Sadelain at the Sloan Kettering Institute in New York and by Carl June at the University of Pennsylvania, building on earlier work by the Israeli researcher Zelig Eshhar.

A few hundred feet from where we had been sitting was the cell therapy unit, a vault-like, enclosed facility with steel doors, sterile rooms, and incubators. There groups of technicians were processing cells collected from dozens of patients enrolled in the clinical studies and then storing them in vat-like freezers. Each freezer bore the name of a character from the animated TV sitcom *The Simpsons;* a fraction of Emily's cells were frozen in Krusty the Clown. Another portion of her T cells had been modified to express a gene that would recognize and kill her leukemia, cultured in the lab to increase their numbers exponentially and then returned to the hospital to infuse them back into Emily.

The infusions, which took place over three days, were largely uneventful. Emily sucked on an ice pop while Dr. Grupp dripped the cells into her veins. In the evenings, she and her parents went to stay with an aunt who lived nearby. The first two nights, she played games and got piggyback rides from her father. On the third day, though, she crashed: throwing up, and spiking an alarming fever. The Whiteheads rushed her back to the hospital. Things rapidly spiraled downward. Her kidneys failed. Emily drifted in and out of consciousness, verging on multi-organ system failure.

"Nothing made sense," Tom told me. His six-year-old daughter was moved to the intensive care unit, where her parents and Grupp kept an all-night vigil.

Carl June, the physician-scientist who was also treating Emily, told me candidly, "We thought she was going to die. I wrote an e-mail to the provost at the university, telling him that one of the first children with the treatment was about to die. The trial was finished. I stored the e-mail in my out-box but never pressed Send."

The lab technicians at Penn worked overnight to determine the cause of the fever. They found no evidence of infection; instead, they found elevated blood levels of molecules called cytokines — signals secreted during active inflammation. In particular, levels of a cytokine known as interleukin 6 (IL-6) were nearly a thousand times normal. As the T cells killed the cancer cells, they were releasing a storm of these chemical messengers, like a rioting crowd disgorging inflammatory pamphlets on a rampage.

By a strange twist of fate, however, June's own daughter had a form of juvenile arthritis, an inflammatory condition. He knew about a new drug, approved by the US Food and Drug Administration (FDA) just four months earlier, that blocks IL-6. As a last-ditch effort, Grupp rushed an application to the hospital pharmacy requesting permission to use the new therapy off-label. The board granted its approval for the IL-6–blocking drug that evening, and Grupp injected Emily with a dose in the ICU.

Two days afterward, on her seventh birthday, Emily woke up. "Boom," Dr. June said, waving his hands in the air. "Boom," he repeated. "It just melted away. We did a bone marrow biopsy twenty-three days later, and she was in a complete remission."

"I have never seen a patient that sick get better so quickly," Grupp told me.

The deft management of Emily's condition — and her startling recovery — saved the field of cell therapy. Emily Whitehead remains in that deep remission to this day. No cancer is detectable in her marrow or her blood. She is considered cured.

"If Emily had died," June told me, it's likely that the whole trial would have been shut down." It would have set back cellular therapy perhaps a decade or even longer.

During a pause in the sessions at the conference, Emily and I joined a tour of the medical campus led by Dr. Bruce Levine, one of Dr. June's colleagues. He is the founding director of the facility at Penn where T cells are modified, quality controlled, and manufactured, and was among the first to handle Emily's cells. The technicians here worked singly or in pairs, checking boxes, optimizing protocols, shuttling cells between incubators, sterilizing their hands.

The facility may as well have doubled as a small monument to Emily. Photographs of her were plastered on the walls: Emily at eight, in pigtails; Emily at ten, holding a plaque; Emily at twelve, with missing front teeth, smiling next to President Barack Obama. At a certain point during the tour, I watched the real Emily looking out the window at the hospital across the street. She could almost see into the corner ICU room where she had been confined for nearly a month.

The rain came down in sheets, streaking the windows with droplets.

I wondered how she felt, knowing that there were three versions of her in the hospital: the one here today, on a break from school; the one in the pictures, who had lived and almost died in the ICU; and the one frozen in the Krusty the Clown freezer next door.

"Do you remember coming into the hospital?" I asked.

"No," she said, looking out into the rain. "I only remember leaving."

As I watched the advance and retreat of Sam's illness, and the remarkable recovery of Emily Whitehead, I knew that I was also observing the birth of a kind of medicine in which cells were being repurposed as tools to fight illness — cellular engineering. But it was also the replay of a centuries-old story. We are built out of cellular units. Our vulnerabilities are built out of the vulnerabilities of cells. Our capacity to engineer or manipulate cells (immune cells, in both Sam's and Emily's cases) has become the basis of a new kind of medicine — albeit a kind of medicine that is still in midbirth. If we knew how to arm Sam's immune cells more effectively against his melanoma without unleashing the autoimmune attack, would he be alive today, spiral notebook in hand, writing sports pieces for a magazine?

Two new humans, examples of cellular manipulation and reengineering. Emily, for whom our understanding of the laws of T cell biology were seemingly sufficient to hold a lethal disease at bay for more than a decade, and, hopefully, for her lifetime. Sam, for whom we still seem to be missing some critical insight of how to balance a T cell's attack on cancer and an attack on the self.

What will the future bring? Let me clarify: I use the phrase "new human" throughout the book, and in its title. I mean it in a very precise sense. I explicitly do not mean the "new human" found in sci-fi visions of the future: an AI-augmented, robotically enhanced, infrared-equipped, blue-pill-swallowing creature who blissfully cohabitates the real and virtual worlds: Keanu Reeves in a black muumuu. Nor do I mean "transhuman," endowed with augmented abilities and capacities that transcend the ones we currently possess.

I mean a human rebuilt anew with modified cells who looks and feels (mostly) like you and me. A woman with crippling, recalcitrant depression whose nerve cells (neurons) are being stimulated with electrodes. A young boy undergoing an experimental bone marrow transplant using gene-edited cells to cure sickle cell disease. A type 1 diabetic infused with his own stem cells that have been engineered to produce the hormone insulin to maintain a normal blood level of glucose, the body's fuel. An octogenarian who, following multiple heart attacks, is injected with a virus that will home to his liver and permanently lower artery-clogging cholesterol, thus reducing his risk of another cardiac event. I mean my father, implanted with neurons, or a neuron-stimulating device, that would have steadied his gait so that he might not have suffered the fall that led to his death.

I find these "new humans" — and the cellular technologies used to create them — vastly more exciting than their imaginary sci-fi counterparts. We've altered these humans to alleviate suffering, using a science that had to be handcrafted and carved with unfathomable labor and love, and technologies so ingenious that they stretch credulity: such as fusing a cancer cell with an immune cell to produce an immortal cell to cure cancer; or extracting a T cell from a young girl's body, engineering it with a virus to weaponize it against leukemia, and then transfusing it back into her body. We will meet these new humans in virtually every chapter in this book. And as we learn to rebuild bodies and parts with cells, we will meet them in the present and in the future: in cafes, supermarkets, train stations, and airports; in neighborhoods; and in our own families. We will find them among our cousins and grandparents, our parents and siblings — and perhaps in our selves.

In a little less than two centuries — from the late 1830s, when the scientists Matthias Schleiden and Theodor Schwann proposed that all animal and plant tissues were made of cells, to the spring of Emily's recovery — a radical concept swept through biology and medicine, touching virtually every aspect of the two sciences, and altering both forever. Complex living organisms were assemblages of tiny, self-contained, self-regulating units — living compartments, if you will, or "living atoms," as the Dutch microscopist Antonie van Leeuwenhoek called them in 1676. Humans were ecosystems of these living units. We were pixelated assemblages, composites, our existence the result of a cooperative agglomeration.

We were a sum of parts.

The discovery of cells, and the reframing of the human body as a cellular ecosystem, also announced the birth of a new kind of medicine based on the therapeutic manipulations of cells. A hip fracture, cardiac arrest, immunodeficiency, Alzheimer's dementia, AIDS, pneumonia, lung cancer, kidney failure, arthritis — all could be reconceived as the results of cells, or systems of cells, functioning abnormally. And all could be perceived as loci of cellular therapies.

The transformation of medicine made possible by our new understanding of cell biology can be broadly divided into four categories.

The first is the use of drugs, chemical substances, or physical stimulation to alter the properties of cells — their interactions with one another, their intercommunication, and their behavior. Antibiotics against germs, chemotherapy and immunotherapy for cancer, and the stimulation of neurons with electrodes to modulate nerve cell circuits in the brain fall in this first category.

The second is the transfer of cells from body to body (including back into our own bodies), exemplified by blood transfusions, bone marrow transplantation, and in vitro fertilization (IVF).

The third is the use of cells to synthesize a substance — insulin or antibodies — that produces a therapeutic effect on an illness.

And most recently, there is a fourth category: the genetic modification of cells, followed by transplantation, to create cells, organs, and bodies endowed with new properties.

Some of these therapies, such as antibiotics and blood transfusion, have been so deeply entrenched in the practice of medicine that we hardly think of them as "cellular therapies." But they arose from our understanding of cell biology (germ theory, as we shall soon see, was an extension of cell theory). Some other therapies, such as immunotherapy for cancer, are twenty-first-century developments. Yet others, such as the infusion of modified stem cells for diabetes, are so new that they are still considered experimental. Yet all these — the old and the new — are "cellular therapies" because they depend critically on our understanding of cell biology. And each advance has changed the course of medicine and, equally, changed our conception of being human and living as humans.

In 1922, a fourteen-year-old boy with type 1 diabetes was resuscitated from a coma — born anew, as it were — by the infusion of insulin extracted from the pancreatic cells of a dog. In 2010, when Emily Whitehead received her infusion of CAR (chimeric antigen receptor) T cells, or twelve years later, when the first patients with sickle cell anemia are surviving, disease-free, with gene-modified blood stem cells, we are transitioning from the century of the gene to a contiguous, overlapping century of the cell.

A cell is the unit of life. But that begs a deeper question: What is "life"? It may be one of biology's metaphysical conundrums that we are still struggling to define the very thing that defines us. Life's definition cannot be captured by a single property. As the Ukrainian biologist Serhiy (or Sergey, as he was commonly known) Tsokolov put it: "Every theory, hypothesis, or point of view adopts

37

life's definitions in accordance with its own scientific interests and premises. There are hundreds of working, conventional definitions of life within scientific discourse, but none has been able to achieve a consensus." (And Tsokolov, who unfortunately died in the prime of his intellectual life in 2009, would know, as it was the particular stone in his shoe. He was an *astro*biologist; his research involved finding life beyond Earth. But how can one find life if scientists are struggling to define the term itself?)

Life's definition, as it stands now, is akin to a menu. It is not one thing but a series of things, a set of *behaviors,* a series of processes, not a single property. To be living, an organism must have the capacity to reproduce, to grow, to metabolize, to adapt to stimuli, and to maintain its internal milieu. Complex, multicellular living beings also possess what I might call "emergent" properties: properties that emerge from systems of cells, such as mechanisms to defend themselves against injury and invasion, organs with specialized functions, physiologic systems of communication between organs and even sentience and cognition. And it is not a coincidence that all these properties repose, ultimately, in the cells, or systems of cells. In a sense, then, one might define life as having cells, and cells as having life.

The recursive definition is not nonsensical.

Had Tsokolov met his first astrobiological being — say, an ectoplasmic alien from Alpha Centauri — and asked whether s/he/it is "living" or not, he might have asked whether this Being fulfills the menu of life's properties. But he might have also queried the Being: "Do you have cells?" It is difficult to imagine life without cells, just as it is impossible to imagine cells having no life.

Perhaps that fact outlines the importance of the story of the cell: we need to understand cells to understand the human body. We need them to understand medicine. But most essentially, we need the story of the cell to tell the story of life and of our selves.

What *is* a cell, anyway? In a narrow sense, a cell is an autonomous living unit that acts as a decoding machine for a gene. Genes provide instructions — code, if you will — to build proteins, the molecules that perform virtually all the work in a cell. Proteins enable biological reactions, coordinate signals within the cell, form its structural elements, and turn genes on and off to regulate a cell's identity, metabolism, growth, and death. They are the central functionaries in biology, the molecular machines that enable life.*

*Genes provide the code to build ribonucleic acid (RNA) that, in turn, is deciphered to build proteins. But aside from carrying the code to make proteins,

39

Genes, which carry the codes to build proteins, are physically located in a double-stranded, helical molecule called deoxyribonucleic acid (DNA), which is further packaged in human cells into skein-like structures called chromosomes. As far as we know, DNA is present inside every living cell. Scientists have hunted for cells that use molecules other than DNA to carry their instructions — RNA, for instance — but so far, they've never found an RNA instruction-carrying cell.

By *decoding,* I mean that molecules within a cell *read* certain sections of the genetic code, like musicians in an orchestra reading their parts of a musical score — the cell's individual song — thereby enabling a gene's instructions to become physically manifest in the actual protein. Or, put more simply, a gene carries the code; a cell deciphers that code. A cell thus transforms information into form; genetic code into proteins. A gene without a cell is lifeless — an instruction manual stored inside an inert molecule, a musical score without a musician, a lonely library with no one to read the books within it. A cell brings materiality and physicality to a set of genes. A cell *enlivens* genes.

But a cell is not merely a gene-decoding

---

some of these RNAs carry out diverse tasks in cells, some of which are yet to be deciphered. RNA can also regulate genes and function in concert with proteins in some biological reactions.

machine. Having unpacked the code by synthesizing a select set of proteins that is encoded in its genes, a cell becomes an integrating machine. A cell uses this set of proteins (and the biochemical products made by proteins) in conjunction with one another to start coordinating its function, its *behavior* (movement, metabolism, signaling, delivering nutrients to other cells, surveying for foreign objects), to achieve the properties of life. And that behavior, in turn, manifests as the behavior of the organism. The metabolism of an organism reposes in the metabolism of the cell. The reproduction of an organism reposes in the reproduction of a cell. The repair, survival, and death of an organism repose in the repair, survival, and death of cells. The behavior of an organ, or an organism, reposes in the behavior of a cell. The *life* of an organism reposes in the life of a cell.

And finally, a cell is a dividing machine. Molecules within the cell — proteins, again — initiate the process of duplicating the genome. The internal organization of the cell changes. Chromosomes, where the genetic material of a cell is physically located, divide. Cell division is what drives growth, repair, regeneration, and, ultimately, reproduction, among the fundamental, defining features of life.

I have spent a lifetime with cells. Every time I see a cell under a microscope — refulgent,

41

glimmering, alive — I relive the thrill of seeing my first cell. On a Friday afternoon in the fall of 1993, about a week after I had arrived as a graduate student in Alain Townsend's lab at the University of Oxford to study immunology, I had ground up a mouse spleen and plated the blood-tinged soup in a petri dish with factors to stimulate T cells. The weekend passed, and on Monday morning, I switched on the microscope. The room was so dimly lit that it was not even necessary to pull down the curtains — the city of Oxford was *always* dimly lit (if cloudless Italy was a land made for telescopes, then foggy, dark England seemed custom-made for microscopes) — and I put the plate under the scope. Wading beneath the tissue culture medium were masses of translucent, kidney-shaped T cells that possessed what I can describe only as an inner glow and a luminous fullness — the signs of healthy, active cells. (When cells die, the glow dims, and they shrivel and turn granular, or pyknotic, to use the jargon of cell biology.)

"Like eyes looking back at me," I whispered to myself. And then, to my astonishment, the T cell *moved* — deliberately, purposefully, seeking out an infected cell that it might purge and kill. It was alive.

Years later, then, it was nothing less than thrilling — mesmerizing — to watch the cellular revolution unfold in humans. When I first met Emily Whitehead, in a fluorescent-lit corridor outside the auditorium at the University of Pennsylvania, it was as if she had allowed me to enter a portal that linked the future and the past. I trained as an immunologist at first, then a stem cell scientist, and, finally, a cancer biologist before I became a medical oncologist.* Emily embodied all these past lives — not just mine, but, more importantly, the lives and labors of thousands of researchers, looking down thousands of microscopes, over thousands of days and nights. She embodied our desire to get to the luminous heart of the cell, to understand its endlessly captivating mysteries. And she embodied our aching aspiration to witness the birth of a new kind of medicine — cellular therapies — based on our deciphering the physiology of cells.

To encounter my friend Sam in his hospital room and watch his remission and relapse whiplash him week upon week, was to experience an opposite chill — not exhilaration but an apprehension of how much there was yet to learn and know. As an oncologist, I focus

*I even had a brief foray into neurobiology between 1996 and 1999, when I worked with Professor Connie Cepko at Harvard Medical School, studying the development of the retina. I studied glial cells long before they were in vogue in neurobiology. Cepko, a developmental biologist and geneticist, taught me the science and art of lineage tracing, a method that we will encounter later in this book.

on cells that have gone rogue; cells that have marauded spaces where they should not exist; cells dividing out of control. These cells distort and overturn the very behaviors that I describe in this book. I try to understand why and how that happens. You might think of me as a cell biologist caught in an upside-down world. And so the story of the cell is one that is stitched into the very fabric of my scientific and personal lives.

As I wrote furiously from the early months of 2020 into 2022, the Covid-19 pandemic continued wildfiring its way throughout the globe. My hospital, my adopted city of New York, and my homeland overflowed with the bodies of the sick and the dead. By February 2020, the ICU beds at Columbia University Medical Center, where I work, were full of patients drowning in their own secretions, with mechanical ventilators forcing air in and out of their lungs. The early spring of '20 was particularly bleak: New York turned into an unrecognizable, windblown metropolis of empty byways and avenues, where people hid from people. India's most lethal surge hit almost a year later in April and May 2021. Bodies were burned in parking lots, back alleyways, slums, and children's playgrounds. At the crematoria, the fires burned so often and so briskly that the metal grids holding the bodies corroded and melted.

I sat in a clinic room in the hospital at first,

and then, when the cancer clinic itself was pared to its bare minimum, isolated with my family at home. Staring out the window at the horizon, I thought yet again about cells. Immunity and its discontents. The Yale University virologist Akiko Iwasaki told me that the central pathology caused by SARS-CoV2 (severe acute respiratory syndrome coronavirus 2) was "immunological misfiring" — a dysregulation of immune cells. I had not even heard the term before, but its immensity hit me: at its core, the pandemic, too, was a disease of cells. Yes, there was the virus, but viruses are inert, lifeless, without cells. Our cells had awoken the plague and brought it to life. To understand crucial features of the pandemic, we would need to understand not just the idiosyncrasies of the virus but also the biology of immune cells and their discontents.

For a while, then, it seemed as if every byway and avenue of my thinking and being led me back to cells. I am not sure how much I conjured this book into life, and how much this book demanded to be written.

In *The Emperor of All Maladies,* I wrote about the aching quest to find cures for cancer or to prevent it. *The Gene* was propelled by the quest to decode and decipher the code of life. *The Song of the Cell* takes us on a very different journey: to understand life in terms of its simplest unit — the cell. This book is not about hunting for a cure or deciphering a code. There is no single adversary. Its protagonists want to understand life by understanding a cell's anatomy, physiology, behavior, and its interactions with surrounding cells. A cell's music. And their medical quest is to seek cellular therapies, to use the building blocks of humans to rebuild and repair humans.

Rather than a chronological unfolding, then, I had to choose a very different structure. Each part of the book takes a fundamental property of complex living beings and explores its story. Each part is a mini history, a chronology of discovery. Each part illuminates a fundamental property of life (reproduction, autonomy, metabolism) that reposes in a particular system of cells. And each contains the birth of new cellular technology (say, bone marrow transplantation, in vitro fertilization [IVF], gene therapy, deep brain stimulation, immunotherapy) that arises from our understanding of cells and challenges our conceptions of how humans are built and how we function. The book is itself a sum of parts: history and personal history, physiology and pathology, past and future — and an intimate chronicle of my own growth as a cell biologist and doctor — spun together into a whole. The organization is cellular, if you will.

When I began this project in the winter of 2019, I chose initially to dedicate it to Rudolf

46

Virchow. I was taken by this reclusive, progressive, soft-spoken German physician-scientist who, resisting the pathological social forces of his times, promoted free thinking, was a champion of public health, despised racism, published his own journal, carved a unique and self-assured path through medicine, and launched an understanding of the study of diseases in organs and tissues based on dysfunctions of cells — "cellular pathology," as he described it.

I returned, in the end, to a patient, a friend, being treated for cancer with a novel form of immunotherapy and Emily Whitehead — patients who had opened new inroads into our understanding of cells and cellular therapy. They were among the first to experience our early attempts to harness cells for human therapy and to transform cellular pathology into cellular medicine — partially successful and partially not. It is to them, and their cells, that this book is dedicated.

PART ONE
DISCOVERY

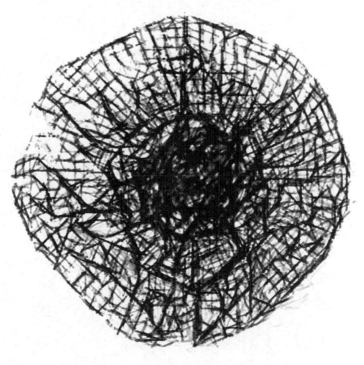

Both of us, you and I, began as single cells.

Our genes are different, albeit marginally. The way our bodies develop are different. Our skin, our hair, our bones, our brains are all built differently. Our life experiences vary widely. I lost two uncles to mental illness. I lost a father to a deadly spiral following a fall. A knee to arthritis. A friend — so many friends — to cancer.

And yet, despite all the yawning gaps between our bodies and experiences, you and I share two features. First, we arose from a single-celled embryo. And second, from that cell came multiple cells — those that populate your body and mine. We are built of the same material units and are akin to two different nuggets of matter built from the same atoms.

What are we made of? Some ancients believed that we were created by menstrual blood that had congealed into bodies. Some believed we came preformed: mini-beings that just expanded over time, like human-shaped balloons blown up for a parade. Some thought humans were sculpted from mud and river water. Some thought we transformed gradually in the womb from tadpole-like beings to fish-mouthed creatures, and, finally, into humans.

But if you looked down a microscope at your skin and mine, or your liver and mine, you would find them strikingly alike. And you'd realize that all of us were, in fact, built out of living units: cells. The first cell gave rise to more cells, and then divided to form even more, until our livers and guts and brains — all the elaborate anatomical architectures in the body — were gradually formed.

When did we realize that humans were, in fact, composites of independent, living units? Or that these units are the basis of all the functions that the body is capable of — in other words, that our physiology reposes, ultimately, in cellular physiology? And conversely, when did we posit that our medical fates and futures were intimately linked to the changes in these living units? That our diseases are consequences of cellular pathology?

It is to these questions — and embedded within them, the story of a discovery that touched and radically transformed biology, medicine, and our conception of humans — that we first turn.

THE ORIGINAL CELL
AN INVISIBLE WORLD

---

*True knowledge is to be aware
of one's ignorance.*
— Rudolf Virchow, letter to
his father, ca. 1830s

Let us give thanks, first, to the softness of Rudolf Virchow's voice. Virchow was born in Pomerania, Prussia (now split between Poland and Germany), on October 13, 1821. His father, Carl, was a farmer and a city treasurer. We know little of his mother, Johanna Virchow, née Hesse. Rudolf was a diligent and bright student — thoughtful, attentive, and clever with languages. He learned German, French, Arabic, and Latin, and earned distinctions for his academic work.

At the age of eighteen, he wrote his high school thesis, "A Life Full of Work and Toil Is Not a Burden but a Benediction," and began to prepare for a professional life in the clergy. He sought to become a pastor and preach to a congregation. But Virchow was anxious about the weakness of his voice. Faith emanated from the strength of inspiration, and inspiration from the strength of elocution. But what if no one could even *hear* him as he tried to project from the pulpit? Medicine and science seemed like more forgiving professions for a reclusive, studious, soft-spoken boy. Upon graduation in 1839, Virchow received a military scholarship and chose to study medicine at the Friedrich-Wilhelms Institute in Berlin.

The world of medicine that Virchow entered in the mid-1800s might have been divided into two halves — anatomy and pathology — one relatively advanced and the other still in a muddle, respectively. In the sixteenth century, anatomists began to describe the forms and structures of the human body with increasing precision. The best-known anatomist of all was the Flemish scientist Andreas Vesalius, a professor at Padua University in Italy. The son of an apothecary, Vesalius arrived in Paris in 1533 to study and practice surgery. He found surgical anatomy in absolute disarray. There were few textbooks and no systematic map of the human body. Most surgeons and their students depended, loosely, on the anatomical teachings of Galen, the Roman physician who lived between AD 129 and 216. Galen's centuries-old treatises on human anatomy were based on animal studies, had become badly outdated, and, frankly, were often incorrect.

The basement of Paris's hospital Hôtel-Dieu,

where decaying human cadavers were dissected, was a dingy, airless, badly lit space with half-feral dogs roaming underneath the gurneys to gnaw on the drippings — a "meat market," as Vesalius would describe one such anatomical chamber. The professors sat on "lofty chairs [and] cackle like jackdaws," he wrote, while their assistants hacked and tugged through the body at random, eviscerating organs and parts as if pulling out cotton stuffing from a toy.

"The doctors did not even attempt cutting," Vesalius wrote bitterly, "but those barbers, to whom the craft of surgery was delegated, were too unlearned to understand the writings of the professors of dissection. . . . They merely chop up the things which are to be shown on the instructions of the physician, who, having never put his hand to cutting, simply steers the boat from the commentary — and not without arrogance. And thus all things are taught wrongly, and days go by in silly disputations. Fewer facts are placed before the spectators in that tumult than a butcher could teach a doctor in his meat market." He concluded, grimly: "Aside from the eight muscles of the abdomen, badly mangled and in the wrong order, no one had ever shown a muscle to me, nor any bone, much less the succession of nerves, veins, and arteries."

Frustrated and confused, Vesalius decided to create his own map of the human body.

He raided charnel houses near the hospital, sometimes twice a day, to haul specimens back to his laboratory. The graves at the Cemetery of the Innocents, often open to the air, with bodies ground to the bone, provided perfectly preserved specimens for skeletal drawings. And walking by Montfaucon, the massive, three-tiered gibbet of Paris, Vesalius spotted prisoners' corpses hanging from the gallows. He would secretly make off with the freshly hung bodies, their muscles, viscera, and nerves intact enough for him to flay them open layer by layer and map the locations of the organs.

The intricate drawings that Vesalius produced over the next decade transformed human anatomy. Occasionally, he cut the brain into horizontal sections, like a melon sliced down from its tip, to create the kind of images that a modern computerized axial tomography (CAT) scan might produce. Other times he overlaid the blood vessels above the muscles or opened the muscles into flaps, like a series of anatomical windows that one could imagine passing through to reveal the surfaces and layers beneath them.

He might draw the human abdomen visualized from the bottom up, like the fifteenth-century Italian painter Andrea Mantegna's perspective of Christ's body in *The Lamentation of Christ,* and cut the picture into slices, in the way that a magnetic resonance imaging

(MRI) scan might visualize it. He collaborated with the painter and printmaker Jan van Kalkar to produce the most detailed and delicate drawings of human anatomy that existed. In 1543, he published his anatomical works in seven volumes entitled *De Humani Corporis Fabrica (The Fabric of the Human Body)*. The word *Fabric* in the title was a clue to its texture and purpose: this was the human body treated like physical material, not mystery; made of fabric, not spirit. It was part medical textbook, with nearly seven hundred illustrations, and part scientific treatise, with maps and diagrams that would lay the foundation for human anatomical studies for centuries to come.

Coincidentally, it was published the same year that the Polish astronomer Nicolaus Copernicus would put out his "anatomy of the heavens," the monumental book *The Revolutions (On the Revolutions of Heavenly Spheres)*, which featured a map of the heliocentric solar system that placed the Earth in orbit and the sun firmly at its center.

Vesalius had put human anatomy at the center of medicine.

But while anatomy, the study of structural elements of the human body, made radical advances, pathology — the study of human diseases, and their causes — had no such center. It was a mapless, dispersed universe.

There was no comparable book of pathology, and no common theory to explain diseases — neither revelations, nor *Revolutions*. During the sixteenth and seventeenth centuries, most diseases were attributed to miasmas: poisonous vapors emanating from sewage or contaminated air. The miasmas carried particles of decaying matter called miasmata that somehow entered the body and forced it to decay. (A disease such as malaria still carries that history, its name created by joining the Italian *mala* and *aria* to form "bad air.")

Early health reformers thus concentrated on sanitary reform and public hygiene to prevent and cure illness. They dug sewage systems to dispense waste, or opened ventilation ducts in homes and factories to prevent the contagious fog of miasmata from accumulating indoors. The theory seemed to be fogged by an indisputable logic. Many cities, undergoing rapid industrialization and unable to deal with the influx of wageworkers and their families, were malodorous arenas of smog and sewage — and disease seemed to track the worst-smelling, most populated areas. Resurgent waves of cholera and typhus stalked the poorer parts of London and its vicinities, such as the East End (now glistening with shops and restaurants selling high-end linen aprons and expensive bottles of single-distillery gin). Syphilis and tuberculosis were rampant. Childbirth was a terrifying event, with a distinct likelihood of ending not in birth but in death — of the infant, the mother, or both. In the wealthier parts of town, where the air was clean, and sewage adequately discarded, health prevailed, while the poor, who lived in miasma-filled areas, inevitably succumbed to illness. If cleanliness was the secret to health, then disease must be a condition of uncleanliness or contamination.

But while the notion of vaporous contamination and miasmata seemed to carry a vague ring of truth — and provided perfect justification to further segregate rich and poor neighborhoods in cities — the understanding of pathology was riddled with peculiar puzzles. Why, for instance, did a woman who gave birth in one part of an obstetrics clinic in Vienna, Austria, have nearly three times the rate of postpartum death compared with a woman who gave birth in the adjacent clinic? What caused infertility? Why would a perfectly healthy young man suddenly succumb to a disease that racked his joints with the most excruciating pain?

Throughout the eighteenth and nineteenth centuries, doctors and scientists searched for a systematic way to explain human diseases. But the best they could achieve was an unsatisfactory surplus of explanations that ultimately relied on gross anatomy: each disease was the dysfunction of an individual organ. The liver. The stomach. The spleen. Was there some deeper organizing principle that connected these organs, and their diffuse and mystifying disorders? Could one even think of human pathology in a systematic manner? Perhaps the answer was not to be found in visible anatomy but rather in microscopic anatomy. Indeed, by analogy, eighteenth-century chemists had already begun to discover that the properties of matter — the combustibility of hydrogen or the fluidity of water — arose from the emergent properties of invisible particles, molecules, and atoms that comprised them. Was biology perhaps similarly organized?

Rudolf Virchow was a mere eighteen years old when he enrolled at the Friedrich-Wilhelms Institute of medicine in Berlin. The institute was designed to train medical officers for the Prussian army, and its work ethic was duly martial: students were expected to attend sixty hours of classes a week by day and memorize facts by night. (At the Pépinière, the surgical institute, senior military doctors often surprised students with "attendance drills." If a student was found missing from class, the entire section was punished.) "It goes on like this every day without a stop from six in the morning until eleven at night, except on Sunday," he wrote glumly to his father, "[. . .] and in this process you get so tired that in the evening you find yourself yearning for a hard bed — on which, having slept in half lethargy, you wake up in the morning almost as tired as before." They ate a daily ration of meat, potatoes, and watery soup, and lived in small, isolated, self-contained chambers. Cells.

Virchow learned facts by rote. Anatomy was taught reasonably: the gross map of the body had been slowly perfected since Vesalius's time by generations of vivisectors and thousands of autopsies. But pathology and physiology lacked fundamental logic. Why organs worked, what they did, and why they fell into dysfunction was pure speculation — spun, as if by martial dictum, from conjecture into fact. Pathologists had long been divided into schools that argued for various sources of disease. There were the miasmists, who thought that diseases originated in contaminated vapors; the Galenists, who believed disease to be a pathological imbalance among four bodily fluids and semifluids referred to as "humors"; and the "psychists," who argued that illness was a manifestation of a frustrated mental process. By the time Virchow entered medicine, most of these theories had become confusing or defunct.

In 1843, Virchow finished his medical degree and joined Berlin's Charité hospital, where he began to work closely with Robert Friorep, a pathologist, microscopist, and the curator of pathological specimens at the hospital. Liberated from the intellectual rigidity

61

of his former institute, Virchow yearned to find a systematic way to understand human physiology and pathology. He delved into the history of pathology. "There is an urgent and far-reaching need to understand [microscopic pathology]," he wrote — but the discipline, he felt, had gone off track. Perhaps the microscopists were right: perhaps this systematic answer couldn't be found in the visible world. What if the failing heart or the cirrhotic liver were mere epiphenomena — emergent properties of a deeper underlying dysfunction invisible to the naked eye?

As he pored through the past, Virchow realized that there had been pioneers before him who had also visualized this invisible world. Since the late seventeenth century, researchers had found that plant and animal tissues were all built out of unitary living structures called cells. Might these cells sit at the heart of physiology and pathology? If so, where did they come from, and what did they do?

"True knowledge is to be aware of one's ignorance," Virchow had written in a letter to this father as a medical student in the 1830s. "[H]ow much and how painfully do I feel the gaps in my knowledge. It is for this reason that I do not stand still in any branch of science. . . . There is much that is uncertain and irresolute about me." In medical science, Virchow had found his footing at last, and it was as if an agitated pang in his soul had been soothed. "I am my own advisor," he wrote with newfound confidence in 1847. If cellular pathology did not exist, he would invent the field from scratch. Having acquired the maturity of a physician and a thorough knowledge of medical history, he could finally stand still and fill the gaps.

THE VISIBLE CELL
"FICTITIOUS STORIES ABOUT THE LITTLE ANIMALS"

---

*In the sum of the parts, there
are only the parts.
The world must be measured by eye.*
— Wallace Stevens

"The world must be measured by eye."

Modern genetics was launched by the practice of agriculture: the Moravian monk Gregor Mendel discovered genes by cross-pollinating peas with a paintbrush in his monastery garden in Brno. The Russian geneticist Nikolai Vavilov was inspired by crop selection. Even the English naturalist Charles Darwin had noted the extreme changes in animal forms created by selective breeding. Cell biology, too, was instigated by an unassuming, practical technology. Highbrow science was born from lowbrow tinkering.

In the case of cell biology, it was simply the art of seeing: the world measured, observed, and dissected by the eye. In the early seventeenth century, a Dutch father and son team of opticians, Hans and Zacharias Janssen, placed two magnifying lenses on the top and bottom of a tube and found that they could magnify an unseen world.* Microscopes with two lenses would be eventually termed "compound microscopes," while those with single lenses were called "simple"; both relied on centuries of innovation in glassblowing that had made its way from the Arabic and Greek worlds to the workshops of Italian and Dutch glassmakers. In the second century BC, the writer Aristophanes described "burning globes": spheres of glass sold as baubles in the market to concentrate and direct beams of light; if you looked carefully through a burning globe, you might see that same miniature universe magnified. Stretch that burning globe into an eye-sized lens, and you get the spectacle — invented supposedly by an Italian glassmaker, Amati, in the twelfth century. Mount it on a handle, and you have a magnifying glass.

The crucial innovation introduced by the Janssens was to fuse the art of glassblowing to the engineering of moving the pieces of glass on a mounted plate. By assembling one or two

*Some historians have argued that the Janssens' competitors, eyeglass makers Hans Lipperhey and Cornelis Drebbel, invented the compound microscope independently. The dates of all these inventions are in dispute, but likely occurred sometime between the 1590s and the 1620s.

perfectly lucid pieces of lens-shaped glass on metal plates or tubes, with systems of screws and cogs to slide them, scientists would soon find their way into an unseen, miniature world — a whole cosmos previously unknown to humans — the obverse of the macroscopic cosmos observable through a telescope.

A secretive Dutch trader had taught himself to visualize this invisible world. In the 1670s, Antonie van Leeuwenhoek, a cloth merchant in Delft, needed an instrument to examine the quality and integrity of thread. Seventeenth-century Netherlands was a booming nexus of cloth merchandising — silks, velvets, wool, linen, and cotton came in swaths and bundles from ports and colonies, and were traded via the Netherlands throughout continental Europe. Building on the Janssens' work, Leeuwenhoek built himself a simple microscope, with a single lens secured on a brass plate, and a tiny stage to mount the specimens. At first, he used it to grade the quality of cloth. But his interest in his handmade instrument soon turned compulsive: he focused his lens on whatever objects he could find.

On May 26, 1675, the city of Delft was inundated by a storm. Leeuwenhoek, then forty-two, gathered some of the water from the drains of his rooftop, let it stand for a day, and then put a droplet under one of his microscopes and held it up to the light. He was instantly entranced. No one he knew had seen anything like it. The water was roiling with dozens of kinds of tiny organisms — "animalcules," he called them. Telescopists had seen macroscopic worlds — the blue-tinged moon, gaseous Venus, ringed Saturn, red-flecked Mars — but no one had reported a marvelous cosmos of a living world in a raindrop. "This was to me among all the marvels that I have discovered in nature the most marvelous among them all," he wrote in 1676. "No greater pleasure has yet come to my eye than these spectacle of the thousands of living creatures in a drop of water."*

He wanted to look more, to build finer instruments to visualize this captivating new universe of living beings. And so Leeuwenhoek purchased the highest-quality beads and globules of Venetian glass and then ground and polished them laboriously into perfectly lenticular shapes (some of his lenses, we now know, were made by stretching a rod of glass into a thin needle on a live flame, breaking the end, and then letting the needle "bubble" into a lens-shaped globule). He mounted these lenses on thin metal plates, crafted of

*Leeuwenhoek had observed the presence of microscopic, single-celled organisms as early as 1674, but his letter to the Royal Society, dated 1676, had the most vivid descriptions of such organisms in standing rainwater.

brass, silver, or gold, each with an increasingly complex system of miniature armatures and screws to move parts of the instrument up and down and attain perfect focus. He made nearly five hundred such scopes, each a marvel of meticulous tinkering.

Were such creatures present in other samples of water as well? Leeuwenhoek entreated a man who was traveling to the seaside to bring him back a sample of ocean water in a "clean glass bottle." And again he found tiny single-celled organisms — "the body of a Mouse Color, clear towards the oval point" — swimming in the water. Eventually, in 1676, he recorded his findings and sent them to the most august scientific society of its time.

"In the year 1675," he wrote to the Royal Society of London, "I discover'd living creatures in Rain water, which had stood but few days in a new earthen pot. . . . When these animalcula or living atoms did move, they put forth two horns, continually moving themselves. . . . The rest of the body was roundish, sharpening a little toward the end, where they had a tail, nearly four times the length of the body."

By the time I'd finished writing that last paragraph, I was similarly obsessed: I wanted to look as well. Suspended in mid-pandemic limbo, I chose to build my own microscope, or at least the closest version that I could create. I ordered a metal plate and a turning knob, drilled a hole, and mounted the plate with the best tiny lens I could buy. It looked as much like a modern microscope as a bullock cart resembles, say, a spaceship. I trashed dozens of prototypes until I finally had one that might work. On a sunny afternoon, I placed a droplet of stagnant rainwater from a puddle on the mounting pin and held the apparatus up to the sunlight.

Nothing. Hazy forms, like shadows from a ghostly world, moved across my field of vision. A blur. Disappointed, I adjusted the focusing knob gently, as Leeuwenhoek would have. The anticipation made me feel each turn of the screw viscerally, as if the knob were, in fact, twisting its way up my spine. And suddenly I could see. The drop came sharply into view, and then a whole world within it. An amoeboid form flashed across the lens. There were branches of an organism I could not name. Then a spiral organism. A round, moving blob, surrounded by a halo of the most beautiful, the most tender filaments that I had ever seen. I could not stop seeing. *Cells.*

In 1677, Leeuwenhoek observed human spermatozoa, "a genital animalcule," in his semen as well as in a sample from a man with gonorrhea. He found them "moving like a snake or an eel swimming in water." Yet despite his ardor and productivity, the cloth merchant was notoriously reluctant to let observers or scientists examine his instruments.

The suspicion was reciprocal, as scientists were often just as dismissive of him. Henry Oldenburg, the secretary of the Royal Society, implored Leeuwenhoek to "acquaint us with his method of observing, that others may confirm such Observations as these," and to provide drawings and confirmatory data, for of the roughly two hundred letters that Leeuwenhoek sent to the society, only about half offered evidence or used scientific methods considered fit for publication. But Leeuwenhoek would provide only vague details of his instruments or his methods. As the science historian Steven Shapin wrote, Leeuwenhoek was "neither a philosopher, a medical man, nor a gentleman. He had been to no university, knew no Latin,

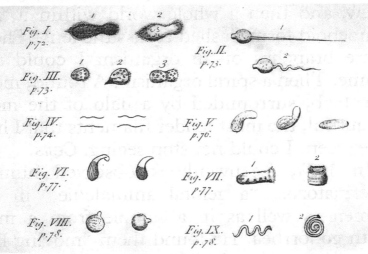

Some of the "animalcules" observed by Leeuwenhoek through his single-lensed microscope. Note the "Fig II" in the lower panel could either be a human spermatozoon or a bacterium with a flagellar tail.

French, or English. . . . His claims [about microscopic organisms existing abundantly in water] strained existing schemes of plausibility, and his identity was of no help in securing credibility for those claims."

He seemed, at times, to revel in the identity of the reticent, guarded amateur — a cloth merchant cajoling a friend to bring him ocean water in a glass bottle. The only way to believe this draper turned microscopist who was also turning biology's vision upside down, proposing a new universe of microscopic organisms, was to trust the testimony of a ragtag group of eight Delft residents that he'd assembled. They swore that the "swimming animals" could, indeed, be observed through his instruments. This was science by affidavit, and Leeuwenhoek's reputation suffered as a result. Suspicious and annoyed, he retreated deeper into a miniature world that seemed visible to him alone. "My work, which I've done for a long time," he wrote indignantly in 1716, "was not pursued in order to gain the praise I now enjoy, but chiefly from a craving after knowledge, which I notice resides in me more than most other men."

It was as if he had been swallowed by his own microscope, shortened in stature. Soon he was almost invisible, diminished, forgotten.

In 1665, nearly a decade before Leeuwenhoek published his letter describing animalcules in water, Robert Hooke, an English scientist and polymath, had also seen cells — although not live ones, and nowhere as diverse as Leeuwenhoek's animalcules. As a scientist, Hooke, perhaps, was quite the opposite of Leeuwenhoek. He had been educated at Wadham College in Oxford, and his intellect ranged widely, foraging through different worlds of science and consuming whole realms as he moved. Hooke was not just a physicist but also an architect, a mathematician, a telescopist, a scientific illustrator, and a microscopist.

Unlike most gentleman scientists of his era — men from wealthy families who could afford to ruminate about the natural sciences without ruing the next paycheck — Hooke came from an indigent English family. As a scholarship student at Oxford, he had survived by apprenticing with the eminent physicist Robert Boyle. By 1662, even as Boyle's subordinate, he had established himself as a powerfully independent thinker and found employment as the "curator of experiments" at the Royal Society.

Hooke's intelligence was phosphorescent and elastic, like a rubber band that glows as it stretches. He would enter disciplines and then expand and illuminate them as if by an internal light. He wrote extensively about mechanics, optics, and material sciences. In the aftermath of the Great Fire of London, which raged for five days in September

1666, destroying four-fifths of the city, Hooke helped the esteemed architect Christopher Wren survey and reconstruct buildings. He built a powerful new telescope through which he could visualize the surface of Mars, and he studied and classified fossils.

In the early 1660s, Hooke began a series of studies with microscopes. Unlike Antonie van Leeuwenhoek's inventions, these were compound microscopes. Two finely ground glass lenses were placed on two ends of a movable tube, which was then filled with water to enhance clarity. As he wrote: "If . . . an Object,

An illustration of the compound two-lensed microscope used by Robert Hooke. Note the brass tube, which held two lenses, a flame with a series of mirrors as a constant source of light, and the specimen mounted at the bottom of the tube.

plac'd very near, be look'd at through it, it will both magnifie and make some objects more distinct than any of the great Microscopes. But because these, though [exceedingly] easily made, are yet very troublesome to be us'd, because of their smallness, and the nearness of the Object; therefore to prevent both of these, and yet have only two refractions, I provided me a Tube of Brass."

In January 1665, Hooke published a book detailing his experiments and observations with microscopy, entitled *Micrographia: Or Some Physiological Descriptions of Minute Bodies Made with Magnifying Glasses with Observations and Inquiries Thereupon.* It was the sleeper hit of the year — "the most ingenious book I read in all my life," wrote the diarist Samuel Pepys. The drawings of minute bodies, never seen before at such magnification, chilled and fascinated his readers. Among the dozens of meticulous illustrations were an enormous rendition of a flea; a gargantuan picture of a louse, its grotesque, parasitic mouth enlarged to one eighth of a page; and the compound eye of a housefly, with its hundreds of lenses, resembling a miniature, multifaceted chandelier. "The Eyes of a Fly . . . appear almost like a Lattice," he wrote. Hooke got an ant drunk on brandy so that he could sketch a detailed image of its antlers. But tucked away among these images of parasites and pests was a relatively prosaic-seeming image that would quietly shake the roots of biology. It was a cross section of a plant stem — a thin slice of cork — that Hooke had placed under his scope.

Hooke found that the cork was not merely a flat, monotonous block of material. "I took a good clear piece of cork," he explained in

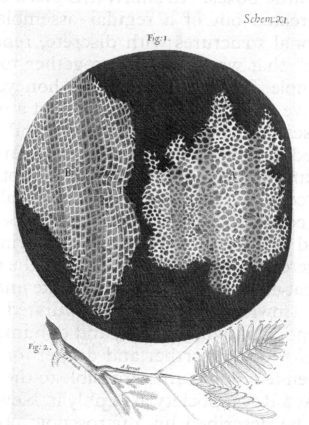

Robert Hooke's drawing of a section of a piece of cork from *Micrographia (1665)*. The book generated enormous and unlikely attention, becoming popular all over England for its magnified images of minute animals and plants. Hooke likely saw cell walls in this specimen, although later, he was also able to visualize actual cells in water.

Micrographia, "and with a pen-knife sharpened as keen as a razor, I cut a piece of it off,
and thereby left the surface of it exceeding
smooth, then examining it very diligently with
a microscope, methought I could perceive it
to be a little porous." These pores or cells
were not very deep but consisted of "a great
many little boxes." In short, this piece of cork
was created out of a regular assemblage of
polygonal structures with discrete, repetitive
"units" that were collected together to form
the whole. They resembled the honeycombs
in a hive — or the living quarters of a monk.

He searched for a name for them and finally
decided on *cells,* from *cella,* a Latin word
meaning "small room." (Hooke had not really
seen "cells" but rather the outlines of walls that
plant cells build around themselves; perhaps,
nestled within them was an actual living cell,
but there's no illustration that proves the point.)
"A great many little boxes," as Hooke imagined
them. Unwittingly, he had inaugurated a new
conception of living beings, and of humans.

Hooke looked further and deeper for small,
independent living units invisible to the naked
eye. At a Royal Society assembly in November
1677, he described his microscopic observations on rainwater. The society recorded his
observations:

The first experiment there exhibited was
the pepper-water, which had been made

with rain-water . . . put whole into it about nine or ten days before. In this Mr. Hooke had all week discovered great numbers of exceedingly small animals swimming to and fro. They appeared the bigness of a mite through a glass, that magnified about a hundred thousand times in bulk; and consequently, it was judged, that they were a hundred thousand times less than a mite. Their shape was to appearance like a very small clear bubble of an oval or egg form; and the biggest end of this egg-like bubble moved foremost. They were observed to have all manner of motions to and fro in the water; and by all, who saw them, they were verily believed to be animals; and that there could be no fallacy in the appearance.

In the decade that followed, Antonie van Leeuwenhoek, having learned of Hooke's earlier work, communicated with him, realizing that the animalcules that he had seen tumbling under his scopes might be analogous to the collection of living units — cells — that Hooke had witnessed in cork, or the organisms tumbling in pepper water. But there is an abject and disappointed tone in these letters, such as this one from November 1680: "As it has often reached my ear that I only tell fictitious stories about the little animals . . ." But in a prescient note, written in 1712, he continued, "Nay, we may yet carry it farther, and

77

discover in the smallest particle of this little world a new inexhausted fund of matter, capable of being spun out into another universe."

Hooke replied only sporadically, but he ensured that Leeuwenhoek's letters were translated and presented to the Royal Society. Yet, although Hooke had likely saved Leeuwenhoek's reputation for posterity, his own influence on cell biological thinking was still rather limited. As the historian of cell biology Henry Harris described it: "Hooke did not for a moment suggest these structures were the residual skeletons of the basic subunits of which all plants and animals were constituted. Nor would he necessarily have imagined, if he had thought of basic subunits at all, that they would have the size and shape of the cork cavities that he had observed." He had seen "the walls of a living cell in cork, but he misunderstood their function, and he clearly had no conception of what, in the living state, occupied the spaces within these walls."* A piece

*In 1671, the Royal Society received two additional communications: one from the Italian scientist Marcello Malpighi and one from Nehemiah Grew, a secretary of the society, both describing cellular forms in various tissues, particularly in plant material. However, even though both Leeuwenhoek and Hooke acknowledged their work, both Malpighi's and Grew's observations on cellular anatomy were largely ignored

of dead cork with pores in it; what more to make of his micrographic drawing? Why was a plant stem built in this manner? How did these "cells" arise? What was their function? Were they universal to all organisms? And what was the relevance of these living compartments to the normal body or to disease?

Hooke's interest in microscopy eventually dwindled. His peripatetic intellect needed to roam widely, and he returned to optics, mechanics, and physics. Indeed, Hooke's interest in virtually *everything* may have been his critical failing. The Royal Society's motto, *Nullius in verba,* translated loosely as "Take no one's word for evidence," was his personal mantra. He loped from one scientific discipline to the next, offering potent insights, believing no one's word, claiming dominion over critical parts of a science, but never asserting complete authority over any one subject. He had built himself on the model of the Aristotelian philosopher-scientist — an inquirer into all matters of the world, an adjudicator of all evidence — rather than the contemporary vision

---

in the seventeenth century. Grew's illustrations of cells in plant stems have been relegated to history, but Malpighi, who went on to explore the microscopic anatomy of animal tissues, lives on through many of the cellular structures that are named after him: among them, the Malpighian layer of the skin and the Malpighian cells in the kidney.

of the scientist as the authority on a single subject, and his reputation suffered as a result.

In 1687, Isaac Newton published *Philosophiae Naturalis Principia Mathematica (Mathematical Principles of Natural Philosophy),* a work so far-reaching in its depth and breadth that it shattered the past and shaped a new landscape for the future of science. Among its revelations: Newton's law of universal gravitation. Hooke, however, argued that *he* had formulated the laws of gravitation earlier, and that Newton had plagiarized his observations.

It was a preposterous claim. Indeed, Hooke, and several other physicists, had suggested that planetary bodies were attracted to the sun through invisible "forces," but none of the prior analysis had anywhere near the mathematical rigor or scientific depth that Newton brought to the puzzle in *Principia.* Hooke's and Newton's argument festered over decades, although Newton, arguably, had the last chortle. In one often-repeated story, likely apocryphal, the sole portrait of Hooke went missing when Newton oversaw the movement of the Royal Society to its new quarters in Crane Court in 1710, seven years after Robert Hooke's death — then neglected to commission a posthumous version. The pioneer of optics, the man who brought whole universes into view, is invisible to us. No definitive likeness, or portrait, of Hooke exists today.

THE UNIVERSAL CELL

"THE SMALLEST PARTICLE
OF THIS LITTLE WORLD"

*I could exceedingly plainly perceive it to be
all perforated and porous,
much like a Honey-comb, but that the
pores of it were not regular. . . .
[T]hese pores, or cells . . . were indeed
the first microscopical pores I ever saw.*
— Robert Hooke, 1665

*As soon as the microscope was applied to
the investigation
of the structure of plants, the great
simplicity of their structure . . .
necessarily attracted attention.*
— Theodor Schwann, 1847

In the history of biology, there are often valleys of silence that follow the peaks of monumental discoveries. Gregor Mendel's discovery of the gene in 1865 was followed by what one historian called "one of the strangest silences in the history of science": genes (or "factors" and "elements," as Mendel loosely called

81

them) were not mentioned for nearly forty years, before being rediscovered in the early 1900s. In 1720, the London physician Benjamin Marten reasoned that tuberculosis — phthisis, or consumption, as it was then called — was a contagious disease of the respiratory system, likely carried by microscopic organisms. He called the potential contagious elements "wonderfully minute living creatures," and *contagium vivum,* or "living contagion." (Note the word *living.*) Marten would almost have become the father of modern microbiology had he deepened his medical discoveries, but it would take nearly a century before microbiologists Robert Koch and Louis Pasteur each linked disease and putrefaction to the microbial cell.

Yet if you zoom into these valleys of history, they are far from silent or inactive. They represent extraordinarily fecund periods when scientists busily try to wrap their minds around the magnitude, generality, and explanatory power of a discovery. Is the discovery a universal, sweeping principle of living systems, or a particular idiosyncrasy of a chicken, an orchid or a frog? Does it explain previously inexplicable observations? Are there further levels of organization that lie beyond it?

Part of the explanation for this valley of silence has to do with the time required to develop instruments and model systems to answer these questions. Genetics had to await the work of the biologist Thomas Morgan, who explored the inheritance of traits in fruit flies in the 1920s to prove the physical existence of the gene, and, eventually, the birth of X-ray crystallography, the technique used to decipher the three-dimensional structure of molecules such as DNA, in the 1950s, to understand what genes look like in physical form. Atomic theory, first enunciated by John Dalton in the early 1800s, had to await the development of the cathode-ray tube in 1890, and the mathematical equations required to model quantum physics in the early twentieth century to elucidate the structure of the atom. Cell biology had to wait for centrifugation, biochemistry, and electron microscopy.

But perhaps an equal answer exists in the conceptual, or heuristic, changes required to switch from the description of an entity — a cell under a microscope, a gene as a unit of heredity — and move toward understanding its universality, organization, function, and behavior. Atomistic claims are the most audacious of all: the scientist is proposing a fundamental reorganization of a world into unitary entities. Atoms. Genes. Cells. You have to *think* of a cell in a different manner: not as an object under a lens but as a functional site for all physiological chemical reactions, as an organizing unit for all tissues, and as the unifying locus for physiology and pathology. You have to move from a continuous organization

83

of the biological world to a description that involves discontinuous, discrete, autonomous elements that unify that world. Metaphorically, we might say that you have to see past "flesh" (continuous, corporeal, and visible) to imagine "blood" (invisible, corpuscular, and discontinuous).

The period between 1690 and 1820 represents such a valley for cell biology. Since Hooke's discovery of cells — or cell walls, to be precise — in a piece of shaved cork, hordes of botanists and zoologists trained their microscopes on animal and plant specimens to understand their microscopic substructures. Right up to his death in 1723, Antonie van Leeuwenhoek kept looking through his microscopes and documenting elements — "living atoms," as he called them — of the invisible world. The thrill of that first encounter with this invisible world never left him (and, I suspect, will never leave me).

In the late seventeenth and early eighteenth centuries, microscopists such as Marcello Malpighi and Marie-Francois-Xavier Bichat realized that Leeuwenhoek's "living atoms" weren't necessarily, or exclusively, single celled; in more complex animals and plants, they organized themselves into tissues. The French anatomist Bichat, in particular, distinguished twenty-one (!) forms of elementary tissues out of which human organs were built.

84

Tragically, he died at thirty from tuberculosis. And although Bichat was occasionally wrong about the structures of some of these elementary tissues, he moved cell biology toward histology: the study of tissues, and of systems of cooperating cells.

More than any microscopist, though, it was François-Vincent Raspail who tried to build a theory of cellular *physiology* out of these early observations. Yes, there were cells, cells everywhere, he acknowledged — in plant and animal tissues — but to understand why they existed, they must be *doing* something.

Raspail believed in doing. A self-taught botanist, chemist, and microscopist, he was born in 1794 in Carpentras, Vaucluse, in southeastern France. He fashioned himself as an enlightened freethinker, refusing to take Catholic vows, and dedicated himself to opposing moral, cultural, academic, and political authority. He deigned to join scientific societies, finding them clubby and old-fashioned, and chose not to attend medical school, either. However, Raspail had no compunction about falling in with secret societies to liberate France during the Revolution of the 1830s, which led to his imprisonment from 1832 to early 1840. While incarcerated, he trained his fellow inmates in antisepsis, sanitation, and hygiene. In 1846 Raspail was tried again for an attempted coup of the government — as well as for dispensing medical advice to prisoners without a formal medical degree. He was exiled to Belgium, although even his prosecutors were apologetic about the trial: "The Court is today confronted with an eminent scientist, a man whom the medical profession would be honored to have as a member if he would but deign join it and accept a diploma from the Medical Faculty." Characteristically, Raspail refused.

And yet, amid all these political diversions, and with no formal training in biology, between 1825 and 1860, Raspail published more than fifty papers on an array of subjects, including botany, anatomy, forensics, cell biology, and antisepsis. Furthermore, reaching beyond his predecessors, he began to investigate the composition, function, and origin of cells.

What were cells made of? "Each cell selects from its surrounding milieu, taking only what it needs," he wrote in the late 1830s, presaging a century of cellular biochemistry. "Cells have various means of choice, resulting in different proportions of water, carbon, and bases which enter into the composition of their cell walls. It's easy to imagine that certain walls permit the passage of certain molecules," Raspail continued, anticipating both the idea of a selective, porous cell membrane, the autonomy of a cell, and the notion of the cell as a metabolic unit.

What did cells do? "A cell is [. . .] a kind of

laboratory," he posited. Pause for a moment to contemplate the scope of that thought. Using no more than basic assumptions about chemistry and cells, Raspail deduced that a cell performs chemical processes to make tissues and organs function. In other words, *it enables physiology*. He imagined the cell as the site for the reactions that sustain life. But biochemistry was in its infancy, and so the chemistry and reactions that occurred within this cellular "laboratory" were invisible to Raspail. He could describe it only as a theory. A hypothesis.

Finally, where did cells come from? Tucked away as an epigraph in an 1825 manuscript, Raspail formulated the Latin aphorism *Omnis cellula e cellula:* "From cells come cells." He did not investigate this further, having no tools or experimental methods to prove his point, but he had already changed the fundamental conception of what a cell is and does.

Unorthodox souls receive unorthodox rewards. Raspail, who thumbed his nose at both society and Societies, was never recognized by the scientific establishment in Europe. But one of Paris's longest boulevards, stretching from the Catacombs to Saint-Germaine, is named after him. As you walk along Boulevard Raspail, you pass behind the Institut Giacometti, with its sculptures of lonely, skeletal men on small islands of pedestals, lost in perennial thoughts. Every time I stroll along the street, I think of cell biology's reluctant, defiant pioneer (although Raspail, I should note, was not particularly skeletal). The concept of the cell as a laboratory for an organism's physiology returns to me: every cell growing in one of my incubators is a lab within a lab. The T cells that I had seen under the microscope in the Oxford lab were "Surveillance labs," swimming in fluid to find viral pathogens hiding within other cells. The sperm cells that Leeuwenhoek had seen under his scope were "Information labs," collecting hereditary information from a male, packaging it in DNA, and attaching a powerful swimming motor to deliver it to the egg cell for reproduction. The cell, as it were, is experimenting with physiology, passing molecules in and out, making chemicals and destroying chemicals. It is the laboratory of reactions that enables life.

At another time, or perhaps in another place, the discovery of unitary, autonomous forms of living matter — cells — might not have created much of a fuss in biology. But at its moment of inception, cell biology happened to collide with two of the most contentious debates about life that were raging through seventeenth- and eighteenth-century European science. Both may seem arcane today, but they represented two of the most serious challenges to cell theory. As the discipline emerged out of its shadowy caul in the 1830s, cell biologists would need to address both these challenges head-on before their discipline could mature.

The first of the debates arose from the vitalists: a group of biologists, chemists, philosophers, and theologians that was convinced that living beings could not possibly be built out of the same chemicals that were pervasive in the natural world. Theories of vitalism had existed since Aristotle's time, but the fusion of vitalism with late-eighteenth-century Romanticism produced an ecstatic depiction of Nature suffused with a special "organic" animus that was irreducible to any chemical or physical matter or force. French histologist Marie-François-Xavier Bichat in the 1790s and German physiologist Justus von Liebig in the early 1800s were both influential proponents. In 1795, the movement found its richest poetic voice in Samuel Taylor Coleridge, who imagined all of "animated nature" trembling into existence as this vital force flowed through it, just as a breeze might resonate through a harp and produce music that is irreducible to its mere notes. As Coleridge wrote: "And what if all of animated nature / Be but organic Harps diversely framed / That tremble into thought, as o'er them sweeps / Plastic and vast, one intellectual breeze / At once the Soul of each, and God of all."

There must be some divine mark that distinguishes the fluids and bodies of living beings, vitalists posited. The wind in the harp. Humans weren't merely an agglomeration of "lifeless," inorganic chemical reactions, and even if we were made of cells, the cells themselves must also possess these vital fluids. Vitalists had no problem with cells per se. As they saw it, a divine Creator fashioning the entire repertoire of biological organisms over the course of six days may well have chosen to construct them out of unitary blocks (how much easier it is to build an elephant and a millipede out of the same blocks, especially if you have a rush order with just six days to deliver the goods). Their concern was the *origin* of cells. Some vitalists alleged that cells were born within cells, like humans inside human wombs; others speculated that cells "crystallized" spontaneously out of vital fluid, like chemicals crystallizing in the inorganic world — except in this case, it was living matter that generated living matter. A natural corollary of vitalism was the notion of "spontaneous generation": that this vital fluid pervading all living systems was necessary and sufficient to create life out of its own. Including cells.

Counterposed against the vitalists was a small, embattled group of scientists who argued that living chemicals and natural chemicals were one and the same, and that living beings came from living beings — not spontaneously, but through birth and development. In the late 1830s, in Berlin, the German

90

scientist Robert Remak looked at frog embryos and chicken blood under a microscope. He was hoping to capture the birth of a cell, a particularly rare event in chicken blood, and so he waited. And waited. And then, late one evening, he saw it: under his scope, he watched a cell quiver, enlarge, bulge, and split in two, giving rise to "daughter" cells. Nothing less than a jolt of euphoria must have shot up Remak's spine, for he had found incontrovertible evidence that developing cells arose from the division of preexisting cells — *Omnis cellula e cellula,* as Raspail had so inconspicuously tucked into an epigraph.* But Remak's trailblazing observation was largely ignored, for as a Jew, he was denied full professorship at the university. (A century later, his grandson, a distinguished mathematician, would perish in the Nazi death camp at Auschwitz.)

Vitalists continued to claim that cells coalesced out of the vital fluids. To prove them wrong, non-vitalists would have to find a way to explain how cells arose — a challenge, vitalists believed, that could never be met.

*The German botanist Hugo von Mohl had also observed the birth of cells from cells in plant meristems. Both Remak and Virchow knew of von Mohl's work, which would later be extended by Theodor Boveri and Walther Flemming, among others, who described the stages of cell division in plant and sea urchin cells.

The second debate that simmered through the early 1800s was preformation: the idea that the human fetus was already fully formed, albeit miniaturized, when it first appeared in the womb following fertilization. Preformation had a long and colorful history: arising, possibly, in folklore and myth, it was adopted by early alchemists. In the mid-1500s, the Swiss alchemist-physician Paracelsus wrote about "transparent" mini-humans, "somewhat like a man," that were already present in a fetus. So convinced were some alchemists about the pre-existence of all human forms in a fetus that they thought that incubating a chicken egg with sperm would generate a fully formed human, since the instructions to build one from scratch were already present in the sperm. In 1694, the Dutch microscopist Nicolaas Hartsoeker published drawings that showed miniature mini-humans in sperm, replete with head, hands, and feet all tucked origami-like into the sperm's head, that he had apparently observed under the microscope. The riddle for cell biologists was to prove how a creature as complex as a human could emerge from a fertilized egg if there *wasn't* a preformed template already present inside it.

It was the demolition of the theories of vitalism and preformation — and their displacement by cell theory — that would firmly establish the new science and usher in the century of the cell.

■ ■ ■ ■

In the mid-1830s, while François-Vincent Raspail was languishing in prison and Rudolf Virchow was still a struggling medical student, a young German lawyer named Matthias Schleiden had become frustrated with his profession. He tried, unsuccessfully, to put a bullet through his head but missed his mark. Chastened by his failure to shoot himself, Schleiden decided to abandon law and turn to his true passion: botany.

He began to study plant tissues under the microscope. The instruments were now vastly more sophisticated than Hooke's or Leeuwenhoek's, with superior lenses and finely tuned knobs to achieve exquisitely sharp focal points. As a botanist, Schleiden was naturally curious about the nature of plant tissues, and when he looked at stems, leaves, roots, and petals, he found the same unitary structures that Hooke had discovered. Tissues, he wrote, were made of agglomerations of tiny, polygonal units: "an aggregate of fully individualized, independent, separate beings, the cells themselves."

Schleiden discussed his findings with zoologist Theodor Schwann, in whom he found a faithful, sympathetic partner and a lifelong collaborator. Schwann, too, had observed that animal tissues had a system of organization visible only by microscope: they were built, unit by unit, out of cells.

"A great portion of animal tissues originates from or consists of cells," Schwann wrote in an 1838 treatise. "The extraordinary diversity in the figure [of organs and tissues] is produced by different modes of junction of simple elementary structure, which, though they present different modifications, are yet essentially the same, namely, *cells*." Complex plant and animal tissues were built out of these living units — skyscrapers made from Lego blocks. They shared the same system of organization. The fiber-like cells of muscle might *look* completely unlike a red blood corpuscle or a liver cell, but "even though they present different modifications," Schwann wrote, they were the same: living units used to build living organisms. In every tissue that Schwann examined meticulously, there were smaller units of life: the "great many little boxes" that Hooke had described.

Neither Schwann nor Schleiden had found something new or unveiled an undiscovered property of the cell. It wasn't novelty that brought them fame; it was the sheer brazenness of their claim. They collated the work of their predecessors — Hooke, Leeuwenhoek, Raspail, Bichat, and a Dutch physician-scientist named Jan Swammerdam — and synthesized it into a radical proposition. What all of these researchers had uncovered, the two men realized, was not some special or idiosyncratic property of certain tissues, or in certain

94

animals and plants, but a sweeping and universal principle of biology.* What do cells *do*? Well, they build organisms. Gradually, as the reach and generality of their claim became evident, Schleiden and Schwann proposed the first two tenets of cell theory:

1. All living organisms are composed of one or more cells.

*As historians of science explore the early years of cell biology with greater depth, Schwann and Schleiden's claim to being the first elucidators of cell theory becomes more clouded. In particular, the groundbreaking work of the scientist Jan Purkinye (or Purkinjě , as he is more commonly known) and some of his students, including Gabriel Gustav Valentin, seems to have been relatively ignored. Part of this may have been the by-product of scientific nationalism: Schwann, Schleiden, and Virchow worked in Germany and wrote their works in German, considered the highbrow language of science, while Purkinjě and his pupils worked in Breslau. Though the city was formally Prussian territory, it was widely considered a backwater outpost mostly populated by Polish citizens. In 1834, having acquired a new microscope, Purkinjě and Valentin made several observations of tissues and sent the Institute of France an essay arguing that some animals and plants were built of unitary components. Unlike Schwann and Schleiden, though, they did not advocate for a sweeping, universal principle uniting all living matter.

2. The cell is the basic unit of structure and organization in organisms.

Yet even Schwann and Schleiden struggled to understand where cells came from. If animals and plants were built out of independent autonomous living units, then where did these units come from? After all, the cells in an animal must have arisen from the first fertilized cell, and then the cell must have expanded millions or billions of folds to build the organism. What, then, was the process by which cells arose and multiplied?

Both Schwann and Schleiden had been star-struck students of the physiologist Johannes Müller, the singularly dominant force in the rarified world of German biological sciences. A "conflicted, enigmatic, transitional figure," as the scholar of science Laura Otis described him to me, Müller was a scientist haunted by contradictions — caught, on one hand, between the vitalist belief that living matter had special properties, but also in constant search for unifying scientific principles that governed the living world.* Influenced

---

*Müller's internal conflict about vitalism was evident through much of his writing. In his introduction to his seminal book, *Elements of Physiology*, for instance, he reflected his uncertainty about life arising out of vital fluids versus "ordinary" inorganic material: "It must at any rate, however, be admitted, that the

by Müller's search for unifying principles, Schleiden turned to the question of the origin of cells. The only mechanism that Schleiden could find to explain his microscopic findings on cells — on how a great many organized units could arise inside tissues — was to relate them to a chemical process that also yields a great many organized units out of a chemical — i.e., *crystallization*. Cells must arise by some sort of crystallization process in a vital fluid, Müller had argued, and Schleiden could not bring himself to disagree.

Yet the more Schwann studied tissues under the microscope, the more he came close to overturning this theory. Where were these so-called living crystals? In his book *Microscopical Researches,* he wrote, "We have, indeed, compared the growth of organisms with crystallization . . . but [crystallization] involves very much that is uncertain and paradoxical." But as paradoxical as it was, even Schwann couldn't move beyond the orthodoxy of vitalism, despite what his eyes were telling him. He proposed: "The main outcome is that a common principle underlies the development . . . much as the same laws govern the formation of crystals." Try as

---

mode in which the ultimate elements are combined in organic bodies, as well as the energies by which the combination is effected, are very peculiar [and] they cannot be regenerated by any chemical process."

he might, he could not understand how a cell might be born.

In the fall of 1845, in Berlin, Rudolf Virchow, then twenty-four years old and barely out of medical school, was called to consult on a medical case involving a fifty-year-old woman with implacable fatigue, a swollen abdomen, and a palpable, enlarged spleen. He drew a drop of blood from her and examined it microscopically. The sample exhibited an extraordinarily elevated level of white cells. Virchow called it *leukocythemia* and then simply *leukemia* — an abundance of white blood cells in blood.

A similar case had been reported from Scotland. On a March evening in 1845, a Scottish physician named John Bennett was called urgently to see a twenty-eight-year-old slate layer who was dying mysteriously. "He is of dark complexion," Bennett wrote, "usually healthy and temperate; [he] states that twenty months ago, he was affected with great listlessness on exertion, which has continued to this time. In June last he noticed a tumor in the left side of his abdomen which has gradually increased in size till four months since, when it became stationary."

Over the next few weeks, Bennett's patient developed massive tumors in his armpits, groin, and neck. At the patient's autopsy a few weeks later, Bennett found that the slate layer's blood was full of white blood cells. Bennett proposed that the patient had succumbed to an infection. "The following case seems to me particularly valuable," Bennett wrote, "as it will serve to demonstrate the existence of true pus, formed universally within the vascular system." A spontaneous "suppuration of blood," he called it — again returning implicitly, as the vitalists did, to spontaneous generation. But there was no other sign of infection or inflammation anywhere, a fact that befuddled physicians.

The Scottish case was treated as a medical curiosity or anomaly, but Virchow, having seen a version of this peculiarity for himself, was intrigued. If Schwann, Schleiden, and Müller were correct about cells forming due to the crystallization of vital fluids, why — or how — had millions of white cells crystallized out of nowhere in blood?

The origin of these cells kept tugging at Virchow. He could not imagine tens of millions of white blood cells developing out of nothing and for no reason. Virchow began to wonder whether these millions of abnormal white cells might have come from other cells. The cells even *looked* like one another, with cancer cells being monotonous and similar in appearance. He knew of Hugo von Mohl's observations about plant cells, showing cells dividing to form two daughter cells. And there had been Remak, of course, waiting patiently

99

by his microscope until he had seen frog and chicken cells arise out of cells. But if that process could happen in plants and animals, then why not in human blood? And what if the leukemia that he'd seen resulted from a physiological process, cell division, gone berserk? What if dysfunctional cells begat dysfunctional cells, and it was this constant, dysregulated birth of cells that caused leukemia?

The themes that had run through Virchow's life had, thus far, been remarkably consistent: a restless, relentless inquisitiveness, and a skepticism about accepted wisdom and orthodox explanations. In 1848, this restlessness acquired a political dimension. Earlier that year, a famine had broken out in Silesia; then a lethal epidemic of typhus had swept through the region. Goaded by the press and public uproar, the Ministries of Interior and Education belatedly formed a commission to investigate the outbreak. Virchow, one of its appointees, traveled to Silesia, bordering the Polish edge of the Prussian Empire (and now largely in Poland). During his several weeks there, he began to realize that the pathology of the state had become the pathology of its citizens. Virchow wrote a furious article about the epidemic and published it in the medical journal he'd recently cofounded, the *Archives for Pathological Anatomy and Physiology and Clinical Medicine* (later retitled *Virchow's*

Archives). The cause of the disease, he concluded, was not just the infectious agent but also decades of political misrule and social neglect.

Virchow's accusatory writings did not escape notice. He was marked a liberal — a dangerous, pejorative term in Germany at the time — and placed on watch. When a fulminant populist revolution swept through Europe in 1848, Virchow took to the streets to protest. He founded yet another publication, *Medical Reform,* in which the confluence of his scientific and political beliefs could be used as a sledgehammer against the state.

These antics of a firebrand activist — even a man who had established himself as one of the most brilliant researchers of his generation — were not perceived favorably by the royalists. The rebellion was quashed, with brutal efficiency in some areas, and Virchow was ordered to resign from his post at the Charité hospital. He was forced to sign a document declaring that he would restrict his political writings and then was shuttled off, with quiet ignominy, to a quieter institute in Würzburg where he could be kept out of limelight — and out of trouble.

It's tempting to speculate on the thoughts spinning around Virchow's mind as he moved from buzzing, effervescent Berlin to dozy, suburban Würzburg. If the revolution of 1848

101

had carried a historical moral, it was that the state and its citizens were reciprocally connected. The sum was made of the parts, and the parts built the sum. Sickness or neglect in just one part could become a diffuse sickness of the whole, just as a single cancer cell could generate billions of malignant cells and precipitate a complex, deadly disease. "The body is a cell state in which every cell is a citizen," Virchow would write. "Disease is merely the conflict of the citizens of the state brought about by the action of external forces."

In Würzburg, isolated from the whirling carnival of Berlin and its politics, Virchow began to formulate two additional principles that would alter the future of cell biology and medicine. He accepted Schwann and Schleiden's belief that all tissues, from animals and plants, were made out of cells. But he could not bring himself to believe that cells arose spontaneously out of vital fluid.

But where did cells come from? As with Schwann and Schleiden, it was time for unifying maxims, and Virchow was ready. Every piece of evidence had been laid out by his predecessors; he merely had to pick up the crown and place it on his head. This feature of cells arising from cells wasn't true just for *some* cells and *some* tissues, Virchow stated, but for *all* cells. It was not an anomaly or an idiosyncrasy, but a universal property of life in plants, animals, and humans. The division of one cell gave rise to two, and two to four, and so forth. *"Omnis cellula e cellula,"* he wrote — "from cells come cells." Raspail's phrase had become Virchow's central tenet.

There was no coalescence of cells from vital fluid or from within the vital fluid of an individual cell. There was no "crystallization." These were fantasies: no one had observed any of these phenomena. By now, three generations of microscopists had been looking at cells. And what scientists *had* observed was the birth of cells from other cells — and that, too, by division. There was no need to invoke special chemicals or divine processes to describe the origin of a cell. A new cell came from the division of a prior cell; that's all there was to it. "There is no life," Virchow wrote, "except through direct succession."

Cells came from cells. And cellular physiology is the basis of normal physiology. If Virchow's first tenet concerned normal physiology, his second was its converse; it reconceived medicine's understanding of abnormality. What if dysfunctions in cells, he began to wonder, were responsible for malfunctions in the body? *What if all pathology were cellular pathology?* In late summer of 1856, Virchow was asked to return to Berlin — his youthful political sins forgiven in light of his growing scientific prominence. Shortly thereafter, he published his most influential book, *Cellular Pathology,* a series of lectures initially delivered at the Pathological Institute of Berlin in the spring of 1858.

Cellular Pathology detonated through the world of medicine. Generations of anatomic pathologists had thought about diseases as the breakdown of tissues, organs, and organ systems. Virchow argued that they had missed the real source of the illness. Since cells were the unit blocks of life and physiology, Virchow reasoned, then the pathological changes observed in diseased tissues and organs should be traced back to pathological changes in the units of the affected tissue — in other words, to cells. To understand pathology, doctors needed to look for essential disruptions not just in visible organs but in the organ's invisible units.*

*Virchow remembered the work of two Scottish surgeons from the previous century, John Hunter and his younger brother, William, as well as that of Giovanni Morgagni, a pathologist in Padua. Autopsies performed by the Hunters, Morgagni, and a host of other pathologists and surgeons had revealed that when a disease beset an organ, there were inevitable, telltale pathological findings in the anatomy of the affected tissue or organ. In tuberculosis, for instance, the lungs filled with white, pus-filled nodules called granulomas. In heart failure, the muscular walls of the heart were typically thin and weary looking. Virchow posited that in each of these cases, there

The words *function* and its opposite, *dys-function,* were crucial: normal cells "did" normal things to ensure the sanctity and physiology of the body. They were not just passive, structural features. They were actors, players, doers, workers, builders, creators — the central functionaries in physiology. And when these functions were somehow disrupted, the body fell into disease.

Once again it was the simplicity of the theory that bore its power and reach. To understand disease, a physician need not look for Galenic humors, psychic aberrations, internal hysterias, neuroses, or miasma — or for God's will, for that matter. The alterations in anatomy or the spectra of symptoms — the slate layer's fevers and lumps, followed by the abundance of white cells in his blood — could all be traced back to alterations and malfunctions in cells.

In essence, Virchow had refined Schwann and Schleiden's cell theory by adding three more crucial tenets to the two founding ones ("All living organisms are composed of one or more cells," and "The cell is the basic unit of structure and organization in organisms"):

---

was a *cellular* dysfunction that was the true cause of disease. At a microscopic level, a failing heart was the consequence of failing heart *cells*. The pus-filled granulomas of tuberculosis were the consequence of *cellular* reactions to the mycobacterial disease.

3. All cells come from other cells *(Omnis cellula e cellula)*.
4. Normal physiology is the function of cellular physiology.
5. Disease, the disruption of physiology, is the result of the disrupted physiology of the cell.

These five principles would form the pillars of cell biology and cellular medicine. They would revolutionize our understanding of the human body as assemblages of these units. They would complete the atomist conception of the human body, with the cell as its fundamental, "atomic" unit.

The final phase of Rudolf Virchow's life bore testimony not only to his theories about the cooperative social organization of the body — cells working with cells — but also a belief in the cooperative social organization of the state: humans working with humans. Immersed within a society that was becoming progressively racist and anti-Semitic, he argued vehemently for equality among citizens. Illness was an equalizer; medicine was not designed to discriminate. "Admission to a hospital must be open to every ill person who stands in need of it," he wrote, "whether he has money or not, whether he is Jewish or heathen."

In 1859, he was elected to the Berlin City Council (and eventually, in the 1880s, to the

Reichstag). And he began to witness in Germany the resurgence of a malignant form of radical nationalism that would eventually culminate in the Nazi state. The central myth of

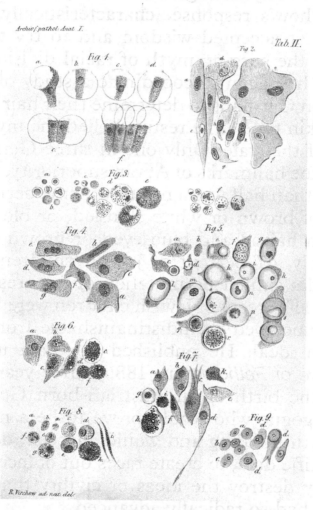

A drawing from Virchow's Archives, *ca. 1847, illustrating the organization of cells and tissues. Note the multiple abutting or adhering cell types in figure 2. Figure 3f shows the various cells found in blood, including ones with granules and many-lobed nuclei (neutrophils).*

what would later be termed "Aryan" racial superiority, and a nation dominated by "clean" *Volk* who were blond, blue eyed, and white skinned, was a pathology already sweeping malevolently through the country.

Virchow's response, characteristically, was to reject accepted wisdom and to try to restrain the surging myth of racial division: in 1876, he began to coordinate a study of 6.76 million Germans to determine their hair color and skin tone. The results belied the mythology of the state. Only one in three Germans bore the hallmarks of Aryan superiority, while more than half was a mixture: some permutation of brown or white skinned, or blond or brown haired and blue eyed or brown eyed. Notably, 47 percent of Jewish children possessed a similar permutation of features, and a full 11 percent of Jewish children were blond and blue eyed — indistinguishable from the Aryan ideal. He published the data in the *Archive of Pathology* in 1886, three years before the birth of an Austrian-born German demagogue who would prove to be a master of myth building and would succeed, despite scientific data, to create races out of faces and utterly destroy the ideas of civility that Virchow had so radically advanced.

Virchow spent much of his final years working on social reform and public health, with a focus on sewage systems and the hygiene of cities. He left a luminous (and voluminous) trail of papers, letters, lectures, and articles as he transitioned from doctor, to researcher, to anthropologist, to activist, to politician. But it is his earlier writings — the musings of a fiercely inquisitive young man searching for a cellular theory of disease — that remain the most timeless. In a prescient lecture in 1845, Virchow defined life, physiology, and embryonic development as the consequences of cellular activity: "Life is, in general, cell activity. Beginning with the use of the microscope in the study of the organic world, far-reaching studies [. . .] have shown that all plants and animals are, in the beginning [. . .] a cell within which other cells develop to give rise again to new cells that together, undergo transformation to new forms, and, finally . . . constitute the amazing organism."

In a letter replying to a scientist who had asked him about the basis of illness, he identified the cell as the locus of pathology: "Every disease depends on an alteration of a larger or smaller number of cellular units in the living body, every pathological disturbance, every therapeutic effect, finds its ultimate explanation only when it's possible to designate the specific living cellular elements involved."

These two paragraphs — the first proposing the cell as a unit of life and physiology, and the second proposing the cell as the unit locus of disease — are pinned on a board in my office. In thinking about cell biology, cellular therapies, and the building of new humans out of cells, I inevitably return to them. They are, as it were, the twin melodies that ring throughout this book.

In the winter of 2002, I saw one of the most complicated medical cases that I have ever encountered, at Massachusetts General Hospital in Boston, where I spent three years as a resident doctor. The patient, M.K., a young man of about twenty-three, was suffering from an unremitting, severe pneumonia that would not respond to antibiotics. Pale and shriveled, he lay curled up in bed under the sheets, moist from a fever that seemed to ebb and rise with no apparent pattern. His parents — Italian American second cousins, I learned — sat by his bedside, wearing dazed, blank expressions on their faces. His body had been so devastated by chronic infections that he looked like he could have been just twelve or thirteen years old. The junior residents and nurses could not find a vein in his hands to insert an intravenous line, and when I was asked to place a large-bore central IV line in his jugular vein to deliver antibiotics and fluids, it was as if my needle were piercing dried parchment. His skin had a papery, translucent quality that nearly crackled as I touched it.

M.K. had been diagnosed with a particular variant of severe combined immunodeficiency (acronymed SCID), in which both B cells (white cells that make antibodies) and T cells (that kill microbially infected cells and help mount an immune response) are dysfunctional. A grotesque English garden of microbes — some common, some exotic — grew out of his blood: *Streptococcus, Staphylococcus aureus, Staphylococcus epidermidis,* weird fungal varieties, and rare bacterial species whose names I could not even pronounce. It was as if his body had been transformed into a living petri dish for microbes.

But there were elements of the diagnosis that made no sense. When we checked M.K., his B cell count was lower than expected, but not alarmingly so. The same was true of his blood levels of antibodies, the immune system's foot soldiers against disease. MRI and CAT scans revealed no lumps or masses that might indicate malignant disease. Further blood tests were ordered. Throughout the entire ordeal, the patient's mother stayed with him, red eyed and silent, sleeping on a cot, and putting him to sleep with his head on her lap every night. Why on earth was this young man so terribly sick?

We were missing some sort of cellular dysfunction. Late on a freezing November evening, sitting at my desk in Boston — a dense mantle of snow had blocked the streets; to drive home was to risk skidding zigzag across the streets — I ticked off the possibilities in my head. What we needed was some systematic dissection, akin to an anatomical dissection, of cellular pathology; a cell atlas of this patient's body. I opened Virchow's textbook of lectures and reread some lines: "every animal presents itself as a sum of vital unities . . . a so-called individual always represents a social arrangement of parts." Every cell, he continued, "has its own special action, even though it derive[s] its stimulus from other parts."

"A social arrangement of parts." *"Every cell . . . derives its stimulus from another cell."* Imagine a cellular network — a social network — in which one node rents the whole net. Think of an actual fisherman's net with a tear in a crucial site. You might find a random sagging point on the edge of that fishing net and conclude that it was the source of the problem. But you would miss the actual source — the epicenter — of the puzzle. You would focus on the periphery, while it was the center that would not hold.

The next week, the pathologists brought his blood and bone marrow to the lab and began to dissect the subsets of cells, part by part, as if performing a surgical dissection — a "Virchovian analysis," as I might describe it. "Ignore the B cells," I urged them. "Let's go through the blood, cell by cell, and look for the center of the sagging net." The neutrophils that traffic through the blood and organs, searching for microbes, were normal, as were the macrophages, another white blood cell with a similar function. But when we counted and analyzed the T cells, the answer jumped out of the plots on the page: they were severely low in number, immature in their development, and virtually nonfunctional. At last, we had found the center of the broken net.

The abnormalities in all the other cells, and the breakdown in his immunity, were just *symptoms* of this T cell dysfunction: the collapse of T cells had cascaded through the entire immune system, causing the whole network to disintegrate. This young man did not have the variant of SCID that he had been initially diagnosed with. It was like a Rube Goldberg machine gone awry: a T cell problem had become a B cell problem, cascading further into a total collapse in immunity.

In the weeks that followed, we attempted a bone marrow transplant to restore M.K.'s immune function. Once the new marrow had engrafted, we reasoned, we might be able to transfuse him with functional T cells from the donor to restore his immunity. He survived the transplant. The marrow cells grew back, and his immunity was restored. The infections abated, and he began to grow again. Cellular normalcy had restored the normalcy of an organism. At a five-year follow-up, he was still free of infections, and with restored immune function, and with B cells and T cells communicating again.

Every time I think of M.K.'s case and my memories of him in his hospital room — his father trudging to Boston's North End in the snow to bring him his favorite Italian meatballs, only to find them untouched by the young man's bedside, and the mystified, befuddled doctors writing medical note after note with multiple question marks crisscrossing the pages — I also think of Rudolf Virchow, and the "new" pathology that he advanced. It isn't sufficient to locate a disease in an organ; it's necessary to understand which *cells* of the organ are responsible. An immune dysfunction might arise from a B cell problem, a T cell malfunction, or a glitch in any of the dozens of cell types that comprise the immune system. For instance, patients with AIDS are immunocompromised because the human immunodeficiency virus (HIV) kills a particular subset of cells — CD4 T cells — that help coordinate an immune response. Other immunodeficiencies arise because B cells cannot make antibodies. In each case, the superficial manifestations of the disease might overlap, but the diagnosis and treatment of the particular immune deficiency is impossible without pinpointing the cause. And pinpointing the cause involves dissecting an organ system in terms of the composition and function of its unit parts: cells. Or, as Virchow reminds me daily: "Every pathological disturbance, every therapeutic effect, finds its ultimate explanation only when it's possible to designate the specific living cellular elements involved."

To locate the heart of normal physiology, or of illness, one must look, first, at cells.

THE PATHOGENIC CELL
MICROBES, INFECTIONS, AND THE ANTIBIOTIC REVOLUTION

Like hermits, microbes need only be concerned with feeding themselves; neither coordination nor cooperation with others is necessary, though some microbes occasionally join forces. In contrast, cells in a multicellular organism, from the four cells in some algae to the thirty-seven trillion in a human, give up their independence to stick together tenaciously; they take on specialized functions, and they curtail their own reproduction for the greater good, growing only as much as they need to fulfill their functions. When they rebel, cancer can break out.

— Elizabeth Pennisi, *Science*, 2018

Rudolf Virchow was not the only scientist to arrive at an understanding of cells by contemplating pathology in the 1850s. The animalcules that Antonie van Leeuwenhoek had visualized tumbling under his microscope almost two centuries earlier were likely autonomous, single-celled living beings: microbes. And although the vast majority of such microbes are harmless, some have the capacity to invade human tissues and initiate inflammation, putrefaction, and deadly diseases. It was germ theory — that microbes are independent, living cells capable, in some cases, of causing human illnesses — that would first bring the cell (in this case, the microbial cell) into intimate contact with pathology and medicine.

The link between microbial cells and human disease emerged from the answer to a question that had preoccupied scientists and philosophers for centuries: What is the cause of rot? Rot wasn't just a scientific problem but also a theological one. In some Christian doctrines, the bodies of saints and kings were supposedly spared putrefaction, especially as they awaited the intermediate state between death, resurrection, and ascension to heaven. Yet when the decomposition rates of saints and sinners seemed no different, there was a theological reckoning to be had: whatever caused putrefaction apparently wasn't behaving according to the laws of God. It was hard, after all, to reconcile a divine corpse ascending to the heavens with decomposing pieces of it falling off like corporeal jetsam.

In 1668, Francesco Redi published a controversial article titled "Experiments on the Generation of Insects." Redi concluded that maggots, one of the first signs of putrefying material, could arise only from eggs laid by flies, not out of thin air, again challenging the vitalist doctrine of spontaneous generation. When Redi covered a piece of veal or fish with a thin muslin veil, allowing in air, but not flies, the flesh remained free of maggots, while the same flesh, exposed to air and flies, grew abundant maggots. Earlier theories of miasmata dictated that the decomposition of flesh arose from within, or from miasma floating in the air. Redi argued that this decomposition arose when living cells (maggot eggs) landed on flesh from the air. *"Omne vivum ex vivo,"* Redi wrote. "All life comes from life." The founder of experimental biology, as he is known, had, in short, enunciated the precursor to Virchow's much bolder statement. Life came from life, he proposed — just a step away from the idea that cells came from cells.

In Paris, in 1859, Louis Pasteur took Redi's experiments further. He placed boiled meat broth in a swan-neck bottle, a round flask with a vertical neck bent into an S shape, like a swan's neck. When Pasteur left the swan-neck bottle open to the air, the broth remained sterile: microbes in the air could not easily travel through the curve in the neck. But when he tipped the flask to expose the broth to the air, or cracked the swan neck, the broth grew out a turbid culture of microbes. Bacterial cells, Pasteur concluded, are carried in air and dust. Putrefaction, or rotting, was not caused by the inner decomposition of living creatures — or some visceral form of interior sin. Rather, decomposition only happened when these bacterial cells landed on the broth.

Decomposition and disease might have seemed superficially very different, but Pasteur made a crucial link between them. He studied infections in silkworms, the decomposition of wine, and the transmission of anthrax in animals. In all these cases, he determined that infections were caused not by the consequence of floating particles of miasma, or divine malefactions, but by invasions by microbes — single-celled organisms that entered other organisms and caused pathological changes and tissue degeneration.

In Wollstein, Germany, Robert Koch, a young, low-ranking but medically trained officer working in a makeshift lab, made the most radical advances to Pasteur's theory. In early 1876, he learned to isolate anthrax bacteria from infected cows and sheep and visualize them under a microscope. They were quivering, transparent, rod-shaped microbes, and, though fragile-looking, potentially lethal. The bacteria could also form round, dormant spores highly resistant to desiccation or heat; add water, or put them in a susceptible host, and the spores would spring back from dormancy to deadly life, generating the rod-shaped anthrax bacilli, multiplying rapidly, and unleashing the disease. Koch took a droplet of blood from an anthrax-infected cow, made a tiny slit in the tail of a mouse with a sterile wooden sliver, and waited. It remains an incredible, if inexplicable, lapse in the history of biology that, until 1876, no

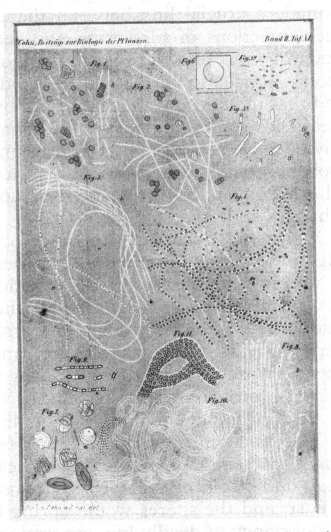

A drawing of the observations on Bacillus anthracis *made by Robert Koch. Note the long, strand-like forms of the bacillus, as well as the tiny circular spores.*

other scientist had experimented with transferring disease from one organism to the next in a systematic, scientific manner.

Anthrax bacteria secrete a poisonous toxin that kills cells. The mouse developed anthrax lesions. Its spleen was dark and swollen with dead cells, and its lungs pitted with similar black lesions. When Koch examined the spleen under a microscope, he found the same quivering, rod-shaped bacteria teeming inside it, surrounded by millions of dead mouse cells. He then repeated the experiment — inoculating a mouse, harvesting the spleen, and transferring a droplet to yet another mouse — a full twenty times. Each time, the recipient mouse developed anthrax. Koch's final experiment was his most ingenious: he created a sterile glass chamber and hung a droplet of liquid extracted from a dead ox's eye inside it. He injected a piece of spleen from an anthrax-infected mouse into the droplet. The same rod-shaped bacteria grew thickly in the fluid, making the clear droplet dusky with microbial cells.

The march of Koch's experiments was steady and systematic — almost drill-like in its precision. Louis Pasteur had assumed causality by association: the rotting of wine was associated with an overgrowth of bacteria; the putrefaction of broth was linked to its contact with microorganisms. Koch, in contrast, desired a more formal architecture of causality. First, he had isolated a microorganism from a diseased animal. Next, he'd demonstrated that introducing the pathogen into healthy animals caused the same disease. Then he'd re-isolated the microbe from inoculated animals, grown the organism again in pure form in a culture, and shown that it could re-create the disease. How could anyone puncture the logic? "In view of this fact," he wrote in his notes, "all doubts as to whether the *Bacillus anthracis* is really the cause and contagium of anthrax just fall silent."

In 1884, eight years after he had concluded his anthrax experiments, Koch used his observations and experiments to postulate four tenets of a theory of causality for a microbial disease. To claim that a microbe causes a particular illness (in the way that, say, *Streptococcus* causes pneumonia or *Bacillus anthracis* causes anthrax), he proposed the following: (1) the organism/microbial cell must be found in a diseased individual, not in a healthy individual; (2) the microbial cell must be isolated and cultured from the diseased individual; (3) the inoculation of a healthy individual with the cultured microbe must recapitulate the essential features of the disease; and (4) the microbe must be re-isolated from the inoculated individual and match the original microorganism.[*]

*Koch's postulates of disease causality, while applicable to most infectious diseases, do not take host factors into account, and are not easily applied

Koch's experiments, and his tenets, reso-
nated deeply through biology and medicine,
deeply influencing Pasteur's thinking as well.
And yet, despite their intellectual proximity
(or perhaps *because* of it), Koch and Pasteur
developed a seething rivalry over the next decades. (Of course, the Franco-Prussian War
of the 1870s did not particularly encourage
scientific comradery between the French and
Germans, either.) Pasteur's papers on anthrax, published nearly contemporaneously
with Koch's, used the French term *bacteridia**

---

to noninfectious diseases. Smoking, for instance,
causes lung cancer — but not all cigarette smokers
get lung cancer. You cannot isolate cigarette smoke
from a cancer patient and transmit the disease to
a second patient, although secondary smoking can
certainly cause lung cancer. HIV indubitably causes
AIDS — but not every individual exposed to HIV
gets infected by the virus and develops AIDS, as
host genetics affect the virus's capacity to enter cells.
You cannot isolate a microbe or cause from patients
with the neurodegenerative disease multiple sclerosis
(MS) or transfer the disease to another human. Over
time, epidemiologists would create a broader criteria
to determine causality for noninfectious diseases.
*The French scientist Casimir Davaine had also
observed rod-shaped microorganisms in anthrax
specimens and called them *bacteridia*. Pasteur's use
of the term was scientific homage to his French
colleague and a snub to the Germans.

with almost vindictive pleasure, referring to Koch's terminology in an obscure footnote: "*Bacillus anthracis* of the Germans." And Koch traded scientific insult for mockery: "Up to now, Pasteur's work on anthrax has led to nothing," he wrote in a French journal in 1882.

Boiled down to its essence, their scientific quarrel was rather minor: Pasteur insisted that, by repeated culture in the lab, bacterial cells could be weakened in their ability to cause disease, or, in the jargon of biology, attenuated. Pasteur intended to use attenuated anthrax as a vaccine: the weakened bacteria would strengthen immunity but not cause disease. According to Koch, however, attenuation was nonsense, as microbes were constant in their pathogenicity. In time, both men would be proved correct: some microbes can be attenuated, while others are difficult to temper. But taken together, Pasteur's and Koch's work pointed to a new direction in pathology. Autonomous, living microbial cells, they had demonstrated, caused both putrefaction and disease — at least in animal models and in cultures.

But what was the association between putrefaction caused by microbial cells and *human* disease? The first hint of a potential link came from a Hungarian obstetrician, Ignaz Semmelweis, who worked as an assistant in a

Viennese maternity hospital in the late 1840s. The clinic was divided into two wards: the first clinic and the second clinic. Childbirth, in the nineteenth century, was almost as much life threatening as it was life giving. Infections — puerperal fever, or, more colloquially, "childbed fever" — caused postpartum death rates that ranged from 5 percent to 10 percent for mothers. Semmelweis noted a peculiar pattern: compared with the second clinic, the first clinic had a significantly higher rate of maternal mortality from childbed fever. News of this discrepancy, spread via gossip and rumor throughout Vienna, was an open secret. Pregnant women would beg, cajole, or manipulate their way to be admitted to the second clinic. Some women, wisely, even opted for so-called street births — outside the clinics — reasoning that the first clinic was a far more dangerous place to have a baby than the street.

"What protected those who delivered outside the clinic from these destructive unknown endemic influences?" Semmelweis mused. It was a rare opportunity to perform a "natural" experiment: two women, with the same condition, entered through two doors of the same hospital. One emerged with a healthy newborn; the other was dispatched to the morgue. Why? Like a detective eliminating potential culprits, Semmelweis made a mental list of causes, crossing them off one by one. It wasn't overcrowding, or the women's ages, or the lack of ventilation, or the length of their labor, or how close the beds were to one another.

In 1847, Semmelweis's colleague Dr. Jacob Kolletschka cut himself with a scalpel while performing an autopsy. He was soon febrile and septic; Semmelweis could hardly help but notice that Kolletschka's symptoms mirrored those of the women with childbed fever. Here, then, was a potential answer: the first clinic was run by surgeons and medical students who shuttled casually between the pathology department and the maternity ward — from performing cadaver dissections and autopsies straight to delivering babies. In contrast, the second clinic was run by midwives, who had no contact with cadavers and never performed autopsies. Semmelweis wondered if the students and surgeons, who routinely examined women without gloves, were transferring some material substance — "cadaverous material," he called it — from the decomposing cadavers into a pregnant woman's body.

He insisted that the students and the surgeons wash their hands with chlorine and water before entering the maternity wards. Semmelweis kept careful records of the deaths in the two clinics. The impact was astonishing, with the mortality rate in the first clinic declining by 90 percent. In April 1847, the mortality rate had been nearly 20 percent:

one in five women died of childbed fever. By August, after rigorous hand washing had been instituted, the mortality among the new mothers had declined to 2 percent.

As stunning as the results were, Semmelweis had no explanation that he could visualize. Was it blood? A fluid? A particle? Senior surgeons in Vienna didn't believe in germ theory and had no interest in a junior assistant's insistence that they wash their hands between the clinics. Semmelweis was harassed and ridiculed, passed over for a promotion, and eventually dismissed from the hospital. The idea that childbed fever was, in fact, a "doctor's plague" — an iatrogenic, physician-induced disease — could hardly sit well with the professors of Vienna. He wrote increasingly frustrated and accusatory letters to obstetricians and surgeons all over Europe, all of whom dismissed Semmelweis as a crank. He eventually packed off to the backwaters of Budapest, only to suffer a mental breakdown. He was admitted to an asylum where the guards beat him, leaving him with broken bones and a gangrenous foot. Ignaz Semmelweis died in 1865, most likely of sepsis caused by the injuries; consumed, possibly, by germs — the very "material" substance that he had tried to identify as a cause of infections.

In the 1850s, not long after Semmelweis had been dismissed to Budapest, an English physician named John Snow was tracking the course of a raging cholera epidemic in the Soho area of London. Snow not only viewed diseases in terms of symptoms and treatments but also regarded geography and transmission as contributing factors: he suspected instinctively that the epidemic was moving in particular patterns throughout particular districts and landscapes, which might provide a clue to its cause. Snow enlisted local residents to pinpoint the time and location of each case. Then he began to track the infection backward in time and space, as if watching a film in reverse — finding origins, sources, and causes.

The source, Snow concluded, was not invisible miasmata floating in the air but rather water from a specific pump on Broad Street, from which the epidemic seemed to spread — or rather, flow — outward, like the ripples from a stone tossed into a pond. When Snow later drew a map of the epidemic, marking every case of death with a bar, the bars surrounded that pump. (A later map, drawn in the 1960s, with dots marking the cases, is now more familiar to most epidemiologists.) "I found that nearly all the deaths had taken place within a short distance of the [Broad Street] pump," he wrote. "There were only ten deaths in houses situated decidedly nearer to another street pump. In five of these cases, the families of the deceased persons informed me that they always sent to the pump in Broad Street, as they preferred the water to that of the pumps which were nearer. In three other cases, the deceased were children who went to school near the pump in Broad Street."

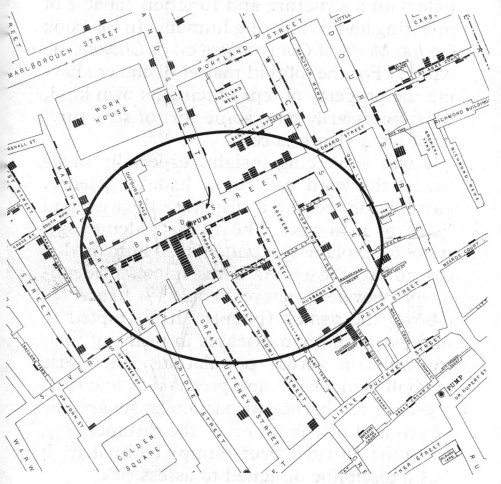

One of John Snow's original drawings from the 1850s of the cases of cholera surrounding the Broad Street pump in London. The arrow shows the pump's location (author's addition) and the number of cases per household is marked by Snow as the height of the bars (note the circle around the area that Snow identified, author's addition).

But what substance was being carried by that contaminated source? By 1855, Snow had begun to examine the water under the microscope. He was convinced that it was something capable of reproduction; some particle with a structure and function capable of infecting and reinfecting humans. In his book *On the Mode of Communication of Cholera,* he wrote: "For the morbid matter of cholera having the property of reproducing its own kind, must necessarily have some sort of structure, most likely that of a cell."

It was a piercing insight, especially in its use of the word *cell.* Snow had, in essence, partially united three disparate theories and fields of medicine. The first, epidemiology, tried to explain the *patterns* of human disease in aggregate. As a discipline, epidemiology "hovered" above people — hence *epi* (above) the *demos* (people). It attempted to understand human diseases in terms of their transmission across populations, their rise and fall in incidence and prevalence, and their presence or absence in particular geographic or physical distributions — the distance, say, from the Broad Street pump. Ultimately, it was a discipline designed to assess risk.

But Snow had also edged a theory of epidemiology toward a theory of pathology, from inferred risk to a material substance. Some *thing* — a cell, no less — in that water was the cause of the infection. The geography, or the map of illness, was just a clue to its root cause; it was the sign of a physical substance moving through time and space, precipitating disease.

Germ theory, the second field, still in its infancy, advanced the notion that infectious diseases were caused by microscopic organisms that invaded the body and disrupted its physiology.

The third was the most audacious of all: cell theory, which held that the invisible microbe causing the disease was, in fact, an independent, living *organism* — a cell — that had contaminated the water. Snow had not seen the cholera bacillus under his microscope. But he had instinctively grasped that the causal elements had to be capable of reproducing in the body, reentering the sewage, and restarting an infectious cycle. The infectious units had to be living entities capable of copying themselves.

It occurs to me, as I write this, how much this framework — germs, cells, risk — still scaffolds the diagnostic art in medicine. Each time I see a patient, I realize, I am probing the cause of his or her disease through three elemental questions. Is it an exogenous agent, such as a bacterium or virus? Is there an endogenous disturbance of cellular physiology? Is it the consequence of a particular risk, be it exposure to some pathogen, a family history, or an environmental toxin?

Years ago, as a young oncologist, I met a previously healthy professor who was suddenly stricken with a relapsing fatigue that was so intense that there were days that he could not lift his limbs to climb out of bed. Over multiple visits to multiple specialists, he had been diagnosed with every conceivable illness: chronic fatigue syndrome, lupus, depression, a psychosomatic syndrome, an occult cancer. The garbled list went on and on.

Every test had come back negative, except for a blood test diagnosing him with chronic anemia. But a low red cell count is a symptom of disease, not a cause. Meanwhile, the weakness advanced relentlessly. A strange rash broke out on his back — another symptom without a cause. A few days later, the man was back in the clinic, with no diagnosis. An X-ray revealed a filmy veil of fluid accumulated in the two-layer pleural sac that surrounds the lungs. I was now sure about the diagnosis. It was cancer, of course, that had been hiding all the while. I inserted a syringe between two ribs, withdrew a small amount of fluid, and sent it to the pathology lab. I was convinced that cancer cells would be found in the fluid, and we would clinch the case.

Yet before sending the patient for further scans and biopsies, I had niggling doubts. My instincts rose against the certitude of my own diagnosis, and so I sent him to the best internist I knew (an odd, otherworldly man who seemed, at times, almost like an anachronistic physician from another century. "Don't forget to smell the patient," this Proust among doctors had advised me once, then went on to list the number of illnesses that could be diagnosed by smell alone; I stood in his office, listening and learning, flummoxed).

A day later, the internist called me.

Had I asked the patient about risks?

I mumbled a vague yes but realized shamefacedly that I had focused my assessment entirely on cancer.

Did I know that my patient had spent the first three years of his life in India? the internist asked. Or that he had traveled there several times since then? It hadn't occurred to me to ask. The man told me that he had lived in Belmont, Massachusetts, since childhood, but I didn't probe further and ask him where he'd been born or when he had moved to the States.

"And have you sent the lung fluid to the bacteriology lab?" wise Dr. Proust asked.

By now, my face was flushed.

"Why?"

"Because it's reactivated tuberculosis, of course."

Thankfully, the lab had reserved half the fluid that I had sent. In three weeks, it grew out *Mycobacterium tuberculosis,* the causal agent of TB. The man was treated with appropriate antibiotics and recovered slowly. In a few months, all his symptoms had vanished.

The whole episode was a lesson in humility. To this day, when I see a patient with an undiagnosable disease, I mumble quietly under my breath, recalling John Snow and my internist friend who liked to sniff patients. *Germs. Cells. Risk.*

The medical application of germ theory was transformative. In Glasgow, Scotland, in 1864, just a few years after Louis Pasteur had completed his experiments on putrefaction (and more than a decade before Robert Koch would prove conclusively that microbes caused disease in animal models), a young surgeon named Joseph Lister chanced upon Pasteur's papers, *Recherches sur la putréfaction*. In an inspired leap, he made a connection between the putrefaction that Pasteur had witnessed in his swan-neck flask and the surgical infections that he saw in his wards. Even in ancient India and Egypt, physicians cleaned their instruments by boiling them. Yet in Lister's time, surgeons paid little attention to the possibility of contamination by microbes. Surgery was an unfathomably unsanitary practice, as if designed intentionally to defy any historical knowledge of hygiene. For example, a pus-covered surgical probe removed from one patient's wound would then be inserted, unsterilized, into another person's body. In fact, surgeons used the phrase "laudable pus" because they thought that the presence of pus was part of the healing process. If a scalpel fell on a blood- and pus-smeared operating room floor, the surgeon would simply wipe it clean on his equally contaminated apron and sanguinely proceed to use the same tool on his next patient.

Lister decided to boil his tools in a solution that would kill the germs that he was convinced were causing the infection. But what solution? Carbolic acid, he knew, was used to remove the rancid stench from sewage and wastewater; if so, it was likely killing germs that created the miasmas surrounding sewage, he thought. And so, taking one inspired leap after another, he began boiling his surgical tools in carbolic acid. The rate of postsurgical infections plummeted. Wounds healed rapidly, and septic shock — the dreaded bane of every surgical procedure — was suddenly diminished in patients. At first, surgeons resisted Lister's theory, but the data became more and more incontrovertible. Like Semmelweis, Lister had turned germ theory into medical practice.

In a little less than a century, from the 1860s to the 1950s, sterility, hygiene, and antisepsis, the only established methods to prevent infections, would be vastly augmented by the invention of antibiotic drugs that killed microbial cells. In 1910, the first among these, an arsenic derivative known as arsphenamine, was discovered by Drs. Paul Ehrlich and Sahachiro Hata, who found it could kill syphilis-causing microbes. Soon there was a seemingly limitless bounty of antibiotics, among them penicillin, an antibacterial chemical secreted by a fungus that was discovered in molding plates by Alexander Fleming in 1928, and the anti-TB drug streptomycin, isolated from bacteria in clods of dirt by Albert Schatz and Selman Waksman in 1943.

Antibiotics, medicines that changed the face of medicine, generally work because they attack something that distinguishes a microbial cell from the host cell. Penicillin kills the bacterial enzymes that synthesize the cell wall, resulting in bacteria with "holes" in their walls. Human cells don't possess these particular kinds of cell walls, thereby making penicillin a magic bullet against bacterial species that rely on the integrity of their cell walls.

Every potent antibiotic — doxycycline, rifampin, levofloxacin — recognizes some molecular component of human cells that is different from a bacterial cell. In this sense, every antibiotic is a "cellular medicine" — a drug that relies on the distinctions between a microbial cell and a human cell. The more we learn about cell biology, the subtler distinctions we uncover, and the more potent antimicrobials we can learn to create.

■■■■

Before we leave antibiotics and the microbial world, let's dwell for a moment on distinctions. Every cell on Earth — which is to say every unit of every living being — belongs to one of three entirely distinctive domains, or branches, of living organisms. The first branch comprises bacteria: single-celled organisms that are surrounded by a cell membrane, lack particular cellular structures found in animal and plant cells, and possess other structures that are unique to them. (It is precisely these differences that are the basis for the specificity of the antibacterial drugs mentioned above.)

Bacteria are disturbingly, ferociously, uncannily successful. They dominate the cellular world. We think of them as pathogens — bartonella, pneumococcus, salmonella — because a few of them cause disease. But our skin, our guts, and our mouths are teeming with several billion bacteria that cause no disease whatsoever. (Science writer Ed Yong's seminal book *I Contain Multitudes: The Microbes Within Us and a Grander View of Life* provides a panoramic view of our intimate and generally symbiotic pact with bacteria.) In fact, bacteria are either harmless or actually helpful. In the gut, they aid digestion. On the skin, some researchers suspect, they inhibit colonization by much more harmful microbes. An infectious disease specialist once told me that humans were just "nice-looking luggage to carry bacteria around the world." He might have been right.

The abundance and resilience of bacteria stagger the mind. Some live in oceanic thermal vents where the water reaches near boiling temperature; they could easily thrive inside a steaming kettle. Some prosper within stomach acid. Yet others live, with seemingly equal ease, in the coldest places on earth, where the land freezes into packed, impenetrable tundra for ten months of the year. They are autonomous, mobile, communicative, and reproductive. They have powerful mechanisms of homeostasis that maintain their internal milieu. They are perfectly self-sufficient hermits, but can also cooperate to share resources.

We — you and me — inhabit a second branch, or domain, called eukaryotes. The word *eukaryote* is a technicality: it refers to the idea that our cells, and the cells of animals, fungi, and plants, contain a special structure called a nucleus (*karyon,* or "kernel," in Greek). This nucleus, as we will soon learn, is a storage site for chromosomes. Bacteria lack nuclei and are called prokaryotes — that is, "before nuclei." Compared with bacteria, we are fragile, feeble, finicky beings capable of inhabiting vastly more limited environments and restricted ecological niches.

And now the third branch: archaea. It may be the singularly most startling fact in the history of taxonomy that this full branch of living beings remained undiscovered until about fifty years ago. In the mid-1970s, Carl Woese, a professor of biology at the University of Illinois at Urbana-Champaign, used comparative genetics — the comparison of genes across various organisms — to deduce that we had misclassified not just some arcane microbe but rather an *entire domain* of life. For decades, Woese fought a spirited but lonely, bitter war that left him ragged at the edges. Taxonomy wasn't just missing the point, he insisted, it was missing a whole living domain. Archaea, Woese argued, were not "almost like" bacteria or "almost like" eukaryotes. ("Almost like" is the taxonomist's version of a parent saying to a child, "Go away, you're bothering me.")

Many prominent biologists ridiculed or simply ignored Woese's work. In 1998, Ernst Mayr, the biologist, wrote an essay on Woese drenched with teacherly condescension ("Evolution is an affair of phenotypes . . . not genes"), getting the story exactly wrong. It wasn't evolution that Woese was contesting, it was taxonomy — which is precisely the question of genes. A bat and a bird may have nearly the same physical characteristics, or phenotypes. It's the difference in their *genes* that gives away the secret: they belong to different taxa. The journal *Science* described Woese as a "scarred revolutionary." But decades later, we have largely accepted, validated, and vindicated his theory, so that archaea are now classified as a distinct, third domain of living creatures.

Superficially, archaea look like bacteria, for the most part. They are tiny and lack some of the structures associated with animal and plant cells. But they are indisputably different from bacteria, or from plant, animal, and fungal cells. In fact, we still know relatively little about them. As Nick Lane, the evolutionary biologist at University College London, puts it in his book *The Vital Question: Energy, Evolution, and the Origins of Complex Life,* they're the Cheshire cats of the living kingdom: absolutely essential to the full story, yet asserting "their presence only by their absence" — in other words, by the fact that they lack the defining features of the other two domains, partly because we've ignored studying them until recently.

This division of life into its principal domains returns us to yet another essential distinction in the trajectory of our story of cells. There are, in fact, two intersecting stories here. The first is the history of cell biology. We have journeyed through vast territory in this first story: from Leeuwenhoek to Hooke visualizing cells in the late 1600s, to the discovery of tissues and organs two centuries later; and from the discovery of bacteria as a cause of putrefaction and disease by Pasteur and Koch to Ehrlich's synthesis of the first antibiotics in 1910. We've moved from the origins of cellular physiology — Raspail's luminously prescient "Every cell is [. . .] a kind of laboratory" — to the young Virchow's brazen proposition that the cell is the locus of both normal physiology and pathology.

But that is the history of cell biology, not the history of the cell. The cell's history dwarfs that of cell biology by billennia. The first cells — the simplest, most primitive of our ancestors — arose on Earth some 3.5 to 4 billion years ago, about 700 million years after the birth of the Earth. (That is a remarkably short period, if you think about it; only about a fifth of the history of the Earth had passed before living beings were already reproducing on it.) How did that "first cell" arise? What did it look like? Evolutionary biologists have grappled with these questions for decades. The simplest cell — call it a "protocell" — had to possess a genetic information system that could reproduce itself. The cell's original replication system was almost certainly made of a strand-like molecule called ribonucleic acid, or RNA. Indeed, in lab experiments, simple chemicals, placed in conditions that resemble the atmospheric conditions on primitive Earth, and trapped within layers of clay, can give rise to precursors of RNA and even strands of RNA molecules.

But the transition from an RNA strand to a *self-replicating* RNA molecule is no small evolutionary feat. Most likely, *two* such molecules were needed — one to act as the template (i.e., the information carrier) and the other to make a copy of the template (i.e., a duplicator).

When these two RNA molecules — template and duplicator — met each other, it was, perhaps, the most important and explosive evolutionary love affair in the history of our living planet. But the lovers had to avoid separation; if the two strands of RNA were to float away from each other, there would be no duplication and, by extension, no cellular life. And so some sort of structure — a spherical membrane — was likely needed to confine these components.

These three components (a membrane, an RNA information carrier, and a duplicator) might have defined the first cell. If a self-replicating RNA system were bound by a spherical membrane, it would make more RNA copies within the confines of the sphere and grow in size by enlarging the membrane.

At some point, biologists believe, the membrane-bound spheroid would split into two, each carrying the RNA duplicating system. (In lab experiments, Jack Szostak and his colleagues have shown that simple spheroidal structures, bound by membranes formed by fat molecules, can absorb more fat molecules, grow, and eventually split into two.) And from that point on, the protocell would launch its long evolutionary march toward the progenitor of the modern cell. Evolution would select more and more complex features of the cell, eventually replacing RNA with DNA as the information carrier.

Bacteria evolved out of that simple progenitor about 3 billion years ago, and they continue to evolve today.* Archaea are probably at least as old as bacteria, arising around the same time — although the precise date is still noisily and vigorously debated — and also continue to exist and to evolve to this day.

But what about nonbacterial, nonarchaeal cells — in other words, *our* cells? About 2 billion years ago (once again the exact date is a matter of debate), evolution took a strange and inexplicable turn. That is when a cell that is the common ancestor of human cells, plant

*This book will not cover this entirely third group of cellular beings, archaea, except for a brief mention. Some biologists argue that features of the modern cell may be explained by some kind of cooperative assembly between bacteria and archaea — but there's debate about the extent to which the evolution of archaea, or some common ancestor, contributes to the evolution of nucleus-bearing cells — that is to say, *our* modern cells. These arguments are essential to evolutionary biologists exploring the early history of life but lie outside the scope of this book.

cells, fungi cells, and amoebal cells appeared on Earth. "This ancestor," as Lane puts it, "was recognisably a 'modern' cell, with an exquisite internal structure and unprecedented molecular dynamism, all driven by sophisticated nanomachines encoded by thousands of new genes that are largely unknown in bacteria." New evidence suggests that this "modern" eukaryotic cell arose *within* archaea. In other words, life has only *two* principal domains — bacteria and archaea — and eukaryotes ("our" cells) represent a relatively recent sub-branch of archaea. We are, perhaps, life-come-lately, the sawdust left over from the carvings of the two main domains of life.

In the parts and chapters that follow, we are about to meet this modern cell. We will encounter its elaborate internal anatomy. We'll discover its "unprecedented molecular dynamism" that enables reproduction and development. We will understand how organized *systems* of cells — multicellular systems with specialized forms and functions — enable the formation and function of organs and organ systems, maintain the constancy of the body, repair broken ankles, and combat decay. And we will contemplate a future where we use this knowledge to develop medicines that attempt to build functional parts of new humans to ameliorate or cure diseases.

But there is one question that we will not and, perhaps, cannot answer. The origin of the modern cell is an evolutionary mystery. It seems to have left only the scarcest of fingerprints of its ancestry or lineage, with no trace of a second or third cousin, no close-enough peers that are still living, no intermediary forms. Lane calls it an "unexplained void . . . the black hole at the heart of biology."

We will soon move to the anatomy, function, development, and specialization of this modern eukaryotic cell. But it is this second story — of the origin of our cells — that neither this book, nor evolutionary science, can yet fully convey.

PART TWO
THE ONE AND THE MANY

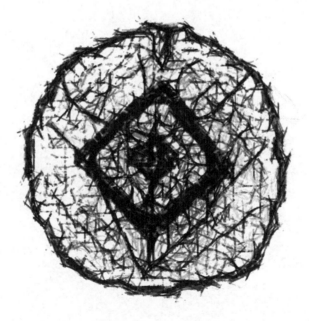

PART TWO

THE ONE AND THE MANY

The words organism *and* organized *share a common root. Both come from the Greek* organon *(later the Latin* organum*), an instrument or tool, or even a method of logic, designed to achieve something. If the cell is the basic unit of life — the living tool that forms the organism — then what is it "designed" to do?*

Well, first it has evolved to be autonomous, to survive as an independent living unit. This autonomy depends, in turn, on organization — on the cell's interior anatomy. A cell is not a blob of chemicals; it has distinct structures, or subunits, within it that allow it to function independently. The subunits are designed to supply energy, discard waste, store nutrients, sequester toxic products, and maintain the internal milieu of a cell. Second, a cell is designed to reproduce, so that one cell can produce all the other cells that populate the organism's body. And finally, for multicellular organisms, the cell (or at least the first cell) is designed to differentiate and develop into other specialized cells, so that various parts of the body — tissues, organs, organ systems — can be formed.

These, then, are among the first and most fundamental properties of the cell: autonomy, reproduction, and development.*

For centuries, we regarded these fundamental features as impregnable. The interior anatomy of the cell and its internal homeostasis were, well, interior and internal — black boxes. Reproduction and development occurred within the womb — another black box. But as we deepen our understanding of the cell, we find ourselves able to pry open these black boxes and alter the fundamental properties of living units. Can we repair a cell's subunit that happens to be defective in function — and if so, to what extent? Can we build a cell with a different kind of interior milieu, different substructures, and therefore different properties? And if we enable human reproduction outside the womb, as we have already, will such an artificially

*In single-celled organisms, one might think of "development" as the maturation of the organism. The maturation of single-celled microbial organisms is now well established. In multicellular organisms, development is more complex. It is a combination of the multiplication of cells, their maturation, their movement to distinct locations, their association with other cells, and their formations of specialized structures with specialized functions to form organs and tissues.

150

created embryo be open to genetic manipulation? What, then, are the permissible limits, and perils, of tampering with the first, fundamental properties of life?

THE ORGANIZED CELL
THE INTERIOR ANATOMY OF THE CELL

*Give me an organic vesicle [cell]
endowed with life, and I will give you
back whole of the organized world.*

— François-Vincent Raspail

*Cell biology finally makes possible a
century-old dream: that of analysis
of diseases at the cellular level, the
first step toward their final control.*

— George Palade

"The cell," Rudolf Virchow proposed in 1852, "is a closed unit of life that bears within itself [. . .] the laws governing its existence." To begin with, a bounded, autonomous living unit — a "closed unit" that bears the laws that govern its existence — must have a boundary.

It is the membrane that defines the boundary; the outer limits of the self. Bodies are bound by a multicellular membrane: the skin. So is the psyche, by another membrane: the self. And so are houses and nations. To define

153

an internal milieu is to define its edge — a place where the inside ends, and the outside begins. Without an edge, there is no self. To *be* a cell, to exist as cell, it must distinguish itself from its nonself.

But what is the boundary of a cell? Where does one cell end and another begin? It also begins and ends with a membrane that surrounds it.

The membrane presents a locus of paradoxes. If it is hermetically sealed, allowing nothing in or out, then it will maintain the integrity of its insides. But how, then, might a cell handle the inevitable requirements — and liabilities — of living? A cell needs pores to permit nutrients to move in and out. It needs docks for signals from the outside to arrive and be processed. What if the organism is starving, and the cell must conserve food and halt metabolism? A cell must excrete waste — but, again, where, or how, to make a hatch to get rid of it?

Every such opening is an exception to the rule of integrity; after all, a doorway to the outside is also a doorway to the inside. Viruses or other microbes might use the routes of nutrient uptake or waste disposal to enter a cell. Porosity, in short, represents an essential feature of life — but also an essential vulnerability of living. A perfectly sealed cell is a perfectly dead cell. But unsealing the membrane through portals exposes the cell to potential harm. The cell must embrace both: closed to the outside, yet open to the outside.

But what are cell membranes made of? In the 1890s, Ernest Overton, a physiologist (and, incidentally, a cousin of Charles Darwin), immersed a variety of cells in hundreds of solutions containing various substances. Chemicals soluble in oil tended to enter the cell, he noted, while those insoluble in oil could not get in. The cell membrane must be an oily layer, Overton concluded, although he could not quite explain how a substance such as an ion or sugar, insoluble in fats, might enter or leave the cell.

Overton's observations deepened the mystery. Was the cell membrane thick or thin? Was it made of one layer of fat molecules (called lipids*) lined in single file, or was it a many-layered structure?

An ingenious study by two physiologists clarified the topological structure of the cell membrane. In the 1920s, Evert Gorter and François Grendel extracted all the fat from the surface of an exact number of red blood cells, spread the molecules in a single layer,

*The constituents were later further subclassified. The most abundant were particular kinds of lipids that had carried a charged molecule — phosphate — as its "head" and a long stretch of carbon as its "tail." Additional molecules, such as cholesterol, were also found embedded within the lipid membrane.

and calculated its surface area. Then they determined the surface area of the cells from which the membranes had been removed. The surface area of the extracted lipids was almost twice the total surface area of the red cells.

That number signaled an unexpected truth: a cell membrane must have two layers of lipids. It is a lipid *bilayer*. Imagine, for a moment, two sheets of paper glued together back-to-back and then shaped into a three-dimensional object — a balloon, say. If the balloon is the cell, then the two sheets of paper form the bilayered cell membrane.

The final piece of the puzzle — how molecules such as sugar or ions pass in and out of the lipid bilayer, and how the cell communicates with its outside — was solved in 1972, nearly fifty years after Gorter and Grendel's experiments. Two biochemists, Garth Nicolson and Seymour Singer, proposed a model in which proteins were embedded, like hatches, or channels, crossing the cell membrane. The lipid bilayer was not uniform or monotonous; it was porous by design. Proteins, floating in the membrane and spanning from inside to outside, allowed molecules to permeate the membrane and allowed other proteins and molecules to bind to the outside of the cell.

Noting the mosaic-like structure of the membrane, with multiple components stippled together, Nicolson and Singer called this the fluid mosaic model of the cell membrane

— a model that electron microscopy then proved to be accurate.

It's simpler, perhaps, to imagine entering and exploring the interior of a cell as an astronaut might imagine exploring an unfamiliar spacecraft. From far away, you might see the spacecraft's/cell's outer contours: the oblong, gray-white sphere of an oocyte, or the crimson disk of a red blood cell.

As you approach the cell membrane, you might begin to see its outer layer more clearly. Bobbing on that fluid surface are proteins. Some might be receptors for signals, while others might function like molecular glue for attaching one cell to another. Some of these might be channels. If you are fortunate, you might watch a nutrient or an ion slip through the pore and into the cell.

And now, you, too, might "board" the craft. You would dive through the hull — that is, into the bilayer membrane, moving rapidly through the space between the two layers, only about ten nanometers thick, or ten thousand times thinner than a strand of human hair — and emerge inside.

Look around and above: now the inner lamella of the cell membrane would hang above you like the fluid surface of the ocean as seen from below. You would also see the inner parts of the proteins dangling above you, like the underbellies of buoys.

157

■ ■ ■ ■

At first, you might swim through the cell's internal fluid, called either protoplasm, cytoplasm, or cytosol. Protoplasm is the "vital fluid" that nineteenth-century biologists discovered in living cells and in living creatures.* Although many cell biologists had noted the existence of a fluid within a cell, Hugo von Mohl was the first to use the term, in the 1840s. The protoplasm is a mind-bogglingly complex soup of chemicals. It is thick and colloidal in some places; watery in others.†

*Indeed, so important is the protoplasm that in the 1850s, a vigorous debate ensued about whether it was the protoplasm — and not the cell — that should be described as the unitary basis for life; the cell was merely a vessel holding it. The German cell biologist Robert Remak was among the strongest proponents of this idea. Eventually the cell theorists won, while the "protoplasmists" admitted a compromise position, maintaining that despite the primacy of the cell, every cell itself contained this vital fluid. The discovery of multiple other organelles within the protoplasm of a cell may have also tempered the idea that the protoplasm was an organism's only necessary and sufficient building block.

†The variations in the physical properties of the protoplasm — watery, semifluid, or like dense jelly — have become an increasing area of focus of recent research. Droplet-like accumulations of chemicals

158

It is the mother jelly that sustains life.

For nearly a half century after von Mohl's work on protoplasm in the 1840s, cell biologists imagined the cell as a liquid balloon filled with a formless, shapeless fluid. But the first thing you might note, once inside the cell, is that the cytoplasm has a molecular "skeleton" that maintains the form of the cell, just as a bony skeleton maintains an organism's form.* This scaffold, termed the cytoskeleton, is composed primarily of filaments of a ropy protein called actin, and tubular structures created by a protein called tubulin.† Unlike bones, though, these ropelike structures crisscrossing the cell are neither

---

suspended within the cell can act as the sites of particular biochemical reactions. The importance of such defined "phases" (as they are called) in many critical reactions is now well established and is being explored for others.

*In 1904, botanist Nikolai Kolstov was among the first to propose that protoplasm had such an organized internal structure. Kolstov would eventually be proved correct when the various elements of the cytoskeleton were observed by powerful microscopes.

†Other proteins also contribute to the cellular skeleton. A third type of protein, called the intermediate filament, is also part of the cytoskeleton in some cells. There are more than seventy different kinds of proteins that make up various intermediate filaments.

static nor merely structural. They form an internal system of organization. The cytoskeleton tethers components of the cell together, and is required for the movement of the cell. When a white blood cell creeps toward a microbe, it uses actin filaments, among other proteins, to push its feelers forward — gelling and un-gelling its front like the ectoplasmic movement of an alien.

Bound to the cytoskeleton, or floating in the protoplasmic fluid, are thousands of proteins that make living reactions (respiration, metabolism, waste disposal) possible. As you swim through the protoplasm, you are certain to encounter one particular molecule of critical importance: a long, strand-like molecule called ribonucleic acid or RNA.

RNA strands are made of four subunits: adenine (A), cytosine (C), uracil (U), and guanine (G). One strand might consist of ACUGGGUUUCCGUCGGGGCCC for thousands of such subunits. The strand carries the message, or code, to build a protein.* You might imagine it as a set of instructions; a Morse code stretched along a tape. One particular RNA, freshly made in the cell's nucleus, may arrive carrying the instructions

---

*RNA has multiple other functions, including regulating the turning of genes on and off, as well as helping in the synthesis of proteins, but we will concentrate on its coding function here.

to build, say, insulin. Other strands, encoding different proteins, might be floating by.

How are these instructions decoded? Look left or right, and you'll spot a massive macromolecular structure called a ribosome, a multipart assemblage first described by the Romanian American cell biologist George Palade in the 1940s. You can't miss it: a liver cell, for instance, contains several million of them. The ribosome captures RNAs and decodes their instructions to synthesize proteins. This cellular protein factory is itself made of proteins and RNA. It is yet another of life's fascinating recursions, in which proteins make it possible to make other proteins.

Building proteins is one of the cell's major tasks. Proteins form enzymes that control the chemical reactions of life. They create structural components of the cell. They are the receptors for signals from the outside. They form pores and channels across the membrane, and the regulators that switch genes on and off in response to stimuli. Proteins are the workhorses of the cell.

You might encounter yet another macromolecular structure, this one shaped like a tubular meat grinder. It is the cell's trash compactor, the proteasome, where proteins go to die. Proteasomes degrade proteins into their constituents and eject the chewed-up pieces back into the protoplasm, completing the cycle of synthesis and breakdown.

■■■

As you keep swimming through the cell's protoplasm, you are bound to come across a multitude of larger, membrane-bound structures. You might imagine each of these as double-walled, enclosed rooms within the spaceship. There's one room for generating power, a room for storage, a room for exporting and importing signals, and another one to discard waste. As microscopists and cell biologists trained their eyes on cells with increasing precision, they found dozens of organized, functional substructures, analogous to organs — kidneys, bones, and hearts — that Vesalius and other anatomists had identified in the body. Biologists called them organelles: mini-organs found inside cells.

Among the first of these structures you're likely to see is a kidney-shaped organelle first described, albeit vaguely, in animal cells in the 1840s by a German histologist named Richard Altmann. These organelles, later renamed mitochondria, were found to be the cell's fuel generators; the furnaces that glow and burn constantly to produce the energy needed for life. There is some debate about the origin of mitochondria. But one of the most intriguing, and widely accepted, theories is that more than a billion years ago, organelles were, in fact, microbial cells that developed the capacity to produce energy via a chemical reaction involving oxygen and glucose. These microbial cells were engulfed or captured by other cells and entered into a working partnership of sorts, a phenomenon termed endosymbiosis.

In 1967, evolutionary biologist Lynn Margulis described this occurrence in a scientific article titled "On the Origin of Mitosing Cells." As Nick Lane explains in *The Vital Question,* Margulis argued that complex organisms "did not evolve by 'standard' natural selection but by an orgy of cooperation, in which cells engaged with each other so closely that they even got inside each other." Too radical, too early. On the streets of San Francisco and New York, it may have been the Summer of Love, with young men and women engulfing each other with ardor, but in scientific halls, Margulis's engulfment theory was met with a barrage of skepticism. For her, the summer of endosymbiotic love turned into a long winter of ridicule and rejection — until decades later, when scientists began to note not just the structural similarities between mitochondria and bacteria but also their molecular and genetic commonalities.

Mitochondria are found in all cells, but they are particularly densely packed in cells that need the most energy or that regulate energy storage, such as muscle cells, fat cells, and certain brain cells. They are wrapped around the tails of sperm, to provide them enough swimming energy to reach an egg. They divide within the cell, but when it's the cell's turn to reproduce, mitochondria are only split between the two daughter cells. In other words, they have no autonomous life; they can live only within cells.

Mitochondria possess their own genes and their own genomes, which, suggestively, bear some resemblance to the genes and genomes of bacteria — again supporting Margulis's hypothesis that they were primitive cells that were engulfed by other cells and then became symbiotic with them.

How does a cell generate energy? There are two pathways: one fast and one slow. The fast route occurs mainly in the protoplasm of the cell. Enzymes serially break down glucose into smaller and smaller molecules, and the reaction produces energy. Because the process doesn't use oxygen, it is called anaerobic. In terms of energy, the end product of the fast pathway is two molecules of a chemical called adenosine triphosphate, or ATP.

ATP is the central currency of energy in virtually all living cells. Any chemical or physical activity that requires energy — for instance, the contraction of a muscle or the synthesis of a protein — utilizes, or "burns," ATP.

The deeper slow burn of sugars to produce energy occurs in mitochondria. (Bacterial cells, lacking mitochondria, can use only the first chain of reactions.) Here the end products of glycolysis (literally, the chemical breakdown of sugar) are fed into a cycle of reactions that ultimately produce water and carbon dioxide. This cycle of reactions involves the use of oxygen (and is therefore called aerobic) and is a small miracle of energy production: it generates a much larger harvest of energy, again, in the form of ATP molecules.

The combination of the fast and slow burn nets about the equivalent of thirty-two ATP molecules from every molecule of glucose. (The actual number is slightly lower, since not every reaction is perfectly efficient.) Over the course of a day, we generate billions of little canisters of fuel, to fire a billion little engines, in the billions of cells in our bodies. "Should all the billions of gently burning little fires cease to burn," the physical chemist Eugene Rabinowitch wrote, "no heart could beat, no plant could grow upward defying gravity, no amoeba could swim, no sensation could speed along a nerve, no thought could flash in the human brain."

Next, you might encounter a maze of winding, tortuous pathways, also bound by membranes, that crisscrosses the body of the cell. It is also an organelle and is called the endoplasmic reticulum, although most biologists abbreviate it ER.

This structure was first described by the cell biologists Keith Porter and Albert Claude,*

*The French cytologist Charles Garnier had first

165

working closely with George Palade, at the Rockefeller Institute in New York in the late 1940s. The experiments to delineate the function of this pathway — and its centrality to the biology of the cell — represent one of the most momentous journeys in science.

Palade's own journey into cell biology had been circuitous. He was born in 1912 in Iasi (then called Jassy), in Romania. His father, a professor of philosophy, wanted his son to become a philosopher, too, but George was drawn to a discipline with more "tangibles and specifics." He studied medicine and began his career as a doctor in the capital city of Bucharest. But he was soon lured by cell biology. Like Rudolf Virchow, Palade also wanted to unify cell biology, cellular pathology, and medicine. "[It] finally makes possible a century-old dream: that of analysis of diseases at the cellular level, the first step toward their final control," he would later write.

In the 1940s, Palade was offered a position as a researcher in New York. His trip to the United States across war-shredded Europe was a harrowing pilgrimage. He traveled through bleak, desolate Poland, where he was detained for weeks awaiting immigration. "He thought of himself like a scientific

observed the endoplasmic reticulum in 1897 using a light microscope, but he did not assign it any particular function.

version of the character Christian in *The Pilgrim's Progress*," a colleague of Palade's told me, "somehow exempt from all the thousands of blockades and pitfalls that might thwart his journey to New York — or, for that matter, into the center of the cell."

In 1946, then thirty-four years old, Palade finally arrived in New York. He launched his research career at New York University and then took a job at the Rockefeller Institute. He was appointed an assistant professor in 1948 and given a lab in an "unattractive dungeon" sunk in a third-floor basement of one of the institute's oldest buildings.

The dungeon, however unbecoming, proved to be a haven for cell biologists. "The new field had virtually no tradition; everybody working in it came from some other province in natural sciences," Palade wrote. And so he pulled and borrowed and stole from every branch and province of science — in essence, creating his own discipline: modern cell biology. Palade launched crucial collaborations with Porter and Claude. The lab would soon become the intellectual basement for the field of subcellular anatomy and function, the plinth on which the towering discipline would be constructed.

Just as Robert Hooke and Antonie van Leeuwenhoek, peering down a microscope, revolutionized cell biology in the seventeenth century, Palade, Porter, and Claude discovered a more abstract way of "looking" inside the cell. First, they burst cells open and spun the contents in a high-speed centrifuge along a gradient of densities. As the centrifuge spun with dizzying velocity, pulling down the cell's heaviest subparts to the bottom and leaving lighter subparts above, different components of the cell appeared at different gradients along the length of a tube.

Each component could then be extracted from a particular part of the tube and assessed separately to identify its structural anatomy and the biochemical reactions contained within it: reactions such as oxidation, synthesis, detoxification, and waste disposal. And then, by sectioning the cell into the thinnest slices and training an electron microscope on them, researchers could retrace these components and reactions to their locations in animal cells.

This, too, was "seeing" — but with two kinds of lenses. There was, on one hand, the abstract lens of biochemistry: the centrifugal separation of subcellular components and the discovery of chemical reactions and components confined to them. And on the other hand there was the physical lens of electron microscopy, which assigned these chemical functions to anatomical structures and locations within cells. Palade described this coalescence of the two ways of seeing as a

168

pendulum that swung from microscopic anatomy to functional anatomy and back again:
"[S]tructure — as traditionally envisaged by
the microscopist — was bound to merge into
biochemistry, and biochemistry of . . . subcellular components appeared to be the best way
to get at the function of some of the newly
discovered structures."

It was a ping-pong match in which both
sides won. Microscopists would see subcellular structures; biochemists would assign
functions to them. Or biochemists would find
a function and then turn to microscopists to
pinpoint the structure responsible for that
function. Using this method, Palade, Porter,
and Claude entered the luminous heart of the
cell.

Let's return to the endoplasmic reticulum,
the winding pathway found in virtually every
cell. There is a voluptuousness to this structure: the sheer excess of it, running lace over
lace, like folded pleats. Looking through an
extraordinarily powerful microscope at cells
from a dog's pancreas revealed the outer
edges of the ER membrane studded with tiny,
dense particles.

An abundance of structures, but what do
they all *do*? Palade asked. From the work of
prior researchers, he knew that the ER was associated with synthesizing and exporting proteins, which carry out virtually all the work
of the cell. Some, like enzymes responsible

for metabolizing glucose, are synthesized within a cell and stay there to perform their functions. But other proteins — insulin, say, or digestive enzymes — are secreted by cells into the blood or the intestines. And yet other proteins, such as receptors and pores, are inserted into the cell membrane. *But how does a protein reach its destination?*

In 1960, Palade and his coworkers, Philip Siekevitz in particular, used radioactivity — a molecular beacon — to label proteins in a cell and then followed their progress over time. He would "pulse" the cell with a high dose of radioactivity, thereby labeling all the proteins being synthesized, and then "chase" the protein's location using the electron microscope to visualize the progress of these proteins.*

Reassuringly, he found the radioactive signal first associated with ribosomes, the site where proteins were initially synthesized (ribosomes were the tiny, dense particles that

---

*By 1961, Keith Porter had left the group to start his own work at Harvard, and Claude had left earlier for the University of Louvain in Belgium. But Palade was joined by a new set of cell fractionators: Siekevitz, Lewis Greene, Colvin Redman, David D. Sabatini, and Yutaka Tashiro, as well as two experts in electron microscopy, Lucien Caro and James Jamieson. By combining forces with these two groups, Palade traced the progress of a protein through the endoplasmic reticulum.

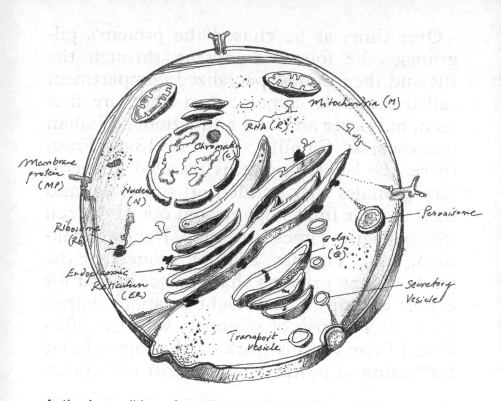

Author's rendition of a cell, showing its various substructures, including the ER (endoplasmic reticulum), N (nucleus), R (RNA), CM (cell membrane), C (chromatin), P (peroxisome), G (Golgi), M (mitochondria), Rb (ribosome), MP (membrane protein). The strands within the cell correspond to elements of the cytoskeleton. Note that the drawing is not to scale.

Palade had seen studding the edges of the ER). Then, to his wonderment, some of the proteins moved from the ribosomes *into* the endoplasmic reticulum.*

*In the years following Palade's discovery, Sabatini and a German immigrant named Günter Blobel made one of the most seminal discoveries about how proteins are targeted either to the ER for secretion out

Over time, as he chased the protein's pilgrimage, he found it moving through the ER and then into a specialized compartment called the Golgi apparatus, a structure first seen, but never ascribed a function, by Italian microscopist Camillo Golgi in 1898. From there, the labeled proteins traveled to secretory granules that budded off the Golgi and then to their final stop: ejected out of the cell (the biologists James Rothman, Randy Schekman, and Thomas Südhof pioneered the study of how proteins that are *not* bound for export end up in their right locations within the cell. The trio of scientists won the 2013 Nobel Prize for this work on the intracellular trafficking of proteins). At almost each point

---

of the cell or for insertion in the cell membrane. In short, a signal directing a protein destined for secretion or destined to the membrane is *already* appended in the protein's sequence, like a postage stamp. Specific cellular pathways recognize this stamp and direct the protein to its predetermined destination. The more detailed version is this: Sabatini and Blobel, a biologist, found that secreted and membrane-resident proteins carry this specific signal — a sequence of amino acids — in their sequences. As the ribosome decodes the RNA and synthesizes a protein, a molecular complex called the signal recognition particle (SRP) recognizes this targeting signal and drags the protein toward the ER. A pore crossing from the cell into the ER enables the transport of the protein into the ER.

in their journey, some of the proteins are modified: they can be clipped short, chemically modified by the addition of a sugar, or twirled around and bound to another protein (the signals to make these modifications are typically contained in the sequence of the protein itself).

The whole process can be imagined as an elaborate postal system. It begins with the linguistic code of genes (RNA) that is translated to write the letter (the protein). The protein is written, or synthesized, by the cell's letter writer (the ribosome), which then posts it to the mailbox (the pore by which the protein enters the ER). The pore routes it to the central posting station (the endoplasmic reticulum), which then sends the letter to the sorting system (the Golgi), and finally brings it to the delivery vehicle (the secretory granule). There are, in fact, even codes appended to proteins (stamps) that enable the cell to determine their ultimate destination. This "postal system," Palade realized, is how most proteins get to their correct locations within the cell.

The pioneering studies of Palade, Porter, and Claude threw open a new world of subcellular anatomy. The twinning of two ways of seeing — microscopy and biochemistry — was synergistic. As biologists used these methods on cells, they found dozens of such functional, anatomically defined subcellular structures. The Belgian biologist Christian

Author's representation of the migration of a secreted protein from the ribosome to the ER to the Golgi and finally to secretory granules. Note the insertion of the protein into the ER as it is synthesized. The protein is modified in the ER, where chains of sugars may be added to it. Its journey continues into the Golgi, where it may be further modified and then routed to a secretory vesicle destined to extrude the protein from the cell, or to other vesicles to bring it to other cellular compartments.

de Duve, yet another Rockefeller Institute scientist, discovered an enzyme-laden structure called a lysosome. Like a cellular "stomach," it digests worn-out cellular parts, as well as invading bacteria and viruses.

Plant cells contain structures called

chloroplasts, the sites of photosynthesis, the conversion of light into glucose. Chloroplasts, like mitochondria, carry their own DNA, again suggesting an origin in microbes that were engulfed by other cells. There is a membrane-bound structure called a peroxisome, another of de Duve's discoveries, where some of life's most dangerous reactions — for instance, the oxidation of molecules — is sequestered, and where hydrogen peroxide, an intensely reactive chemical, is generated. Were the peroxisome to open up and release its internal poisons, the cell would be attacked by its own reactive contents. It is the chalice, filled with poisons to metabolize other poisons, that the cell keeps carefully closed.

I've reserved for last the most essential, and still the most mysterious, of organelles: the nucleus. Bacteria don't possess nuclei, but in cells that do — all plant and animal cells, including human cells — the nucleus is where the bulk of the cell's genetic material, the instruction manual for life, is stored. It is the storage bank for DNA, for the genome.

The nucleus is the command center; the captain's bridge of the cell. It is the place that both receives and then disseminates most of life's signals. RNA, the code to build proteins, is copied from the genetic code here and then exported out of the nucleus. We might imagine the nucleus as the center of the center of life.

The cellular anatomist Robert Brown observed the nucleus in orchid cells in 1836. Noting its central position in the cell, he named the structure after the Greek word for *kernel*. Yet its function, or its vitality to the function of the cell, would remain unknown for an entire century. Like all cells, the nucleus is surrounded by a porous two-layered membrane, although its pores are far less characterized or known.

The nucleus, as I mentioned before, houses the organism's genome, made of long stretches of deoxyribonucleic acid. The DNA double helix is elaborately folded and packaged around molecules called histones, and tightened and wound further into structures called chromosomes. If a single cell's DNA could be stretched out straight, like a wire, it would measure six and a half feet. And if you could do that for every cell in the human body and laid all of that DNA end to end, it would stretch from the Earth to the sun and back again more than sixty times. String together all the DNA in every human being on the planet, and it would reach the Andromeda galaxy and back nearly two and half times.

The nucleus, like the cell's inner liquid, the cytoplasm, is also organized, although we still know little about its organizational structure. Scientists studying the nucleus believe that it contains its own skeleton made of molecular fibers. Proteins, traversing the cytoplasm, enter through the pores of the nuclear membrane and bind to the DNA and turn genes on and off. Hormones, bound to proteins, traffic in and out. ATP, the universal source of energy, moves swiftly through the pores.

The process of switching genes on and off is vital, giving the cell its identity. The set of on/off genes instructs a neuron to be a neuron, and a white cell to be a white cell. During the development of an organism, genes — or rather proteins encoded by genes — tell cells about their relative positions and command their future fates. Genes are turned on and off by external stimuli such as hormones, which also signal changes in a cell's behavior.

When a cell divides, every chromosome is copied, and the two copies separate in space. In human cells, the nuclear envelope dissolves, one full set of chromosomes migrates to each of the two newborn daughter cells, and the nuclear membrane reappears around them — in essence, regenerating a daughter cell with a new nucleus and chromosomes lodged inside it.

Yet much of the nucleus remains a mystery: the doors to the command center of the cell are still partially closed. As one biologist put it, "We can only hope that what the geneticist J. B. S. Haldane posited on the cosmos will prove not to be true for the nucleus: " 'Now, my suspicion is that the universe is not only queerer than we suppose, but queerer than we *can* suppose. If we appropriately bear in mind that the nucleus may be more complicated than we may have once thought, and yet just may be knowable, then this very belief may empower us and our students and successors to penetrate the subject's awaiting depths, the next of which now beckon. There is every reason to believe in this program. So let us be of good cheer.'"

Membrane. Protoplasm. Lysosome. Peroxisome. Nucleus. The subunits of the cell that we've met are vital to its existence; they perform specialized functions that allow a cell to possess and maintain an independent life. Their location, organization, and orchestration are crucial. In short: *a cell's autonomy lies in its anatomy.*

And that autonomy, in turn, enables an essential feature of living systems: the capacity to maintain the fixity of an internal milieu — a phenomenon termed "homeostasis." The concept of homeostasis (the word is derived from the Greek words *homeo* and *stasis,* loosely meaning "related to stillness") was first described by the French physiologist Claude Bernard in the 1870s and further developed by the Harvard University physiologist Walter Cannon in the 1930s.

For generations before Bernard and Cannon, physiologists had described animals as assemblages of machines, sums of dynamic parts. Muscles were motors; the lungs a pair of bellows, the heart a pump. Pulsing, swiveling, pumping; physiology's emphasis was on movement, on actions, on work. *Don't just stand there, do something.*

Bernard inverted that logic. *"La fixité du milieu intérieur est la condition de la vie libre, indépendante":* the constancy of the interior environment is the condition of free and independent life, Bernard wrote in 1878. In shifting physiology's focus from action to the maintenance of fixity, Bernard changed our conception of how an organism's body works. A major point of physiological "activity," paradoxically, was to enable stasis. *Don't just do something, stand there.*

Bernard and Cannon studied homeostasis in organisms and organs, but it is increasingly recognized as a fundamental feature of cells — and, indeed, of life. To understand cellular homeostasis, we begin, again, with the membrane that separates the cell from its external milieu, so that its internal reactions can be sequestered and self-contained. The membrane has also evolved pumps to move unwanted substances out of the cell — again, to maintain a constancy in the cell's internal space. The protoplasm contains chemical buffers so that the acidity or the alkalinity of the cell does not change, even when the chemical environment outside the

179

cell changes. A cell needs energy, and the mitochondria deliver. The proteasome disposes of unwanted, or misfolded, proteins. Specialized storage organelles in some cells ensure that a supply of nutrients can act as a backup repository should there be a shortage externally. Toxic by-products of metabolism are shepherded to the peroxisome to be destroyed.

We will soon move from autonomy and homeostasis to other fundamental features of the cell — reproduction, functional specialization, and the cell's capacity to divide and form multicellular organisms. But linger with me, for a moment, to mark the extraordinary discoveries that this chapter covers. The two decades between 1940 and 1960 may rank as the most fertile and productive period for cell biologists seeking to dissect the functional anatomy of the cell's interior. There's a majesty and mastery in these decades — similar to the majesty and mastery that dates to almost exactly a century prior, when Schwann, Schleiden, Virchow, and others laid the foundations of cell biology. If the insights revealed during this period seem "routine" today (some version of the phrase "The mitochondria are the cell's energy factory" can inevitably be found in every high school science textbook), it's because we've forgotten, as we often do, the spine-tingling awe that each of these discoveries generated in its time. I don't think it's hyperbolic to describe the transitions from the discovery of the cell to the revelation of its structural anatomy and, finally, to the elucidation of its *functional* anatomy as one of science's most inspiring achievements.

The discovery of functional anatomy enabled an integrated view of the cell and, by extension, of the defining features of life. A cell, as noted before, is not just a system of parts sitting next to parts, just as a car is not a carburetor sitting next to an engine. It is an integrating machine that must amalgamate the functions of these individual parts to enable the fundamental features of life. Between 1940 and 1960, scientists began to *integrate* the separate parts of the cell to understand how an autonomous living unit might function and become "living."

Inevitably, these fundamental discoveries eventually launched the development of new medicines. If gross anatomy and physiology had launched a new era for surgery and medicine in the eighteenth and nineteenth centuries, functional cellular anatomy and physiology announced new loci for disease and therapeutic intervention in the twentieth century. The functional breakdown of an organ, we have long known, causes a disease: the kidney fails, the heart weakens, bones fracture. But what of the functional break-down of a cellular *organelle*?

In the summer of 2003, an eleven-year-old hockey player named Jared began to lose vision in both his eyes. The world was dimming, slowly, and Jared, trying his best to keep playing sports, struggled to find the lines on the hockey ice surface. His parents brought him to an ophthalmologist at the Mayo Clinic in Rochester, New York, to obtain a diagnosis.

A week later, the clinic uncovered the source: Jared had a condition called Leber hereditary optic neuropathy (LHON). "I'm so very sorry, but Jared will be blind," the ophthalmologist at the Mayo informed Jared's parents gently. Typically, the inherited illness arises out of a mutation in a gene named mtND4 that is found in the mitochondria. (The culprit gene was found and mapped in 1988, just two years before the launch of the Human Genome Project.) For reasons still unknown, it specifically affects the function of the eye's retinal ganglion cell, which transmits information from the retina to the optic nerve, and onward to the brain.

In affected children, the disease progresses inexorably. At first, the nerve fibers along the optic disk begin to bloat. Then the optic nerve atrophies, and the retinal nerves become thin and dull in appearance. Jared had inherited the most common LHON mutation: in nucleotide position 11778 in the mitochondrial genome, which is about sixteen thousand bases long in total.*

"1 1 7 7 8," Jared wrote in his diary. "I wish this had been the combination to my hockey locket, or bike lock, or even to my school locker. Instead, it was to be the combination to a genetic mutation at nucleotide position 11778 that would unlock the disease in my body at the age of eleven and would ultimately change my life forever. . . . Blind, what the hell was blind? I'm eleven years old. I'm a hockey player. I dig chicks, and they dig me. I've got a lot of friends and no worries. Blind? What do they mean I won't be able to see? Not be able to see what? . . . Just fix it, Dad, and let me go play with my buddies."

*Mitochondrial mutations are special because they can be inherited only from your mother, while most other mutations can come from either parent. Mitochondria don't have an autonomous existence; they can live only inside cells. They divide when a cell divides and then are apportioned to the two daughter cells. When an egg cell forms in the mother, all its mitochondria are from her cells. Upon fertilization, the sperm cell injects its DNA into the egg — but not any mitochondria. Therefore, every mitochondrion that you are born with is maternal in origin. The mutation in the mtND4 gene that Jared inherited had to come from his mother. It is likely that it occurred by chance, during the creation of the egg, because his mother did not have the disease.

But Dad, try as he might, could not fix it. Jared's ganglion cells began to decay. Cannily, the parents turned Jared's attention to playing the guitar. He learned to play by touch and sound alone. And while the blindness advanced — gradually but relentlessly — so did his music. "So here I am at the Musicians Institute in Los Angeles, California, eight years after playing my first ear-piercing concert for my mother and father in Guitar Center. I believe I'm the first blind student that has attended this remarkable music institution, which is pretty cool. I guess they believed I was good enough to keep up with all the other students who are required to read music." Jared had lost sight but found sound.

In 2011, a group of ophthalmologists in Hubei, China, modified a virus called AAV2 to carry the normal version of the ND4 gene. The virus infects human and primate cells but doesn't cause any overt or acute disease, and it can be modified to carry a "foreign" gene such as ND4. Millions of gene-modified virus particles were suspended in a droplet of liquid. A tiny needle pierced the edge of the patient's cornea and deposited the drop of dense viral soup into the vitreous layer, just above the retina.

The scientists knew they were treading on dangerous, trip-wired territory: in September 1999, Jesse Gelsinger, a teenager with a mild metabolic disease that affected his liver's ability to metabolize by-products of protein degradation, resulting in near toxic levels of ammonia in his blood, had been infused with a gene-modified adenovirus. The boy's doctors had hoped that the transfusion of the virus, an experimental therapy, would cure Jesse's illness. Tragically, though, Jesse suffered a catastrophic immune reaction to the virus, quickly resulting in fatal organ failure. The fallout from his death was immediate. For the first decade of the twenty-first century, the field of gene therapy was forced into a deep winter. Few researchers attempted to deliver gene-modified viruses into humans, and regulators bound the field with stringent rules.

But the retina is a special site. Not only is a raindrop's worth of virus sufficient to infect the cells, but also the retina is uniquely immune privileged: along with a few other places in the body — the testes, among them — it is not actively surveyed by an immune response and therefore highly unlikely to generate a severe reaction to an infectious agent. Furthermore, the gene therapy vectors had improved vastly since the Gelsinger saga, bolstering scientists' confidence that the gene could be delivered without provoking an adverse reaction.

In 2011, Chinese physicians enlisted eight patients with LHON for a small clinical trial. There were early signs of success: the virus bore the genes into the retinal ganglion cells, and the cells synthesized the correct ND4 protein, which found its way to the mitochondria. Over the next thirty-six months, visual acuity improved in five of the eight patients.

The studies are continuing as I write this, with the researchers refining the nature of the patients enrolled and extending the observation period. The viral product, now called Lumevoq, is currently in late-phase clinical trials for LHON patients with early vision loss. In May 2021, trialists reported the completion of the RESCUE trial, in which gene therapy was used to halt the progressive loss of vision in patients with the mutation within six months of their visual impairment. The placebo-controlled, double-blinded, multicentered, and randomized trial — a gold standard study — involved thirty-nine subjects. (One patient received a lower dose of the virus, leaving thirty-eight evaluable patients.) One eye was injected with the virus, while the other was treated with a sham injection (containing no virus). At twenty-four weeks, both the treated and control (untreated) groups continued to demonstrate the inevitable decline in visual acuity. The vision loss plateaued in both eyes at forty-eight weeks. But at ninety-six weeks, surprisingly, *both* the treated and untreated eyes in about three-quarters of the treated subjects showed significant improvements in acuity. The trial was thus both a success and a mystery: while the eye treated with gene therapy was expected to improve, why did the untreated eye also improve? Is there some interconnection between retinal ganglion cells, or some other mechanism of connection, across two eyes that we don't know about? Did the virus leak into the circulation and affect the other eye?

Unfortunately, for patients such as Jared, who have lost their vision completely, the replacement of ND4 is unlikely to be of benefit: for them, it is too late to restore vision. When the responsive cells have died, the replacement of the function of an organelle can no longer be of benefit. An organelle can function only in the context of the right cell.

If the trials continue apace, and the benefits prove to be long term — still a big "if" — Lumevoq will eventually find its place in the medical pharmacopeia. But the launching of a cell-modifying therapy that attempts to change mitochondrial function has already signaled a new direction for medicine.

In the 1950s and 1960s, medicine and surgery witnessed an explosion of *organ*-directed therapies: rerouting blood vessels in a heart to bypass a blockage, or replacing a diseased kidney with a transplanted organ. A new universe of drugs emerged — antibiotics, antibodies, chemicals to prevent blood clots or reduce cholesterol. But this is *organelle*-directed therapy: the replenishment of a functional

187

deficiency in the mitochondrion of a retinal ganglion cell. It represents the culmination of decades of study of cellular anatomy, the dissection of subcellular compartments, and the characterization of their dysfunction in diseased states. It is gene therapy, of course, but also cell therapy in situ — in other words, the restoration of function of a diseased cell in its native anatomical location in the human body.

THE DIVIDING CELL
CELLULAR REPRODUCTION
AND THE BIRTH OF IVF

There is no such thing as reproduction. . . .
When two people decide to have a baby,
they engage in an act of production.
— Andrew Solomon, *Far from the Tree:*
Parents, Children,
and the Search for Identity

A cell divides.

Perhaps the most monumental event in the life cycle of a cell is the moment it gives birth to daughter cells. Not every cell is capable of reproducing: some cells, such as some neurons, have undergone permanent or terminal division and will never divide again. But the converse is not true: *every* cell is the product of birth from another cell — *Omnis cellula e cellula*. As the French biologist François Jacob once put it, "The dream of every cell is to become two cells" (except, of course, those that have opted out of the dream altogether).

Conceptually speaking, cell division in animals might be broadly divided into two

189

purposes or functions: production and reproduction. By *production,* I mean the creation of
new cells to build, grow, or repair an organism. When skin cells divide to heal a wound,
when T cells divide to produce an immune
response, the cells are giving birth to new
cells either to *produce* a tissue or an organ, or
to fulfill a function.

But it is a completely different matter when
sperm or eggs are generated in the human
body. Here they are being generated to undergo *re*production — dividing to produce
not a new function or an organ but rather a
new organism.

In humans and multicellular organisms,
the process for the production of new cells to
build organs and tissues is called mitosis —
from *mitos,* the Greek word for "thread." In
contrast, the birth of new cells, sperm, and
eggs for the purpose of *re*production — to
make a new organism — is called meiosis,
from *meion,* the Greek word for "lessening."

The German scientist who discovered mitosis was a disillusioned, short-sighted military doctor seeking a new vision for biology.
Walther Flemming, the son of a psychiatrist,
trained in medicine in the 1860s. Like Rudolf
Virchow, he also attended a military medical school, and, like Virchow again, he found
the discipline regimented and inflexible and
soon turned to studying cells. Humans — all

multicellular creatures — were made of cells, and yet the building of an organism out of cells, from one cell to several billion, was a mysterious process. In the 1870s, Flemming became particularly intrigued by cellular anatomy, and he began to use aniline dyes and their derivatives to stain tissues, hoping to illuminate subcellular structures.

At first, Flemming saw very little. The dye revealed only a wispy, threadlike substance located almost exclusively within the nucleus, the typically spherical, membrane-bound structure within the cell first discovered by the Scottish botanist Robert Brown in the 1830s.

Flemming, following his colleague Wilhelm von Waldeyer-Hartz, christened the nucleus-dwelling, threadlike substances chromosomes — "colored bodies," a neutral name. He wondered about their function, and their dynamics during cell division. His curiosity aroused, Flemming kept looking at dividing cells under the scope. Looking but not *seeing*. Sight — real sight — requires insight. Other scientists, such as von Mohl and Remak, had observed cells divide but concluded very little about the orchestration or phases of the process. Flemming realized that they had been looking *at* cells, not *inside* cells. His critical insight appeared in 1878: it involved staining chromosomes with blue dye during cell division and then following the whole process of division under the microscope, thereby capturing the activity of chromosomes and the nucleus within the cell.

What did chromosomes *do*? And how was the nucleus, or the chromosomes inside it, related to cell division? "What forces act during cell division?" he asked in a two-part paper written in 1878 and 1880. "Do the shifts in the positions of the visible formed structures in the cell [the nucleus and the chromosomes, during cell division] follow a scheme, and if so, what scheme?"*

*Theodor Boveri and Walter Sutton would make the next logical link: connecting chromosomes to inheritance. In short, they would link *genetic* inheritance to the *anatomical/physical* inheritance of chromosomes, thereby situating genes (and inheritance) on chromosomes. In his experiments with peas, Gregor Mendel could identify genes only in the abstract as "factors" that would move across generations and carry traits, or features, from parents to their offspring; he had no means to identify the physical location of these factors. Sutton and Boveri, among others, would provide the first evidence that the inheritance of characteristics (that is, genes) occurs through the inheritance of chromosomes. Work by the fruit fly geneticist Thomas Morgan and others would build on this theory, finally placing the locus of genes on chromosomes. Decades later, studies by Frederick Griffith, Oswald Avery, James Watson, Francis Crick, and Rosalind Franklin,

192

The scheme, he found, was startlingly systematic.* It was staged as precisely as a military drill. In the larvae of salamanders — in the dividing cells of mammals, amphibians, and fish — Flemming found a common rhythm of cell division that ran through virtually every organism. It was an exhilarating result: no scientist before him had even faintly imagined that the cells of such diverse organisms would follow a nearly identical and rhythmic scheme during the division of their cells.

The first step, Flemming found, was the condensation of the threadlike chromosomes into thickened bundles — "skeins," he called them. The dye now bound strongly; the chromosomes glowed under the scope, like reels of thread dyed with deep indigo. Then the condensed chromosomes doubled and split

---

among others, would identify DNA — the molecule that lies at the center of the chromosome — as the carrier of genetic information. And further research by Marshall Nirenberg and colleagues at the National Institutes of Health (NIH) would identify how genes are decoded to create proteins that ultimately provide forms and features to organisms.

*The botanist Karl Wilhelm von Nägeli regarded Flemming's experiments as an anomaly — but then, he also dismissed Mendel's paper as the work of a crank. It was only decades later that the universal principles of cell division were elucidated in all organisms.

along a defined axis, creating structures that reminded him of two starbursts splitting. The "nuclear figures began to organize themselves into successive stages during division," he wrote. The nuclear membrane dissolved, and the nucleus, too, began to split. At last, the cell itself divided, its membrane segmented, giving rise to two daughter cells.

Once in the daughter cells, the chromosomes uncondensed slowly and returned to their wispy "resting stage," back in the nuclei of the daughter cells — as if reversing the process that had initiated cell division. Since the chromosomes doubled at first and then halved upon cell division, the number of chromosomes in the daughter cells was conserved. Forty-six became ninety-two, and was halved to forty-six. Flemming called it homotypic, or "conservative," cell division: the parent cell and the daughter cells ended up with the same, conserved number of chromosomes.* Between the 1880s and the early 1900s, the biologists Theodor Boveri, Oscar Hertwig, and Edmund Wilson would contribute a great many details to this initial sketch of cell division, delving further into each of the individual steps that Flemming had initially described.

*Two other cytologists, Eduard Strasburger and Édouard van Beneden, also observed the separation of chromosomes, followed by the division of the cell membrane into two daughter cells (mitosis).

Walther Flemming's drawings of successive stages of mitosis, or cell division. At first, the chromosomes are present in loose threadlike forms in the nucleus. Two abutting cells are shown, each with a nucleus and uncondensed chromosomes. Then the threads tighten into dense bundles. The nuclear membrane dissolves, and the chromosomes separate into two sides of the cell, as if drawn by some forces. When they've fully separated (second-to-last figure), the cell splits, generating two new cells.

Flemming drew the process as a cycle: the threadlike chromosomes condensed into skeins, split, and then returned to a resting state. And then they compacted and dilated again as the cell edged toward its next division cycle — condensing, splitting, and decondensing, almost as if a breath of life was moving through them.

But there had to be a different kind of cell division, the kind that leads to reproduction. It is easy, in retrospect, to understand that the dynamics of this form of cell division could not possibly be the same as that of mitosis: it's a matter of elementary mathematics. In mitosis, you'll recall, the parent cell and the daughter cells end up with the same number of chromosomes. You start, say, with forty-six (the number of chromosomes in human cells);

the chromosomes duplicate (ninety-two), and then each daughter cell gets half: back to forty-six.

But how might those numbers work for reproduction? If sperm and eggs had the same number of chromosomes as their parent cells, 46, then the fertilized egg would contain twice the number, 92. That number would double in the next generation to 184, and then double again to 368, and so forth, increasing exponentially generation upon generation. Soon the cell would explode with chromosomes.

The genesis of sperm and eggs, then, must require first *halving* the number of chromosomes, twenty-three each, and then restoring them back to forty-six upon fertilization. This variant of cell division — reduction, followed by restoration — was observed in sea urchins by Theodor Boveri and Oscar Hertwig in the mid-1870s. In 1883, Belgian zoologist Édouard van Beneden also observed meiosis in worms, confirming the commonality of the process in more complex organisms.

The life cycle for a multicellular organism, in short, could be reconceived as a rather simple back-and-forth game between meiosis and mitosis. Humans, starting with forty-six chromosomes in every bodily cell, produce sperm cells in the testes and egg cells in the ovaries via meiosis, each ending up with twenty-three chromosomes. When sperm and egg meet to form a zygote, the number of chromosomes is restored to forty-six. The zygote grows through cell division, mitosis, to produce the embryo, and then develops progressively mature tissues and organs — heart, lungs, blood, kidneys, brain — with cells that have forty-six chromosomes each. As the organism matures, it eventually develops a gonad (testes or ovaries), with forty-six chromosomes in each cell. And here the game shifts again: when the cells in the gonads make male and female reproductive cells, they undergo meiosis, generating sperm and eggs with twenty-three chromosomes each. Fertilization restores the number to forty-six. A zygote is born, and the cycle repeats. Meiosis, mitosis, meiosis. Halve, restore, grow. Halve, restore, grow. Ad infinitum.

What controls the division of a cell? Flemming had witnessed the systematic stages of mitosis. But who, or rather what, conducts this staging? In the decades after Flemming published his seminal work on cell division, cell biologists noted that the life cycle of a dividing cell could be divided into phases.

Let us begin with cells that opt out of the cycle altogether. They are permanently or semipermanently resting — *quiescent,* to use biology jargon. This phase is now termed G-zero, the G0 referring to the "gap," or resting cycle. In fact, some of these cells will *never* divide; they are post-mitotic. Most mature neurons are good examples.

197

When a cell makes a decision to enter the cycle of division, it moves into a new gap period, termed G1. It is almost as if it is dipping its toe into the waters of cell division, contemplating its decision. Few changes are visible under a microscope during G1 but in molecular terms, this first gap is monumental: the proteins that coordinate cell division are synthesized. Mitochondria are duplicated. The cell gathers molecules, summoning and synthesizing those that are crucial to metabolism and sustenance, and increasing them in number before they will be apportioned to the two daughter cells. It is also the first critical checkpoint in a cell's decision whether or not to commit to the enormity of cell division. Go? Or not go? If certain nutrients are absent, or if the hormonal milieu is not appropriate, the cell might choose to remain in G1. It is a point *before* the point of no return.

The phase that follows Gap 1 is distinct and unique: the duplication of chromosomes — and therefore, the synthesis of new DNA. It demands energy, commitment, and a drastic shift in focus. It is termed the S phase, from *synthesis* — synthesis of duplicate chromosomes. If you inhabited the interior of the cell, swimming, as you once were, in the protoplasm, you might sense a shift in its hub of activity away from the cytoplasm and toward the nucleus. Enzymes that duplicate DNA latch on to chromosomes. Yet other enzymes

198

begin to unwind DNA. The building blocks of DNA are shuttled to the nucleus. A complex assemblage of DNA-replicating enzymes strings along the chromosomes, synthesizing a duplicate copy. And an apparatus to pull the duplicated chromosomes apart begins to form within the cell.

The third phase is perhaps the most mysterious and least understood: a second resting phase, called G2. Why stop a cell from dividing once it has synthesized a duplicate chromosome? Why waste a freshly synthesized strand of DNA? G2 exists as a final checkpoint before cell division because cells cannot afford chromosomal catastrophes such as translocations, broken arms of DNA, drastic mutations, deletions. This is a time when the cell checks and double-checks the fidelity of DNA replication, guarding against damage to DNA, or a devastating event in a chromosome. A cell showered with DNA-damaging radiation or chemotherapy might halt at this stage. Proteins termed the Guardians of the Genome — among them the p53 tumor suppressor — scan the genome and the cell to ensure its health before generating new cells.*

*As checkpoints go, G2 seems like a perfectly simple solution until you realize that it has to perform a rather delicate balancing act. G2 "arrest," as far as we know, is reserved largely for detecting *catastrophic* mutations in a cell. Mutations are generated in the S phase. Like

The final phase is M — mitosis itself — the splitting of the cell into two daughter cells. The nuclear membrane dissolves. The about-to-separate chromosomes tighten even farther into the dense structures that Flemming had stained with his dyes. The molecular apparatus to pull apart duplicated chromosomes is fully assembled. And now the duplicated chromosomes, lying side by side, like twins in a cot, begin to get tugged away from one another, until half occupy one side of the cell, while the other half is pulled to the opposite side. A furrow appears between the cells, and the cytoplasm of the cell is halved. The mother cell generates two daughter cells.

I met Paul Nurse in 2017 on a car ride across the plains of the Netherlands. He was a compact man with an English accent and a broad, open smile who reminded me of an elderly, wizened version of Bilbo Baggins. We were both giving talks at the Wilhelmina Children's

---

any copying machine with an intrinsic error rate, the molecular machines that produce new copies of DNA during the synthesis phase make mistakes. Some of these are instantly repaired, but some remain. If G2 were to arrest *every* mutation, catch *every* mistake, and rectify *every* error, mutants would never be generated, and evolution would come to a grinding halt. G2, then, must be a discerning guardian, knowing when to look and when to look away.

Hospital in Utrecht, so we shared a car from Amsterdam to the campus. Nurse was friendly, humble, and kind, the sort of scientist I took an instant liking to. The landscape around us was flat and homogeneous: dry, tilled fields of hay and straw, punctuated with a few windmills that moved cyclically with the occasional gusts of wind.

Cycles. The mechanics of energy — the rise and fall of the wind — that drove the cycles of a machine. Was the dividing cell such a machine, cycling through division and rest? When Nurse was a postdoc in Edinburgh, he began to wonder about the coordination of a cell's cycle. Which factors govern whether, or when, a cell decides to divide? In the 1870s and 1880s, Flemming and Boveri, among others, had observed the distinct stages of cell division. The question was: Which molecules and signals conducted and regulated these phases? How did a cell "know" when to move from, say, the G1 phase to S phase?

Nurse came from a working-class family. "My dad was a blue-collar worker," he told a journalist in 2014. "My mum was a cleaner. My siblings all left school at fifteen. I was different. I passed exams, and I somehow got into university, got a scholarship, and did a PhD." Decades after his university days, Nurse would learn that his "sister" was actually his mother. Born out of an unmarried liaison, Nurse was raised by his grandmother, who acted as his mother until the secret arrangement was finally revealed to him many years later, when he was in his sixties. He told me the story matter-of-factly as we neared Utrecht. His eyes twinkled. "Reproduction is never as simple as it seems," he added drily.

Nurse's mentor at the University of Edinburgh, Murdoch Mitchison, had been studying the cell cycle in a particular strain of yeast called fission yeast — *fission* because they reproduce much like human cells, by splitting in the middle. The more common yeast cells divide by "budding," a process in which a smaller nub of a daughter cells appears as the cell divides.

In the 1980s, Nurse began to make mutants of yeast that would not divide properly. Nearly five thousand miles away, in Seattle, the cell biologist Lee Hartwell had also arrived at a similar strategy: he was also going to hunt for genes that affected the cell cycle and cell division by producing mutants in a different strain: baker's yeast, the budding variety.

Both Hartwell and Nurse hoped that the mutants would lead them to discover the normal genes that control cell division. It was an old biological trick: disrupting a physiological function in order to illuminate normal physiology. An anatomist might cut, or ligate, an artery in an animal and then track the body part that was no longer perfused and thereby learn the artery's function. Or a geneticist might mutate a gene to disrupt a genetic process — cell division, for example — and thereby uncover the functional master regulators that govern the process of mitosis.

In the summer of 1982, Tim Hunt, a cell biologist from Cambridge University, traveled to the Marine Biological Laboratory in Woods Hole, Massachusetts, on scenic Cape Cod, to help teach a course on embryology. Tourists in whale-printed shorts and linen shirts came to the Cape to eat fried clams and lounge on its expansive sand beaches. Scientists, in turn, came there to scour the shallow, rocky tidepools for clams and, more often, for sea urchins.

Urchin eggs, in particular, offered a precious resource, for they were large, and easy experimental models. Inject a female urchin with a simple salt solution, and she will burst forth with an efflorescence of dozens of orange eggs. Fertilize the eggs with male urchin sperm, and the zygote will develop and begin dividing, with clocklike regularity, to start forming a new, multicellular animal. From Flemming in the 1870s, to Hunt in the 1980s, scientists had used these spiny, globular creatures, with their erotic tongues of flesh (Who ever thought of *eating* them?) as model systems to study fertilization, cell division, and embryology. What the fruit fly had been to early genetics, the urchin would be to the study of the cell cycle.

Hunt wanted to study how the synthesis of proteins was controlled after fertilization, but it was frustrating, halting work. "By 1982," he wrote, "work on the control of protein synthesis in sea urchin eggs had almost ground to a halt; every idea that my students and I tested proved to be false, and the very basis of the system was essentially flawed."

But as dusk fell on the evening of July 22, 1982, Hunt noted a remarkable phenomenon: exactly ten minutes before a fertilized urchin cell divided, one abundant protein would peak in concentration, then disappear. It was rhythmic and regular, the precise churn of a windmill's vane. At the evening seminar, followed by the wine and cheese hour that night, he learned that other scientists, including Marc Kirschner at Harvard, had also been puzzling over how cells transitioned from one phase to the next during the genesis of sperm and eggs, or meiosis. The idea that the rise and ebb of a protein could signal the transition from one phase to the next mesmerized Hunt. He may have hardly finished his glass of wine before he was back in the lab.

Over the next decade, Hunt returned to the Cape year after year with a lab-in-a-suitcase — "tubes and tips and gel plates, and even a peristaltic pump" — to try to decipher the mechanisms that enabled transitions in the cell cycle. By the winter of 1986, Hunt and his students had found more such proteins that increased and decreased precisely in conjunction with phases of mitotic cell division. One might peak and drop in perfect step with the S phase (the stage at which chromosomes are duplicated). Another might rise and fall with the G2 phase (the second checkpoint before cell division occurs). Hunt called these proteins cyclins because he was a cycling enthusiast. He would soon realize that he had given them a befitting name: these proteins seemed to be preternaturally coordinated with phases of the cycles of cell division. The name stuck.

Nurse and Hartwell, meanwhile, were also closing in on the cell cycle controlling genes using their mutant-hunting approach in yeast cells. They, too, had found several genes associated with different phases of cell division. In the late 1980s, they named these cdc, and later cdk, genes.* The proteins encoded by them were called CDK proteins.

*These were initially called cdc genes (for "cell division cycle"), but the terminology was changed to cdc/cdk and then to cdk. The K refers to an enzymatic activity of the proteins encoded by these genes — a kinase — that adds a phosphate group to its target protein and typically activates it. For simplification, I have used cdk for the gene and CDK, in caps, for the protein. The same applies to the cyclin family: the genes are denoted in lowercase, while the proteins ("Cyclins") begin with capital letters.

■ ■ ■ ■

But there was an unsettling mystery in the separate lines of discovery. Despite the obvious convergences in their questions, they had not found the same proteins, with one notable exception: one of Nurse's mutants was, indeed, in a cyclin-like gene.

Why? Why was Hunt finding Cyclin proteins in his hunt for regulators of the cell cycle? And why were Hartwell and Nurse finding a (mostly) different set of proteins that coordinated the division of cells? It was as if two sets of mathematicians, having solved the same equation, had emerged with two different answers — and yet, at least in method, both seemed right. What, in short, had the Cyclins to do with the CDKs?

In the 1980s and 1990s, working with teams of researchers, Hunt, Hartwell, and Nurse discovered a synthesis of all the observations — in essence, reconciling the role of the Cyclins and the CDK proteins in the cell cycle. The proteins act in concert to regulate the transitions in the phases of cell division. They are partners and collaborators — functionally, genetically, biochemically, *physically* linked. They are the yin and yang of cell division.

A particular Cyclin protein, we now know, binds to a particular CDK protein, and activates it. That activation, in turn, unleashes a cascade of molecular events in the

206

cell — pinging from one activated molecule to another, like a pinball — that ultimately "commands" the cell to transition from one phase of the cell cycle to the next. Hunt had solved half the puzzle; Nurse and Hartwell had solved the other half. Or, in diagrammatic form:

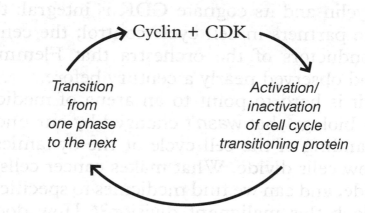

As Nurse told me on the ride to Utrecht, "We were just looking at the same thing from two different sides. If you stepped back, it was really the same thing. It was as if we had caught two different shadows of the same object." The mills whirled around us, completing yet another cycle.

Cyclins and CDKs work together, but different pairs signal the different transitions. One particular Cyclin-CDK association might act as the master regulator of the transition from G2 to M. The Cyclin activates the CDK, which then activates more proteins to facilitate the transition. When the Cyclin is degraded, the activity of the CDK ceases, and the cell awaits the next signal for its next phase.

Another Cyclin-CDK combination regulates the transition from G1 to S. Dozens of other proteins join the coordination of cell division, but intimate association between a Cyclin and its cognate CDK is integral: they are partners in cell cycle control; the central conductors of the orchestra that Flemming had observed nearly a century before.

It is hard to point to an arena of medicine or biology that *wasn't* changed by our understanding of the cell cycle or the dynamics of how cells divide. What makes cancer cells divide, and can we find medicines to specifically block this malignant division?* How does a

---

*It is intriguing, given the central role of Cyclins and CDK proteins in cell division, that few cancer therapies that can block Cyclins or CDKs have emerged or been successful. For the most part, this is because cell division is a universal phenomenon essential for life, and it represents far too degenerate a target for cancer therapy: kill a dividing cancer cell, and you'd also kill a dividing normal cell, thereby unleashing intolerable toxicities. In the late 1990s, a family of drugs was found to inhibit CDK 4/6, two particular members of the CDK family. Almost two decades later, trials proved that new generations of these drugs, in low doses and in combination with

blood stem cell divide to produce a copy of it-self (termed "self-renewal") under certain cir-cumstances, versus producing mature blood cells ("differentiation") under other circum-stances? How does an embryo grow out of a single cell? In 2001, in recognition of the uni-versal importance of their work elucidating the mechanism by which cells control their division, Hartwell, Hunt, and Nurse shared the Nobel Prize for Physiology and Medicine.

Perhaps no arena of medicine is conceptu-ally closer to cell division — to mitosis and meiosis — than artificial or medically assisted human reproduction, or in vitro fertilization (IVF). (The word *artificial* seems odd here. Isn't all of medicine "artificial"? Should we call the use of antibiotics to treat a pneumo-nia "artificial immunity"? Or the delivery of a baby "the artificial externalization of a fetus"? So, I'll use "medically assisted" reproduction, even though "artificial reproduction" is the more commonplace term.)*

---

other medicines, such as Herceptin, an antibody drug against breast cancer, would extend survival in certain patients with breast cancer. The search for cancer-specific Cyclin and CDK inhibitors continues, although the specter of toxicity inevitably hangs over such drugs.

*By "medically assisted reproduction," I am speaking of the body of medicine that seeks to enhance

Let us begin with a fact that is both strikingly self-evident to a cell therapist and startling to someone outside the field: in vitro fertilization (IVF) *is cell therapy*. It is, in fact, among the most common cell therapies in human use. It has been a reproductive option for more than four decades and has produced roughly eight to ten million children. Many of those IVF babies are now adults with children of their own — typically produced without any need for in vitro fertilization. It has become so familiar, indeed, that we don't even imagine it as cellular medicine, although, of course, it is precisely that: the therapeutic manipulation of human cells to ameliorate an ancient and aching form of human suffering: infertility.

---

human reproduction using drugs, hormones, surgical intervention, and ex vivo (outside the body) manipulation of human cells. The range of the discipline is broad: it might involve enhancing the production of human sperm and eggs and enabling their extraction and storage. It might include methods to fertilize sperm and eggs outside the body, or cultivating living human embryos and then implanting them into a female womb to produce a baby. To this list, we might add new technologies that are rapidly intersecting with reproductive strategies: the genetic engineering of human sperm, eggs, and embryos to produce new kinds of cells and, by extension, new kinds of humans.

The technology had an unsteady birth; in fact, it nearly died of prematurity. The scientific animosity, personal rivalry, the public dissent — and even the medical dissent — that accompanied the birth of IVF has largely been airbrushed by its success, but the inception of the technology was nothing but intensely turbulent and controversial.

In the mid-1950s, an unorthodox, secretive professor who taught obstetrics and gynecology at Columbia University, Landrum Shettles, launched a project to create an in vitro fertilized human baby. He wanted to cure infertility. Shettles, who had seven children, rarely went home to rest. His lab was furnished with a large, overgrown fish tank and a series of clocks. He slept on a makeshift cot amid the constant ticktocking, and the medical residents would often find him, in his wrinkled green scrubs, wandering the halls late at night.

At first, Shettles performed his experiments in petri dishes and test tubes. He harvested human eggs from a female donor, then fertilized them with human sperm and managed to keep the primordial embryo alive for six days. He published frequently and won awards for his work, including the Markle Prize from Columbia.

But then his career took an odd turn. In 1973, Shettles agreed to help a Florida couple, Dr. John Del Zio and Doris Del Zio, conceive a child. Shettles did not report the extension of the work — from fertilization in petri dishes to the implantation of an embryo — to the hospital's regulatory or experimental committees. Nor did he inform the chief of obstetrics.

On September 12, 1973, a gynecologist at New York University Hospital harvested eggs from Doris. John carried the eggs and a vial of his semen in a taxi uptown to Shettles's lab. The trip, which I imagine would have taken about an hour through uptown traffic, may have ranked as among the tensest cab rides in New York history.

Dr. Shettles's supervisor, in the meantime, had learned of the experiment, and he was livid. The creation of a human embryo in vitro — a test tube baby — for implantation into a real womb was unheard of, and the medical and ethical implications were obviously uncharted. The story, perhaps apocryphal, is that the supervisor barged into the lab, threw open the incubators with the fertilized eggs, and destroyed the experiment. The Del Zios sued the hospital and won $50,000 for emotional damage.

Unsurprisingly, Shettles — fish tank, sleeping cot, clocks, green scrubs at midnight — was fired by his department and thrown out of the university soon after. He moved to a Vermont clinic, where his unorthodox manners got him into trouble again, and finally settled into his own clinic in Las Vegas, where he promised to continue his dreams of making human babies using IVF.

In England, meanwhile, a duo of scientists, Robert Edwards and Patrick Steptoe, were also attempting in vitro fertilization. Unlike Shettles, they were not tone-deaf to the scientific and moral crossings required to produce a human embryo in a glass jar. They dutifully wrote protocols and papers, presented their work at conferences, and informed hospital committees and departments of their intentions. They worked slowly and methodically, overturning orthodoxy upon orthodoxy. Mavericks they were, but, in the words of science historian Margaret Marsh, "careful mavericks."

Edwards, the son of a railroad worker father and a milling machinist mother, was a geneticist and physiologist interested in cell division and chromosomal abnormalities. His career was derailed temporarily by four years in the British army during World War II as well as by the time spent earning an undergraduate degree in zoology, which he described as "a disaster. My grants were spent, and I was in debt. Unlike some of the students, I had no rich parents. . . . I could not write home, 'Dear Dad, please send me £100, as I did badly in the exams.'"

But Edwards eventually found a spot

studying animal genetics at the University of Edinburgh, where his interests began to swerve toward studying reproduction. He undertook experiments with mouse sperm and then switched to eggs. Collaborating with his wife, Ruth Fowler, an accomplished zoologist, Edwards demonstrated that injecting mice with ovulation-inducing hormones could generate dozens of eggs that were at a similar stage of their life cycles and could therefore, in principle, be harvested and fertilized in vitro, in a dish. In 1963, after a wayfaring career through various universities, Edwards came to Cambridge University to study the maturation of human egg cells. He, Ruth, and their five daughters settled into a modest house on Gough Way, off the Barton Road, and into a lab atop the Physiological Laboratory that was a warren of seven poorly heated rooms.

The field of reproductive biology, and especially the connection between egg and sperm maturation and the cell cycle, was in its infancy. Tim Hunt's work on sea urchins, which would lay the foundation for the cell cycle, was decades away from being published, and the cell division genes that would make Paul Nurse and Lee Hartwell famous were yet undiscovered.

Edwards knew of the work of Harvard scientists John Rock and Miriam Menkin, who, in the mid-1940s, had extracted nearly eight hundred eggs from women undergoing gynecological surgery and tried to fertilize these with human sperm. Success was hit and miss. "[W]e have made numerous attempts to initiate in vitro fertilization of human ovarian eggs," Menkin wrote in a journal article. But the project turned out to be more complicated than either Rock or Menkin had expected; more often than not, the eggs could not be fertilized.

In 1951, Min Chueh Chang, a little-recognized scientist working on reproduction at the Worcester Institute in Massachusetts, realized that the sperm, and not only the egg, might contribute equally to the problem to achieve IVF. Working with rabbits, he proposed that a sperm cell had to be activated — "capacitated," he called it — before it could fertilize an egg. This capacitation, Chang reasoned, was achieved by exposing the sperm to particular conditions and chemicals in the female fallopian tubes.

Edwards sat in the hushed, reverential silence of the library of the National Institute of Medical Research in Mill Hill, London, for months on end, intensely scrutinizing all these prior experiments. It was like studying a litany of failure, but he wanted to try fertilizing human eggs outside the body again. Initially, he worked with a gynecologist, Molly Rose, at Edgware General Hospital, to "ripen" the eggs — in essence, to make them receptive to fertilization. But unlike rabbit and mouse

215

eggs, human eggs wouldn't ripen. "[T]hree, six, nine, and twelve hours, none of them changed their appearance in any way whatsoever. They gazed back at me," he wrote. The eggs seemed, well, impenetrable.

And then, one morning in 1963, Edwards had a crucial insight, as simple as it was profound. What if, he wondered, the "ripening programme in the eggs of primates such as humans might simply take longer than in rodents?" Again, Edwards obtained a small bushel of eggs from Rose and ripened them, but this time he chose to wait.

"I must not look at them too early," he wrote, reprimanding himself for his impatience. "After 18 hours exactly, I looked and saw, alas, the nucleus unchanged, no sign of ripening at all." Failure again. Now he had just two eggs left, staring at him obdurately, unfazed, in the dish. At twenty-four hours, Edwards pulled one out and thought he saw the barest hint of ripening: something was changing in the nucleus.

One egg left.

At twenty-eight hours, he pulled out the last egg and stained it.

"Excitement beyond belief," he wrote. "The chromosomes were just beginning their march through the center of the egg." The cell had ripened; it was ready for fertilization. "There, in one egg in the last of the group, lay the whole secret of the human programme."

216

The moral? We don't reproduce like rabbits. Our eggs need a little more seduction.

Edwards was nearing the end of his lonely decade. But there was yet another circular quandary that he had to countenance: the eggs that Rose had supplied were from women undergoing extensive gynecological surgery and, therefore, highly unlikely to undergo in vitro fertilization. And so, while most convenient as experimental material, the eggs that came from Rose's surgeries were the least amenable for reimplantation. To finish his experiment, Edwards needed human eggs from another source.

The eggs came from patients of Dr. Patrick Steptoe: women with ovarian conditions who had agreed to donate eggs. Steptoe was a consultant obstetrician at the General Hospital in Oldham, a declining, fog-blurred, textile-manufacturing town near Manchester. His particular interest was ovarian laparoscopy, a procedure that entails operating on the ovary and its surrounding tissues using a flexible scope inserted through small incisions in the lower abdomen. The minimally invasive technique was often derided by gynecologists because they imagined it as imprecise compared with invasive open surgery. At one medical conference, a distinguished gynecologist stood up and announced imperiously, "Laparoscopy is of no use whatsoever.

It's impossible to visualize the ovary." Steptoe, soft spoken and reticent, had to rise up to defend his practice. "[You] are hopelessly wrong," he replied. "The whole abdominal cavity can be inspected."

Robert Edwards happened to be in attendance. While gynecologists dismissed Steptoe, Edwards's ears pricked up, for laparoscopic extraction, he realized, was crucial to his success. Unlike the eggs that he had obtained from invasive surgical procedures, laparoscopic extraction would make the procedure vastly more tractable for women — and perhaps precisely for a woman who might want a fertilized egg reimplanted in her womb.

At the end of the presentations, while the audience bickered and quarreled, Edwards ambled up to Steptoe in the foyer.

"You are Patrick Steptoe," he said gently.

"Yes."

"I am Bob Edwards."

They exchanged notes and ideas on in vitro fertilization. On April 1, 1968, Edwards traveled to Oldham to meet with Steptoe. They drew up an experimental plan, and Steptoe agreed to send Edwards some human eggs harvested through his laparoscopic operations. That Oldham was a full five hours away from Cambridge did not deter either of them. The round-trip to bring an egg from Steptoe's clinic to Edwards's lab might involve the better part of a day on a train trundling, slowly, through smoky, rain-drenched Lancashire County towns. The experimental protocol seemed simple, but the details were complicated: Which culture solution would keep the egg and sperm alive? How many hours after egg harvest should the sperm be introduced? How many cell divisions were necessary before a fertilized egg might become viable in a human body? And how might one know which embryo to pick?

From Dr. Barry Bavister, a colleague at Cambridge, Edwards learned that the rate of fertilization increased vastly if you increased the alkalinity of the solution; this was part of the sperm capacitation that had stymied Min Chueh Chang. Edwards picked up additional tricks to activate sperm. And he learned to ripen eggs in culture, waiting for the exact moment of ripening before adding the sperm to it. There were ratios to be determined — How many sperm per egg? — and the exact composition of the fluid for culturing embryos. But piece by piece, Edwards and Steptoe solved the problem of in vitro fertilization. One afternoon, late in the winter of 1968, Jean Purdy, a scientist and nurse working with Edwards, set up the crucial experiment. "Those eggs," she wrote, "were soon ripening in mixtures of culture medium . . . to which some of Barry [Bavister]'s fluid had been added. Thirty-six hours later, we judged that they were ready for fertilization."

That evening, Bavister and Edwards drove to the hospital and studied the culture under the microscope. An awe-inspiring event was unfolding under the lens: the first steps in the conception of human life. According to Purdy, "A spermatozoon was just passing into the first egg. . . . An hour later, we looked at the second egg. Yes, there it was, the earliest stages of fertilization. A spermatozoon had entered the egg without any doubt — we had done it. . . .We examined other eggs and found more and more evidence. Some ova were in the early stages of fertilization, with the sperm tails following the sperm heads into the depths of the egg; others were even more advanced, with two nuclei — one from the sperm and one from the egg — as each [sperm and egg cell] donated its genetic component to the embryo." They had achieved in vitro fertilization.

The paper by Edwards, Steptoe, and Bavister, "Early Stages of Fertilization *in Vitro* of Human Oocytes Matured *in Vitro*," was published in the journal *Nature* in 1969. Unfortunately, Jean Purdy, who had performed the experiment, was not credited, consistent with the conventional practice of cutting women out of science. Later, both Edwards and Steptoe made several attempts to acknowledge her contributions, for IVF was born in Purdy's hands. In the lab, she created the first human embryo produced through IVF; in the hospital, she would later cradle the first IVF baby. In 1985, she died of melanoma at just thirty-nine years of age, never able to fully garner the scientific recognition due to her.

The study set off a public, scientific, and medical furor almost immediately. Attacks came from all sides at once. Some gynecologists did not consider infertility a disease. Reproduction, they argued, was not a requirement for wellness, so why define its absence as an "illness"? As one historian wrote: "It is perhaps difficult now to comprehend the complete absence of infertility from the consciousness of most gynaecologists in the UK at the time, of whom Steptoe was a remarkable exception. . . . Overpopulation and family planning were seen as dominant concerns, and the infertile were ignored as, at best, a tiny and irrelevant minority and, at worst, as a positive contribution to population control." Much of the gynecological research in the United Kingdom and the United States focused on contraception — that is, bringing *fewer* babies into the world. In America, one scientific paper noted, "contraceptive development research increased over 6-fold between 1965 and 1969, and private philanthropic funding went up 30 times."

Religious groups, in turn, pointed to the special status of the human embryo: to produce one in a laboratory petri dish, intended for transfer into a human body, was to violate the most inviolable laws of "natural" human reproduction. And ethicists were hyperconscious of the legacy of Nazi experiments from the 1940s, in which humans were subjected to horrific risks but with little benefit; what if the babies produced by this method, or the mothers who would carry these babies, turned out to bear unknown risks?

It took almost a decade after the publication of "Early Stages of Fertilization" to convince the medical community that infertility *was,* in fact, a "disease." Working with teams of obstetricians and laboratory technicians in the mid-seventies, they launched their first efforts to make a living baby via IVF.

On November 10, 1977, a tiny cluster of living embryonic cells, about twenty-five times smaller than a grain of rice, was transferred into the womb of Lesley Brown. The thirty-year-old British woman and her husband, John, had been trying to conceive naturally for nine years but failed after many attempts. Lesley's fallopian tubes were blocked, and her eggs, albeit functionally normal, were anatomically obstructed in their movement from her ovary to the site of fertilization in the tubes or in the uterus. During the procedure, carried out at Oldham General, her eggs were harvested directly from the ovary, ripened using Edwards and Purdy's protocol, and then fertilized with John's sperm. Purdy was the first to see the embryonic cells begin their division with tiny jerks of movement — a cellular quickening, as it were, captured in a glass jar.

About nine months later, on July 25, 1978, the hospital's operation theater was packed with researchers, doctors, and a team of government administrators. It was close to midnight, when Dr. John Webster, an obstetrician, delivered a baby girl by cesarian section. The operation was completed under a veil of absolute secrecy. Steptoe had originally announced that the birth would occur the next morning but quietly changed it to midnight the night before, in part to outwit journalists who were thronging outside the hospital. Earlier that evening, he had driven off from the hospital in his white Mercedes, in an elaborately planned smoke screen to convince the journalists that the team was packing up for the night. He had crept back surreptitiously at nightfall.

The delivery was spectacularly unremarkable. "[The baby] didn't have to be resuscitated at all, and the pediatrician who examined her for any defects didn't find any," Webster recalled. "We had all been a little concerned that if, by chance, she had been born with a cleft palate or another minor defect that we couldn't pick up beforehand . . . that would have effectively killed off the research — because people would have said that it was due to the technique [of IVF]." Every fingernail, every eyelash, every toe, every joint, every inch of skin was examined. The baby was angelically perfect.

There were no "wild celebrations," said Webster. After the delivery, the obstetrician went to bed to a quiet night of sleep. "I felt quite whacked, really," he recalled. "I simply went back to the house where I was staying and had some supper. I don't think there was even any booze in the cupboard."

The baby was christened Louise Brown. Her middle name is Joy.

The following morning, news of Brown's birth exploded in the press. Over the next week, the hospital was barraged by journalists with flashing cameras and notebooks in hand, straining to get a picture of the mother and her daughter. Louise Brown was called a "test-tube baby" — an odd term, since test tubes were hardly used during the fertilization. (The large glass jar in which she was actually conceived is displayed at the Science Museum in London.) Her birth provoked a tsunami of fury, celebration, relief, and pride. In an indignant letter to *Time* magazine, a woman from Michigan fumed, "[T]he Browns have . . . degraded and institutionalized the child, and for that act, not for their act of medically assisted birth, the Browns should be viewed as symbols of the degeneration of

Western morals." An anonymous package from America arrived at the Browns' home in Bristol, containing a broken test tube splashed with a grotesque spray of fake blood.

Others, however, called her a miracle baby. The July 31 cover of *Time* borrowed the famous detail from *The Creation of Adam,* the Michelangelo painting that adorns the ceiling of the Sistine Chapel, in which God's finger is poised to touch Adam's. Except here, suspended between the two fingers is a test tube, and, in it, a graphic of an embryo: Louise Brown in utero. For men and women who'd been unable to have children, this breakthrough brought extraordinary hope: infertility had been cured, at least for those who still had viable sperm and eggs.

Louise Joy Brown is now forty-three years old. She has her mother's soft, rounded features, her father's open smile, and blond-brown hair, once a torrent of curls, now straightened and golden. She works for a freight company and lives near Bristol. When she was four, she was told that she had been "born in a slightly different way than everybody else." That sentence may be one of the pivotal understatements in the history of science.

Robert Edwards received the Nobel Prize in 2010 for his work. Unfortunately, he died before he could attend the ceremony that December. Steptoe, who was twelve years older than Edwards, had died in 1988. And Landrum Shettles passed away in Las Vegas in 2003, insisting, till the end, that he would have been the first to develop IVF had his efforts not been scuttled by the orthodoxies of his superiors.

This book is about the cell and the transformation of medicine. And while in vitro fertilization may rank among the most commonly used cell therapies in medicine, there is a peculiarity in its history that must be confronted: it was the perfect storm of advances in reproductive biology and obstetrics — not cell biology — that brought the procedure to life.

While the birth of Louise Brown signaled the rebirth of reproductive medicine, the procedural aspects of IVF maintained a rather frigid indifference to the rapidly moving fronts of cell biology. Even Edwards, whose initial interest in reproduction had been sparked by abnormal chromosomal division during egg maturation (a 1962 scientific paper of his was titled "Meiosis in Ovarian Oocytes of Adult Mammals"), wrote virtually nothing further about the discoveries of cell cycle, chromosome segregation, and the molecular control of meiosis and mitosis once those insights had arrived from Nurse, Hartwell, and Hunt in the 1980s. Stranger still was the fact that Hunt was his colleague at Cambridge, while

Nurse worked less than fifty miles away. And the aspects of cellular physiology that you might expect fertilization and embryo maturation to have the most naturally affinity with — the dynamics of cell division, the production of sperm and eggs, the mitotic stages of the zygote — remained in the distant periphery of the field's vision.

IVF, in short, was perceived primarily as a hormonal intervention followed by an obstetric procedure. Eggs and sperm were extracted and went in; a human baby came out. The lab in between, where fertilization occurred and the embryo matured, was merely a link in the chain. The incubator was, quite literally, a black box, albeit a moist, warm one. And how an egg, or sperm, could be made more fecund, or how the best embryos might be selected for implantation — both questions that intimately invoke cell biology, and chromosomal and cellular assessment — remained open and unanswered.

But the insights of Nurse, Hartwell, and Hunt are, at last, beginning to enter the field and transform it. It is now increasingly evident that questions that have furrowed through human reproduction can be answered only by understanding *cellular* reproduction — again bringing to mind Rudolf Virchow's tenet that all disease is cellular disease. IVF is thus learning the vocabulary of Cyclins and CDKs. Why, for instance, is it sometimes tough to harvest eggs from some women, despite hormonal stimulation? In 2016, a group of researchers demonstrated that the very molecules that Nurse, Hartwell, and Hunt had discovered — Cyclins and CDKs — are involved. As long as one such combination, CDK-1 and a Cyclin, remains inactive in egg cells, the cell remains dormant. Quiescent. In G-zero. Release these molecules and activate them, and the egg cells begin to mature. If the eggs mature "prematurely," as it were, they are progressively lost over time. Even with hormonal stimulation, they may be depleted to start. Under such circumstances, the animal is infertile.

Interestingly, this release from quiescence (or cellular "sleep"), and the consequent premature maturation, can be targeted by a newly synthesized drug. This experimental molecule works, predictably, by blocking the Cyclin-CDK activation. In principle, such a drug should be able to put human eggs back to "sleep," potentially enabling higher success rates for IVF in certain groups of women with recalcitrant infertility.

In 2010, a group of researchers at Stanford University School of Medicine took an even simpler approach to developing a toolbox for IVF that relies more intimately on the dynamics of the cell cycle. A perpetual frustration of medically assisted reproduction is that only one in three fertilized embryos reaches a stage where they are likely to generate viable fetuses. To increase the odds, multiple embryos are implanted — but that, in turn, leads to an increased frequency of twins and triplets, which carry their own medical and obstetric complications.

Is it possible to identify one-cell zygotes that are most likely to give rise to healthy, mature embryos? Can one identify such zygotes *prospectively* — in other words, before implanting them, thereby increasing the success rate of a singleton human birth? The Stanford group took 242 human embryos and filmed their maturation from single cell zygotes into hollow, multicellular embryonic balls called blastocysts — an early sign of a healthy, viable embryo. The blastocyst is made of two parts. Its outer shell gives rise to the placenta and the umbilical cord, the developing baby's support system, while the inner mass of cells, hanging on to the wall of the fluid-filled cavity, becomes the embryo. Both the outer shell and the inner mass form out of the first fertilized cell, through the rapid division of cells, mitosis upon mitosis.

The fact that only about one-third of single-cell embryos form blastocysts reflects the one-third success rate of IVF that is found clinically. By playing the film backward, and using software to measure various parameters, the Stanford group identified just three factors that were predictive of future blastocyst formation: the duration of time that it takes the first cell to divide for the first time; the time between that first division and the second; and the synchronicity of the second and third mitosis. By relying on this trio of parameters, the odds of predicting blastocyst formation (and, subsequently, the chance of viable implantation) increased to 93 percent. Imagine IVF performed with a single embryo — no high-risk pregnancies with twins and triplets — and with a 90 percent success rate.

We might also note bemusedly that it was precisely measurements such as these — synchronicity, mitotic time, and the fidelity of cell division — that had enabled Paul Nurse and his students to dissect the cell cycle in yeast cells nearly three decades earlier.

THE TAMPERED CELL
LULU, NANA, AND THE
TRANSGRESSIONS OF TRUST

Do First, Think Later
— An inversion of the proverb

On June 10, 2017, a biophysicist turned geneticist named He Jiankui, who also goes by the nickname JK, met two couples on the campus of the Southern University of Science and Technology in Shenzhen, China. The meeting took place in a nondescript conference room, with swiveling faux leather chairs and a blank projector screen. Two other scientists, Michael Deem, a professor at Rice University and JK's former mentor, and Yu Jun, a cofounder of the Beijing Institute of Genomics, were in attendance, although Yu would state later that they were merely sitting on the side, minding their own business. Perhaps they were discussing the intricacies of the silkworm genome that Yu had sequenced. "Deem and I were chatting about something else," he would say later.

We know very little about the meeting. It

231

was recorded on a grainy video, and a few scattered screenshots remain. The couples had come to JK to give consent for a medical procedure. It was IVF — but with a crucial twist. JK intended to permanently alter the embryo's genes — in essence, creating "transgenic," gene-edited babies — before implanting the embryos back into the womb.

A little more than two years later, on December 30, 2019, He Jiankui would be sentenced to three years in prison for violating fundamental protocols of informed consent and the improper use of human subjects. It would be impossible to tell the story of reproductive biology, or the birth of cellular medicine, without the fable of JK — of the seduction to alter human babies, of scientific aspirations gone wrong, and of the future of gene therapy for embryos left hanging in a fragile limbo.

But in order to tell that story, we must begin nearly a half century earlier. In 1968, the ever-prescient Robert Edwards, of IVF fame, published a paper on a seemingly obscure topic: sex determination in rabbit embryos. Before becoming interested in medically assisted reproduction, Edwards's initial interest in reproductive biology had been sparked by the possibility of detecting chromosomal abnormalities in embryos. In the genetic disorder Down syndrome, for instance, an extra chromosome — number 21 — is left over in the egg or sperm cell. Edwards wondered whether such chromosomal problems could be detected in the embryo — perhaps at the blastocyst stage, the hollow ball of cells — and whether such embryos with chromosomal abnormalities could be selected and discarded before implantation. That way, he reasoned, a couple could choose not to implant a baby with Down syndrome, or any such chromosomal alteration. They could, in effect, select the "right" embryos to implant.

In 1968, Edwards fertilized rabbit eggs and grew them into blastocysts. He held the blastocyst still with a suction pipette — a task akin to immobilizing a water balloon with a vacuum cleaner — and then, with miraculous dexterity, used minuscule surgical scissors to remove about three hundred cells from the outer shell of the blastocyst. He then dyed the extracted cells with chromatin to determine which ones had both X and Y chromosomes, denoting male blastocysts. (Female blastocysts possess two X chromosomes.) In a paper published in *Nature* in April 1968, Edwards and coauthor Richard Gardner reported that by selectively implanting male or female rabbit embryos, they could control the biological sex of mammalian offspring, a task impossible in nature. "Control of the Sex Ratio at Full Term in the Rabbit by Transferring Sexed Blastocysts" began and ended with Edwards's penchant for understatement: "Numerous attempts have been made to control the sex of offspring of various mammals, including man. . . . Now that we can correctly sex rabbit blastocysts, it might be possible to detect other differences in male and female embryos." Edwards had invented a method for embryo selection based on genetic assessment.

By the 1990s, IVF and genetic techniques had advanced to a point that Edwards's technique could be tried on human embryos. At Hammersmith Hospital in London, the scientist Alan Handyside worked with couples who had a history of X-linked diseases for which only male children were at risk of having. By "sexing" the embryos before implantation — as Edwards had done with rabbits — Handyside and his colleagues demonstrated that they could ensure that only female embryos were implanted, thereby eliminating the risk of birthing an infant with the X-linked illness. The technique was called preimplantation genetic diagnosis (PGD), or, in common parlance, embryo selection. PGD was soon extended to screen for embryos with Down syndrome, cystic fibrosis, Tay-Sachs disease, and myotonic dystrophy, among others.

But embryo selection, to be clear, is essentially a *negative* process. By removing only male embryos, you can select embryos that have acquired a particular genetic endowment. However, you cannot fundamentally change the genetic roulette that endows embryos with their genes. In other words, you can cull, or remove, embryos from a set of permutations, but you cannot make embryos with new (de novo) sets of genes. You get what you get (and you don't get upset): permutations of genes from both parents, but nothing outside those predetermined combinations.

But what if we wanted to produce human embryos with genetic features (and futures) that neither parent has? Or what if you wish to alter some information from an embryo's genome — to disable a gene, say, that might cause a lethal disease? In 2012, for instance, I was approached by a woman with a tragic history of breast cancer in her family. The heightened risk of cancer arose from a mutation in the BRCA-1 gene, a mutation that crisscrossed through the family. She carried the harmful variant herself, as did one of her two daughters. Could I help her find a medical strategy to restore the mutated gene in her daughter's embryos? I had little to offer, except the future possibility that she or her daughters might be able to use embryo selection to eliminate (cull) those carrying the BRCA-1 mutation.

Or what if both parents carry mutations in *both* copies of a disease-linked gene? Two copies in the father, two in the mother. A man with cystic fibrosis wants to conceive a baby with the woman he loves, who also happens to suffer from cystic fibrosis. *All* their children would inevitably carry the mutations in both copies, and thereby be inevitably susceptible to the disease. Could a scientist do something to ensure that a child from that union would have at least one corrected copy of the gene? In other words, could a human embryo be the target not just of a negative process — embryo selection — but of a *positive* process: the addition or alteration of a gene, or gene editing?

For decades, scientists had tried with animal embryos. In the 1980s, they had succeeded in introducing genetically modifying cells into mouse blastocysts. After multiple steps, they had generated live "transgenic" mice with genomes that had been deliberately and permanently altered. Transgenic cows and sheep soon followed, all created using somewhat similar techniques. When these animals produced sperm and eggs, they carried the genetic alteration forward into future generations.

But the methods used to create such animals were not easy to apply to humans. The technical hurdles were high. And the ethical concerns about genetic intervention, with the accompanying questions about human eugenics, were just as dissuasive. The dream of generating transgenic humans — those with permanently altered genomes that they would pass on to their children — remained suspended.

236

In 2011, however, a startling new technology crashed onto the scene. Scientists stumbled on a gene-altering method that would be vastly easier to use on cells and, potentially, on early-stage human embryos.* The technique, called gene editing, derives from a bacterial defense system.

Gene editing — making directed, deliberate, and specific changes in a genome — can be deployed through multiple strategies, but the most commonly used form relies on a bacterial protein called Cas9. This protein can be introduced into human cells and then "guided," or directed, to a specific part of a

*It is impossible to name every scientist who contributed to this field — the number is enormous — but some investigators stand out. In the 1990s, a Spanish scientist, Francis Mojica, was the first to perceive that an antiviral defense system was encoded in the bacterial genome. Between 2007 and 2011, Philippe Horvath, working at the Danisco yogurt plant in France, and Virginijus Syksnys, in Vilnius, Lithuania, deepened the understanding of this form of immunity. And between 2011 and 2013, Jennifer Doudna, Emmanuelle Charpentier, and Feng Zhang genetically manipulated the system to make programmable cuts in DNA. This list is necessarily abbreviated; a fuller history can be found online at "CRISPR Timeline," Broad Institute online, https://www.broadinstitute.org/what-broad/areas-focus/project-spotlight/crispr-timeline.

cell's genome to make a deliberate alteration: typically a cut in the genome that usually disables the targeted gene. Bacteria use this system to chop up the genes of invading viruses, thereby inactivating the invader. Pioneers of gene editing, including Jennifer Doudna, Emmanuelle Charpentier, Feng Zhang, and George Church, among others, have adapted this bacterial defense system and turned it into a way to make deliberate edits in the human genome.

Imagine, for a moment, the entire human genome as a vast library. Its books are written in an alphabet containing just four letters: A, C, G, and T, the four building-block chemicals of DNA. The human genome has more than 3 billion such letters — 6 billion per cell if you count the genomes of both parents. Reframed as a library of books, with about 250 words per page and 300 pages per book, we might think of ourselves — or rather the instructions to build, maintain, and repair ourselves — as written in about 80,000 books.

Cas9, when combined with a piece of RNA to guide it, can be directed to make a deliberate change in the human genome. You can analogize it to finding and erasing *one* word in *one* sentence on *one* page in *one* volume of that eighty-thousand-book library. It errs occasionally and erases an unintended word as well, but its general fidelity is remarkable. More recently, the system has been modified not just to erase words but also to implement a vast array of potential changes in a gene, such as adding *new* information or making more subtle alterations. Cas9 is a search-and-destroy eraser. To continue the analogy, it can change *Verbal* to *Herbal* in the preface to volume one of *Samuel Pepys' Diary* in a college library containing eighty thousand books. Every other word, in every other sentence, in every other book in the library is, for the most part, left alone.

In March 2017, according to JK, the Medical Ethics Committee of the Shenzhen Harmonicare Women's and Children's Hospital approved his study to edit a gene in human embryos. "The committee consists of seven people," he wrote. "We were told that the committee held a comprehensive discussion of risks and benefits before reaching the approval conclusion." The hospital would later deny ever having read or approved the protocol. Nor is there any documentation of the "comprehensive discussion" that led to the approval. In addition, the seven people who supposedly approved the protocol are yet to be identified.

The gene that JK was proposing to edit in human embryos was CCR5, an immune-related gene that is a known method of entry for the HIV virus. Previous studies had shown that humans who happen to have two disabled copies of the CCR5 gene with a natural mutation called delta 32 are resistant to HIV infection.

But the logic of He Jiankui's experiment begins to crumble here. First, the couples were chosen because the *father* — not the mother — had a chronic but controlled HIV infection. The risk of HIV transmission from sperm, after the sperm has been washed for IVF processing, is zero. These embryos, in short, were at no greater risk of becoming infected by HIV than an embryo conceived by an HIV-negative couple. Worse, there is evidence that disabling CCR5, which coordinates critical aspects of the immune response, may increase the severity of infection caused by other viruses, such as West Nile and influenza (the latter especially common in China). JK had chosen to edit a gene with no obvious benefit to a human embryo, and with a potentially life-threatening future risk. And whether the couples had been informed of the possible adverse effects of the procedure, and whether informed consent had really been obtained, remain in doubt. In his rush to be the first to make gene-edited humans, JK had, in essence, inverted virtually every principle that governs the ethical use of humans as subjects in clinical studies.

It is difficult to reconstruct what happened next and when, but sometime in early January

2018, twelve eggs were collected from one of the women and the eggs were fertilized with her husband's washed sperm. From JK's slides, it seems as if he injected a single sperm into an egg using a microneedle, a procedure called intracytoplasmic sperm injection (ICSI). At the same time, he must have injected the egg with the Cas9 protein along with the RNA molecule to make a cut in the CCR5 gene.

After six days, JK wrote, four of the single-cell zygotes grew into "viable blastocysts." Not long afterward, he must have biopsied the outer shell of the blastocyst to determine if the edits had been performed.

"Two of the blastocysts were successfully edited," the geneticist wrote. In one of them, both copies of the CCR5 gene had been edited, while in the other one, only one copy was edited. But the gene edits that JK had obtained were not the same as the natural delta 32 mutation found in humans. He had produced a different mutation in the gene, possibly with the effect of conferring HIV resistance but possibly not — it's impossible to know, since no one had performed such a gene edit before. And only one of the embryos had both copies deleted; the other one still had one intact copy. Cells biopsied from the blastocyst were apparently scanned for the possibility that gene edits had been performed inadvertently on other parts of the genome

— off-target edits. One potential unintended edit was found in a sample of the biopsied cells, but the team concluded, without much supportive evidence, that it was "irrelevant."

Despite these multiple caveats, JK's team implanted the two edited embryos into the womb of the mother in early 2018. Soon after, he wrote an e-mail to Steve Quake, his former postdoctoral advisor at Stanford, headlined "Success." It read: "Good news! The woman is pregnant, the genome editing success!"

Quake was immediately concerned. At a previous meeting with JK at Stanford in 2016, he had cajoled him repeatedly, and then urged him sternly, to seek appropriate permission from ethics committees and to obtain informed consent from the patients. So had Matt Porteus, a Stanford professor of pediatrics whom JK had approached for advice. As Porteus recalls: "I spent the next half hour, forty-five minutes, telling them about all the reasons that was wrong, that there was no medical justification; he was not addressing an unmet medical need that, you know, he had not talked about this publicly." JK had sat through the meeting silently, his face flushed, because he had not anticipated such a vehement critique.

Quake forwarded JK's e-mail to an unnamed colleague who was a bioethicist. "FYI, this is probably the first human germ line editing. . . . I strongly urged him to get IRB [institutional review board] approval, and it is my understanding that he did. His goal is to help HIV-positive parents conceive. It's a bit early for him to celebrate, but if she carries to term, it's going to be big news, I suspect."

The colleague wrote back: "I was only telling someone last week that my assumption was that this had already happened. It will definitely be news . . ."

Yes it was. On November 28, 2018, at the International Summit on Human Genome Editing in Hong Kong, JK walked on stage carrying a leather briefcase and wearing dark trousers and a striped, button-down shirt. He was introduced by Robin Lovell-Badge, a geneticist from England. Lovell-Badge had learned only recently that He Jiankui was about to announce the birth of gene-edited human babies in his talk, and he anticipated a media storm. Word of the impending bombshell had already leaked to the press, and journalists, ethicists, and scientists in the audience were staring hungrily at the podium to ask questions. Lovell-Badge introduced JK hesitatingly:

Just to remind everyone here that . . . uh . . . we want to give Dr. He a chance to explain what he's done . . . um . . . in terms of science, in particular, but also . . . um . . . um . . . uh . . . in terms of other aspects of what he's done. So please, can you allow him to speak without interruptions. As I said, I have the right to just cancel the session if there's too much noise or interruption. . . . We didn't know this story beforehand. In fact, he had sent me the slides he was going to include in this session, and it did not include any of the work that he is now going to talk about.

JK's presentation was stilted and vague — almost as if he were a Soviet diplomat reading from a prepared transcript. He flipped through the slides blandly, offering equally banal descriptions of the experiment, often as if he had been a mere onlooker. Cells biopsied from one blastocyst, he said, carried two "likely" disabled copies of the CCR5 gene — although, as I mentioned earlier, neither of the variants was the same as the natural delta 32 mutation found in humans.* The other

*To understand the exact nature of the mutations introduced into the genomes of the babies by JK's method, we need to start with the composition of genes. Genes are "written" in DNA, which is composed of a chain of four subunits: A, C, T, and G. A gene such as CCR5 is composed of a sequence of these subunits: say, ACTGGGTCCCGGGG, and so forth. For most genes, the string of letters can stretch into several thousand such subunits. In the natural human mutation CCR5–delta 32, thirty-two continuous letters are deleted in the middle of the gene, inactivating

244

embryo had one intact copy, and one copy with yet a new mutation not found in nature — possibly conferring resistance to HIV, but possibly not. The mother, JK said, opted to have the two modified embryos implanted, but not the other two unmodified ones. How did she arrive at that decision, given that the route she chose was far riskier? And who had provided her ethical and medical guidance in making that choice? It was as if the questions had not even been considered.

"Gene-edited" twins were born to the woman in October 2018, JK reported — although, bizarrely, in his submitted manuscript about the experiment, which has never been published in a peer-reviewed medical journal but merely made public online, the date was changed to November. The two baby girls, apparently healthy, were named Lulu and Nana. JK refused to divulge their real identities. Some

---

the gene. However, JK didn't re-create that exact thirty-two-letter deletion. With gene editing, it's quite simple to target a gene and erase part of it. But re-creating an *exact* mutation is far more challenging technically. Instead, JK took a shortcut. As a result, one twin has fifteen (not thirty-two) letters missing from one copy of the CCR5 gene, while the other copy is intact. The other twin has four letters missing in one copy, and one extra letter added in the second copy. Neither has the CCR5–delta 32 mutation that occurs naturally in humans.

desultory results had been obtained from the twins' cells — from umbilical cord blood and the placenta — to confirm the presence of the mutation, but crucial questions remained unanswered. Did all cells in their body carry the mutations, or just some?* Were any new off-target mutations observed? Were their CCR5-deleted cells resistant to HIV?

JK repeated the word *successfully* multiple times in his manuscript. But as Hank Greely, a legal scholar and bioethicist at Stanford, wrote: "*Successfully* is iffy here. None of the embryos got the 32-base-pair deletion to CCR5 that is known in millions of humans. Instead, the embryos/eventual babies got novel variations, whose effects are not clear. As well, what does 'partial resistance' to HIV

---

*There are some fundamental scientific questions that He Jiankui hadn't answered and that still remain unanswered. When he had used the CRISPR system to make changes in the embryos, had every cell in the embryo been genetically altered, or just some cells? And if just some cells, which ones? The phenomenon in which some cells of an organism have been genetically altered while others have not been is called mosaicism. Are Lulu and Nana genetic mosaics? The second set of questions derives from the off-target effects of the genetic manipulation. Were other genes altered? Were single cells sequenced to determine whether only CCR5 was changed? If so, how many cells were assessed? We simply do not know.

mean? How partial? And was that enough to justify transferring the embryo, with a CCR5 gene never before seen in humans, to a uterus for possible birth?"

The question-and-answer session that followed JK's presentation can be described only as among the more surreal moments in medical history. At the end of his talk, summoning immense professional restraint, Lovell-Badge and Porteus pushed JK toward a civil, structured discussion of the data. They asked him about the potentially deleterious effects of the gene editing on Lulu and Nana, the nature of informed consent, and the methods used to recruit the couples to the study.

The answers were desultory; it was as if JK were sleepwalking through his experiment and its ethical fallout. "Outside my team . . . um . . . about four people read the informed consent," he stuttered, declining to name any of them. He admitted that he had obtained consent himself, and that two professors — presumably Michael Deem and Yu Jun — had watched him obtain consent from a few patients. (But weren't Deem and Jun supposedly discussing silkworm genetics on one side of the room?) More probing questions elicited answers that seemed archly diffuse: about the global pandemic of HIV and the need for new medicines, but little about the actual gene edits that had been performed on the twins. The panel ended with Dr. David Baltimore, one of the organizers of the summit and a Nobel laureate, appearing onstage and shaking his head in exasperation while delivering one of the most withering reviews of JK's clinical study. "I don't think it has been a transparent process. We've only found out about it. . . . I think that there has been a failure of self-regulation by the scientific community because of a lack of transparency."

And then to the audience. Straining at their leashes through the talk, the listeners erupted with questions. One scientist stood up to ask what "unmet medical need" had been addressed by the experiment: after all, the twins' risk of HIV infection had been *zero,* was it not?

He Jiankui referred vaguely to the possibility that Lulu and Nana might have been HIV negative but still exposed to HIV — a condition called HEU (HIV exposed but uninfected). But this, too, rested on inconceivably weak logic: the mother, after all, did not carry HIV, and sperm washing and in vitro fertilization would have ensured that the embryos were totally unexposed to the virus. He then told the audience that he felt "proud" to have conducted the experiment, eliciting an audible gasp. Other interviewers dug deeper into the issue of consent. Yet others questioned the veil of secrecy around the experiment: Why had virtually no one in the public or the scientific community been informed of the choice?

In the end, JK's presentation — meant, perhaps, to seal his reputation as the first scientist to perform gene editing on a human embryo — dissolved into pandemonium. Journalists, armed with pointed microphones, queued outside the auditorium to heckle him. He was escorted out of the talk by a thicket of organizers, almost as if they were part of a security detail for a political prisoner.

Biochemist Jennifer Doudna, one of the pioneers of the gene-editing system, and cowinner of the 2020 Nobel Prize with her collaborator, Dr. Emmanuelle Charpentier, remembers being "horrified and stunned" by JK's talk. The Chinese biophysicist had tried to contact Dr. Doudna before his talk — perhaps to recruit her support — but she was aghast. By the time she landed in Hong Kong, her in-box was full of desperate e-mails seeking advice. "Honestly, I thought, *This is fake, right? This is a joke,*" she recalled. "'Babies born.' Who puts that in a subject line of an e-mail of that kind of import? It just seemed shocking, in a crazy, almost comedic, way." The talk confirmed Doudna's instincts: JK had crossed the line with little ethical compunction. "Having listened to Dr. He," bioethicist R. Alta Charo stated, "I can only conclude that this was misguided, premature, unnecessary, and largely useless."

Late in 2019, JK was sentenced to three years' imprisonment in China; he has also been barred from performing any IVF research in the future. Meanwhile, as I write this in June 2021, a stocky, passionate geneticist from Russia, Denis Rebrikov, who works in one of Russia's largest government-funded IVF facilities, has announced that he plans to edit a gene for hereditary human deafness. Inheriting two mutated copies of the gene, GJB2, causes deafness. Cochlear implants can restore some hearing of speech, but, oddly, not music; what's more, patients with implants typically require months of rehabilitation.

Rebrikov promises, in the footsteps of Steptoe and Edwards, to be the "careful maverick." But careful or not, he wants to be a maverick nonetheless: he says that although he will seek regulatory approval and obtain informed consent based on strict standards, he will still move forward with genetic manipulations of an embryo. According to Rebrikov, he will take the process ahead step by step: publishing data, sequencing genomes deeply for on-target and off-target effects. And his therapies, he asserts, will be exclusively for deaf couples who carry the mutation in both copies of the gene, who grant full consent, and want a child who is not deaf. He has identified five such couples, and one in particular — a

Moscow husband and wife with mutations in the GJB2 and a deaf daughter — are seriously considering his proposal.

Medical and scientific societies around the world are currently scrambling to establish rules and standards to govern human gene editing in embryos. Some have called for an international moratorium but lack the authority to enforce one. Others would allow the use of gene editing to treat diseases of extraordinary suffering — but does inherited deafness qualify? Although international scientific and bioethics organizations can certainly choose to answer this question, there is no governing body with the power or authority to allow or disbar gene-editing experiments on human embryos.

In vitro fertilization, as I described it earlier, is a cellular manipulation that enables profound forms of human manipulation. Embryo selection, gene editing, and, potentially, the delivering new genes into the genome depend, critically, on cellular reproduction (the meeting of sperm and egg) and the first burst of cellular production (the growth of the early embryo) in a petri dish. Once the creation of the human embryo has been moved out of the womb — once the embryo, at various stages, can be microinjected, cultivated, frozen, culled, genetically modified, grown, biopsied — then an entire array of transformative genetic technologies can be unleashed upon it.

He Jiankui made terrible choices at every level: wrong gene, wrong patients, wrong protocol, wrong purpose. But he was also responding to the inevitable seduction of new technology: he wanted to be the "first." He spoke frequently of his research being his ticket to the Nobel Prize. He compared himself to Edwards and Steptoe, but he reminded me, in fact, of a modern-day Landrum Shettles: fiercely ambitious and restive, passionate about science, yet seemingly unable to discriminate between human experimental subjects and fish in an aquarium.

This is not to excuse his choices; other scientists, armed with the same technology, managed to restrain themselves. But whether by embryo selection or by gene editing, the genetic manipulation of the human embryo to arrest diseases (or, perhaps, to enhance human abilities) seems, every day, to become an inevitable destination for medicine. What began as a treatment for human infertility is now being repurposed as a therapy for human vulnerability. And at the center of this therapy lies an increasingly malleable and increasingly precious cell: the fertilized egg cell, the human zygote.

We are about to move from the cloistered world of the single-celled zygote and approach the developing embryo. But we might pause here and ask the question: Why did we *ever* leave the single-celled world? Why did "we" become "we" — that is, multicellular organisms? Take a yeast cell or some species of single-celled algae. These single cells, or modern cells, as biologist Nick Lane calls them, possess virtually all the features of the cells of vastly more complex organisms, including humans. They are abundant, fiercely successful in their environments, and can thrive in diverse places on Earth. They communicate with one another, reproduce, metabolize, and trade signals. They possess nuclei, mitochondria, and most of the cellular organelles that make an autonomous cell function with extraordinary efficiency. Which begs another question: Why on earth did they choose to form multicellular organisms?

When evolutionary biologists explored this question in the early 1990s, they reasoned that among eukaryotes (nucleus-bearing cells), the transition from single-celled existence to multicellularity may have involved scaling a towering evolutionary wall. A yeast cell, after all, couldn't just wake up one morning and decide that it's better to function as a many-celled organism. In the words of László Nagy, a Hungarian evolutionary biologist, the transition to multicellularity has "been viewed as a major transition with large genetic [and therefore evolutionary] hurdles to it."

But evidence from a series of recent experiments and genetic studies suggests a very different story. First, multicellularity is ancient. Spiral fossils, shaped like the first fronds sent out by ferns, began to appear in blue-green and green algae about two billion years ago; they were all collections of cells that seem to have accreted for a reason. Leaflike "organisms" with radiating structures resembling small veins (veinules) and containing multiple cells, appeared about 570 million years ago and flourished on ocean floors. Sponges agglomerated out of individual cells. Colonies of microorganisms organized themselves into novel "beings," heralding a new kind of existence.

But perhaps the most astonishing feature of multicellularity is that it evolved *independently,* and in multiple different species, not just once, but many, many times. It is as if the drive to become multicellular was so forceful and pervasive that evolution leapt over the fence again and again. Genetic evidence suggests this incontrovertibly. Collective existence — above isolation — was so selectively advantageous that the forces of natural selection gravitated repeatedly toward the collective. The transformation from single cells into multicellularity was, as the evolutionary biologists Richard Grosberg and Richard Strathmann wrote, a "minor major transition."

To some extent, the "minor major transition" from unicellularity to multicellularity can be studied and reproduced in the laboratory. In one of the most intriguing attempts, carried out at the University of Minnesota in 2014, a group of researchers led by Michael Travisano and William Ratcliff made a multicellular being evolve from a unicellular organism.

Skinny and boundlessly enthusiastic, with wire-framed glasses, Ratcliff looks like a perpetual graduate student, although he's a highly cited professor with a large laboratory in Atlanta. One morning in 2010, Ratcliff, on the verge of finishing his PhD in ecology, evolution, and behavior, was chatting with Travisano about the evolution of multicellularity. They both knew that different single-celled organisms had evolved into different multicellular forms for different reasons, and using different pathways.

Ratcliff laughed as he described the experiment, while paraphrasing the famous first line of the classic Tolstoy novel: "All happy families are alike; each unhappy family is unhappy in its own way." For multicellular evolution, he told me, the logic is turned around: every single-celled organism that evolved toward multicellularity took a unique path. It became "happy" — or, rather, more evolutionarily fit — in its distinctive way. Single-celled organisms remained, well, similarly single-celled. It is, in Ratcliff's words, "a reverse Anna Karenina situation."

Travisano and Ratcliff worked with yeast. And so, in December 2010, over Christmas break, Ratcliff set up one of the most magnificently simple evolutionary experiments. He allowed yeast cells to grow in ten separate flasks and then let the flasks stand for forty-five minutes such that single-celled yeast remained afloat, while heavier multicellular aggregates ("clusters") fell to the bottom. (After a few iterations, they found that spinning the broth at low speed in a centrifuge made the selection more efficient.) Ratcliff picked up the multicellular clusters that had been brought down by gravity, cultured them, and repeated the process more than sixty times for each of the ten original cultures, each time selecting the aggregates that fell to the bottom. It was a simulation of multiple generations of selection and growth — Darwin's Galápagos Islands captured in a bottle.

It was snowing heavily when Ratcliff returned to the lab on the tenth day. "Big, heavy blobs of Minnesota snowflakes," he recalled. He dusted off his shoes and anorak, looked at the flasks, and immediately knew that something had happened: the tenth culture was clear, with a sediment at the bottom. And what he saw under the microscope was a mirror of the outside: the sediment in all ten cultures had converged on the selection of a new kind of multicellular aggregate — a crystal-like, many-branched accumulation of several hundred yeast cells. *A living snowflake.* Once they had aggregated, the "snowflakes" continued to survive in these clusters. Cultured again, they did not become single-celled but retained their configuration. Having leapt into multicellularity, evolution refused to go back.

Ratcliff realized that the aggregates ("snowflakeys," he called them) formed because the mother cells and daughter cells clung together even after cell division. This pattern

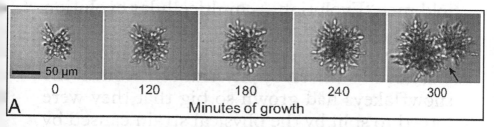

A

The life cycle of a snowflake yeast. The snowflake forms evolved from single-celled yeast cells by selecting for larger clusters. Over time, they maintain these large cluster forms and do not revert to single cells—i.e., they have been evolutionarily selected for multicellularity. New cells are added to the growing branches, increasing the size of the cluster. Initially, the snowflakes were split by the physical strain of their size, like a branch of a tree that has grown too long to remain attached. Over generations, though, specialized cells have now evolved that commit deliberate, programmed suicide to create a cleavage site that fractures to facilitate the division of one cluster into another.

was repeated generation upon generation, like a joint family in which fully grown children forever refuse to leave their ancestral home.

As the experiment continued, and larger and larger snowflake clusters were created, the researchers were puzzled by another question. How did the aggregates propagate? A simple model might suggest that a single cell separates from one cluster and then grows fingerlings to form a new, multicellular starburst form. Instead, they found that the clusters reproduced by splitting down the middle into new clusters after they had reached a certain size. The joint family split into two joint families. "It was breathtaking," Ratcliff told me. "Evolution — multicellular evolution — in a flask."

At first, the propagation of the multicellular clusters was driven by physical constraints: the snowflakeys had grown so big that they were forced to split by the physical strain caused by their size. But then there was an additional surprise: as the clusters kept evolving, a subset of cells in the middle committed a form of deliberate, programmed suicide, thereby enabling a cleft — a cleavage line, a furrow — between the two aggregates, enabling detachment of one cluster from the mother.

I asked Ratcliff what might happen if he kept culturing the snowflakes generation upon generation. He is already up to several thousand and wants to continue to fifty thousand or even a hundred thousand over the course of his life. "Oh, we've already seen the emergence of new properties," he replied with a faraway look, as if imagining the future of this new Being. "The clusters are now twenty thousand times larger than the single cells. And the cells have evolved a kind of entanglement with each other. Now it's hard to break them apart, until the furrow of dead cells forms. And some of them have started dissolving the walls between them. We're trying to see if they're beginning to form some sort of channel of communication to deliver nutrients, or signals, across these large clusters. We've added hemoglobin genes to see if they'll create a mechanism for oxygen transfer. We've started adding genes that might make them convert light into energy, like plants do."

Evolutionary scientists have performed variations of this experiment for a number of different unicellular organisms — yeast, slime molds, algae — and a general principle emerges from them. Under the right evolutionary pressure, single cells can become multicellular aggregates over a mere few generations. Some do take longer, however: in one experiment, unicellular algae became a multicellular agglomerate over 750 generations. That is no more than a blink, a ticktock, in evolutionary time, but 750 lifetimes for an algal cell.

We can only generate theories and lab experiments about *why* single cells are so singularly drawn to form multicellular clusters. To see the real forces of natural selection in action, we would need to rewind time. But the reigning theories suggest that specialization and cooperativity conserve energy and resources while allowing new, synergistic functions to develop. One part of the collective can handle waste disposal, for instance, while another acquires food — and thus the multicellular cluster acquires an evolutionary edge. One prominent hypothesis, bolstered by experiments and mathematical modeling, suggests that multicellularity evolved to support larger sizes and rapid movement, thereby enabling the organism to escape predation (it's hard to swallow a snowflake-sized body) or to make faster, coordinated movements toward weak gradients of food. Evolution raced toward collective existence because "organisms" could race away from being eaten — or, equally, race toward eating. The answer may be unknowable, or perhaps there are many answers. What we do know is that the evolution of multicellularity was not an accident, but purposeful and directional. As I described above in Ratcliff's yeast experiment, certain cells acquire the ability to perform a programmed form of cell death, or self-sacrifice, to split one cluster from another — a sign of cellular specialization at particular, defined locations.

And as Ratcliff has found, as his multicellular aggregates grow generation upon generation, they may be in the process of developing channels to deliver nutrients into the depths of their anatomy.

Notice the words: *specialization, anatomy,* and *location*. At some point in time, perhaps, Ratcliff will begin to describe his clusters as "organisms." He has already started to dissect how they acquire their anatomy. He's wondering about how cells divide to create specialized structures, about what makes them acquire specialized functions, and how those structures determine their locations within the clusters. How might one envision the newly forming channels? Cellular vessels? Nutrient delivery systems? A primeval signaling apparatus? A cell biologist might be tempted to use a word to describe the formation of organized and functional anatomies, and the emergence of specialized cells as these "organisms" grow in size and complexity. She might call it "development."

THE DEVELOPING CELL
A CELL BECOMES AN ORGANISM

Life not so much "is" but "becomes."
— Ignaz Döllinger, nineteenth-century
German naturalist,
anatomist, and professor of medicine

Pause, for a moment, to consider the birth of a human zygote. A sperm swims its way*

---

*The principal mechanism by which sperm swim is a long, whipping tail called the flagellum. At its base is a series of protein molecules that interact with one another to create a minuscule but powerful motor to which the tail is attached, enabling its constant whipping motion. Rings of mitochondria surround the molecular motor, providing all the energy necessary for the sperm's frantic effort to reach the egg. In contrast to the large, whipping flagellum, similar proteins can also form much smaller, mobile, hairlike projections or filaments called cilia, which are central to cell biology. Cilia enable multiple types of cells to move around the body by wiggling their filaments in a constant, often unidirectional motion. Let me

262

across a seemingly oceanic distance and penetrates an egg. A special protein on the surface of the egg and its cognate receptor on sperm clicks the two cells together. Once a single

give you several examples: Cilia that are attached to cells that line the intestines allow nutrients to travel through the body, while those on white blood cells enable them to race through blood vessels to defend the body from infection. Cilia in cells of the fallopian tube are thought to propel a newly released egg toward its site of fertilization, while those in cells lining the respiratory tract beat constantly to expel mucus and foreign particles. And during an organism's development, cilia facilitate cell movement within the embryo. Without properly functioning cilia, it would be virtually impossible to reproduce, develop, or repair a human body. Some children are afflicted with a rare genetic syndrome called primary ciliary dyskinesia, which impairs the cilia's ability to keep the body's highways and byways humming. This can lead to multiple systemic abnormalities such as chronic nasal congestion and frequent respiratory infections from accumulations of phlegm and foreign matter in the airways. Further complicating the picture, about half of PCD patients suffer from congenital organ displacement, on account of cell dysfunction during development; for instance, their heart might be on the right side of the chest instead of the left. Women with PCD tend to be infertile because the cells in the reproductive tract cannot move the egg cells into position for fertilization.

sperm has penetrated an egg, a wave of ions diffuses out from within the egg, initiating a host of reactions that prevent other sperm from entering.

We are, after all, monogamous in the cellular sense.

Aristotle imagined the next steps of the formation of the fetus as a kind of menstrual sculpting. He proposed that the "form" of the fetus was the menstrual blood, which came from the mother. The father supplied the sperm — the "information" — to fashion the blood into the fetal form, and to breathe life and warmth into it. There *was* a logic to this, albeit contorted: conception leads to the loss of menses, and where might that blood go, Aristotle reasoned, but to shape the fetus?

It was a wholly incorrect scheme, but it contained a kernel of truth. Aristotle broke from the ancient idea of preformation, which proposed that the mini-human, called a homunculus, came already premade — eyes, nose, mouth, ears intact — but shrunk into microscopic size and folded tight in the sperm, like a toy that expands to full size when you add water. The preformation theory would preoccupy many scientific minds from ancient times all the way up to the early eighteenth century.

The Aristotelian proposal, in contrast, posited that fetal development occurred through a *series of distinct events* that ultimately led to its form. Genesis occurred by, well, genesis — not mere expansion. As the physiologist William Harvey would write in the 1600s: "There are some [animals] in which one part is made before another, and then from the same material, afterwards, receive at once nutrition, bulk and form." This latter theory would later be called epigenesis, loosely reflecting the idea that genesis occurred through a cascade of embryological alterations that impinged upon or above (*epi*) the developing zygote.

In the mid-1200s, a German friar, Albertus Magnus, whose interests ranged from chemistry to astronomy, studied animal and bird embryos. Like Aristotle, he believed, incorrectly, that the very first steps of fetal formation was a sort of corporeal congealing — like cheese — between the sperm and the egg. But Magnus radically advanced the theory of epigenesis: he was among the first to identify the formation of distinct organs in the embryo: the bulge of an eye where no protrusion had existed before, and the extensions of a chick's wings out of barely discernible bulges on the embryo's two sides.

In 1759, nearly five centuries later, a twenty-five-year-old son of a German tailor, Caspar Friedrich Wolff, wrote a doctoral dissertation titled "Theoria Generationis" in which he advanced Magnus's observations further by describing the series of continuous changes that occurred during embryonic development. Wolff devised an ingenious method to study bird and animal embryos under a microscope. And he was able to watch the stage-by-stage development of organs: the fetal heart beginning its first pulsating motions, and the intestines forming their convoluted tubes.

It was the *continuity* of development that struck Wolff: he could track the formation of new structures that were derived from earlier ones, even though their final morphology bore little physical resemblance to anything in the early embryo. "New objects must be described and explained," he wrote, "and simultaneously their history must be given, even though they have not achieved their firm, lasting form, but are still *continually changing*" (italics my own). For the German poet Johann Wolfgang von Goethe, the serial — and miraculous — metamorphosis of an embryonic form into a mature organism was a sign of Nature at "play." "It is becoming aware of the form with which Nature, so to speak, always plays," he wrote in 1786, "and playing brings forth manifold life." The fetus did not passively balloon into life; Nature "played" with the early forms of an embryo, as a child might play with clay — molding it, sculpting it — into the form of a mature organism.

The observations by Albertus Magnus and, later, Caspar Wolff, of the continual change of fetal organs — Nature's play — would finally demolish preformationism. It would be replaced by a *cell biological* theory of embryological development, in which all the anatomic structures of a developing embryo are formed by dividing cells, creating different structures and performing various functions. As the naturalist Ignaz Döllinger would write in the 1800s, "Life not so much 'is' as 'becomes.'"

But let's return to our zygote floating in the womb. The fertilized cell soon divides into two, then two into four, and so forth, until a small ball of cells is formed. The cells keep dividing and moving — the quickening that nurse-scientist Jean Purdy had observed in Robert Edwards's lab — until the initial mass of cells hollows out within, like a water balloon with a fluid-filled center and the newly formed cells creating the walls of the balloon — a structure called the blastocyst. And a tiny furl of cells divides further and begins to hang off the inner wall of the hollow ball. The outer walls of the cave — the lining of the balloon — will attach to the maternal womb and become part of the placenta, the membranes that surround the fetus, and the umbilical cord. The little bat-like lump of cells hanging inside the ball will develop into the human fetus.*

---

*This is somewhat of a simplification, and I have tried to avoid a lot of the jargon of embryology. For those

The next series of events represents the true marvel of embryology. The tiny cluster of cells hanging from the walls of the cellular balloon, the inner cell mass, divides furiously and begins to form two layers of cells — the outer one called the ectoderm, and the inner called the endoderm. And about three weeks after conception, a third layer of cells invades the two layers and lodges itself between them, like a child squeezing into bed between her parents. It's now the middle layer, called the mesoderm.

This three-layered embryo — ectoderm, mesoderm, endoderm — is the basis of every

---

who want a deeper dive: the wall of the blastocyst, called the trophoblast, gives rise to membranes that house the early embryo — the chorion and the amnion — and a nutrient-supplying structure called the yolk sac. As the chorion invades the womb, forming the placenta, the yolk sac degenerates, thus making the placenta the main source of nutrients. An umbilicus, containing blood vessels and a stalk, connects the embryo to the maternal blood circulation, enabling exchanges of gases and nutrients. For a thorough review of the development of the trophoblast, I suggest Martin Knöfler et al., "Human Placenta and Trophoblast Development: Key Molecular Mechanisms and Model Systems," *Cellular and Molecular Life Sciences* 76, no. 18 (September 2019): 3479–96, doi:10.1007/s00018-019-03104-6. Source: https://pubmed.ncbi.nlm.nih.gov/31049600/.

organ in the human body. The ectoderm will give rise to everything that faces the outer surface of the body: skin, hair, nails, teeth, even the lens of the eye. The endoderm produces everything that faces the inner surface of the body, such as the intestines and the lungs. The mesoderm handles everything in the middle: muscle, bone, blood, heart.

The embryo is now ready for the final sequence of activities. Within the mesoderm, a series of cells assemble along a thin axis to form a rodlike structure called the notochord, which spans from the front of the embryo to its back. The notochord will become the GPS of the developing embryo, determining the position and axis of the internal organs as well as secreting proteins called inducers. In response, just above the notochord, a section of the ectoderm — the outer layer — invaginates, folding inward and forming a tube. This tube will become the precursor of the nervous system, made up of the brain, spinal cord, and nerves.

In one of embryology's many ironies, having set up the framework of the embryo, the human notochord will lose its prominence and function between embryonic development and adulthood. Its only cellular remnant in the adult human body is the pulp that remains stuck between the skeletal bones. In the end, the master maker of the embryo is trapped inside the bony prison of the very creature it has created.

Once the notochord and the neural tube have been generated, individual organs begin to the form out of the three layers (four, if you count the neural tube): the primitive heart, the liver bud, the intestines, the kidneys. About three weeks after gestation, the heart will generate its first beat. A week later, one part of the neural tube will begin to protrude out into the beginnings of the human brain. All of this, remember, emerged from a single cell: the fertilized egg. As physician Lewis Thomas wrote in his collection of essays *The Medusa and the Snail: More Notes of a Biology Watcher,* "at a certain stage there emerges a single cell which will have as all its progeny the human brain. The mere existence of that cell should be one of the great astonishments of the earth."

But what I've written above is descriptive. What about the mechanisms that drive embryogenesis? How do these cells and organs *know* what to become? It is impossible, in a few paragraphs, to capture the immense complexity of the cell-cell and the cell-gene interactions that allow the developing embryo to create each of its parts — organs, tissues, and organ systems — at the right time and in the right place in the body. Each of these interactions is a virtuoso act, an elaborate, multipart symphony perfected by millions of years of evolution. What we *can* capture here is a very basic theme of that symphony — the fundamental mechanisms and processes that enable the developing cell to transform into a developed organism.

In the 1920s, in perhaps one of the most captivating experiments in embryology, a burly, brusque German biologist named Hans Spemann and his student Hilde Mangold began to solve the riddle. Just as Antonie van Leeuwenhoek had learned to grind glass globules into exquisitely lucid lenses, Spemann and Mangold learned to sharpen glass pipettes and needles by heating them on Bunsen burners and pulling the tip gently until the tube — half melted — stretched and thinned, nearly to an invisible point. (Indeed, perhaps a history of cell biology can be written through the lens of the history of glass.) By using these pipettes, needles, suction devices, scissors, and micromanipulators, Spemann and Mangold could extract tiny chunks of tissue from particular parts of frog embryos while the embryos were still globular — long before complex structures, organs, and layers had been formed.

Spemann and Mangold mined one such chunk of tissue from one very early frog embryo. From earlier experiments, in which they had followed the fates of various parts of the embryo, Spemann and Mangold knew that this cluster of cells was already destined to bear the front end of the notochord, parts of the gut, and some of its adjacent organs. This chunk would later be called an "organizer."

They transplanted the tissue under the surface of another frog embryo and waited for the tadpole to grow. What emerged under the microscope was a Janus-like monster. As expected, the chimeric tadpole had two notochords and two guts — one of its own and one from the donor. But the embryo grew even more monstrous, developing into a tadpole with two fully conjoined upper bodies lying side by side, two fully formed nervous systems, and two heads. The tissue extracted from the second tadpole embryo had not just organized itself, but also had commanded the host cells above and around it to adopt fates to *its* specifications. It had "induced," to use Spemann's word, a full second head to grow.*

It would take decades for scientists to identify the precise proteins that were being secreted to "impel" the cells to form a new nervous system and a new head. But Spemann and Mangold had uncovered a basis for the stage-by-stage development of different structures of

*In this case, the transplanted cells happened to come from the front end of the notochord, and so two heads with two nervous systems were formed. The experiments to get the posterior part of the frog embryo to develop from the posterior end of the notochord and mesoderm are much more difficult, for anatomical reasons.

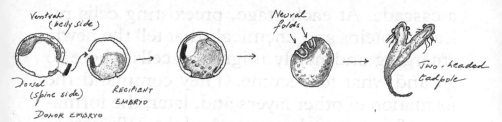

Ventral
(belly side)

Dorsal
(spine side)

DONOR EMBRYO

RECIPIENT
EMBRYO

Neural
folds

Two-headed
tadpole

A representation of an early diagram from Spemann and Mangold's paper describing their experiment. Note the transfer of tissue from the dorsal lip of one embryo to the other induces an embryo with two neural folds, resulting in a tadpole with two heads. A part of the dorsal lip from a very early frog embryo (before any organs or structures have formed) is transplanted into a recipient embryo. The recipient now has two such lips: one of its own and one from the donor. Spemann and Mangold found that the transplanted organizer cells from the donor frog would generate their own neural tubes, guts, and—ultimately—a second fully formed head of a tadpole. In other words, signals from the cells of the dorsal lip induce cells above and around them to form structures of the embryo, including the head and the nervous system. The organizer cells, then, must have an inherent capacity to determine the destinies of their neighbors.

the embryo. Early-developing cells such as the organizer cells secrete local factors that make late-developing cells fix their fates and forms, and these cells, in turn, secrete factors that create organs and the connections between organs.* The growth of an embryo is a *process,*

---

*This begs the question: How do the organizers assume *their* fates? Well, from signals emerging from earlier developing cells — all the way to the single-celled fertilized egg. The fertilized egg already contains

a cascade. At each stage, preexisting cells release proteins and chemicals that tell the newly emerging and newly migrating cells where to go and what to become. They command the formation of other layers and, later, the formation of tissues and organs. And the cells within these layers themselves turn genes on and off, in response to location and their intrinsic properties, to obtain their self-identities. One stage builds upon signals emerging from a prior stage — the tumble of epigenesis that early embryologists had captured so vividly.

Since the 1970s, embryologists have come to discover that the process is even more complex. There is an interplay between intrinsic signals, encoded by genes within cells, and extrinsic signals induced by surrounding cells. The extrinsic signals (proteins and chemicals) reach the recipient cells and activate or repress genes in them. They also interact with one another: cancelling or amplifying their actions, ultimately leading cells to adopt their fates, positions, connections, and locations.

This is how we build our cellular house.

In 1957, a German company called Chemie Grünenthal developed what it considered a

---

protein factors that are distributed in gradients. As soon as it begins to divide, these preset gradients send signals and begin to determine the futures of cells in various parts of the embryo.

marvelous sedative and antianxiety medicine called thalidomide. Marketing was aggressive. The drug was targeted especially to pregnant women, who, given the casual misogyny of the time, were often considered "anxious" and "emotional" and therefore needed to be sedated. Thalidomide was soon approved in forty countries and was being prescribed to tens of thousands of women.

That thalidomide was poised to be a block-buster drug in the United States, where doctors were even more eager to sedate and where it would face even less regulation than in Europe, was obvious to its German manufacturer from the start. In the early sixties, Grünenthal began to search for a partner to bring the drug stateside. Its only hurdle was winning clearance from the US Food and Drug Administration (FDA), typically considered a simple, if somewhat onerous, paper-chasing task. They found a perfect partner in the Wm. S. Merrell Company, which had merged to form pharma conglomerate Richardson-Merrell.

Meanwhile, in early 1960, the FDA had appointed a new commissioner, Frances Kelsey. The Canadian-born forty-six-year-old had earned a PhD and a medical degree from the University of Chicago. After a stint as a teacher of pharmacology (where she learned to how to evaluate the safety of drugs) and a general practitioner in South Dakota (where

275

she learned that even "safe" drugs could produce severe side effects when administered at the wrong dose or given to the wrong patient), she began a long career at the FDA. Eventually she had ascended to the position of chief of the Division of New Drugs, and deputy for scientific and medical affairs in the Office of Compliance. A midlevel bureaucrat. A gatekeeper. One inconsequential flagstone, among many, in the journey of a shiny new medicine developed by a pharmaceutical behemoth and marketed by another.

Merrell's application to bring thalidomide to the United States wound its way through the FDA and eventually landed on Kelsey's desk. But as Kelsey read about the drug, she found herself unconvinced about its safety. The data looked too good. "It was just too positive," she recalled. "This couldn't be the perfect drug with no risk."

In May 1961, as Merrell's executives pressured the Food and Drug Administration to release the drug for general use, Kelsey fired off a response that might well represent one of the most significant letters written in the history of the FDA: "The burden of proof that the drug is safe . . . *lies with the applicant*" (italics my own). She stayed up nights, reading case report upon case report. In February 1961, she noted, a doctor in England had reported severe peripheral nerve numbness after treating some patients; a nurse with

276

access to the drug had given birth to a child with severe limb defects. She pounced on the doctor's case. "In this connection, we are much concerned that apparently evidence [of] peripheral neuritis in England was known to you but not forthrightly disclosed."

Merrell's executives threatened legal action, but Kelsey dug in deeper. She had begun to hear reports of birth defects; now she wanted proof that the drug was safe — not just for peripheral neurons but for pregnant women. When Merrell tried again to apply for a license, Kelsey insisted that the company either prove thalidomide safe or withdraw its application.

As the increasingly acrimonious Merrell-Kelsey battle was unfolding in Washington, DC, more ominous reports began to trickle in from Europe. Women who had been prescribed the drug in Britain and France during their pregnancies began to notice severe congenital malformations in their babies. Some had malformed urinary systems. Some had heart problems. Some had intestinal defects. The most visibly horrific manifestation was that some babies were born with severely shortened limbs, while some had no limbs at all. All in all, about eight thousand malformed babies would be reported in the next few years, and another seven thousand may have died in utero — both likely severe underestimates of the actual degree of harm.

Yet even as one alarming case report after another streamed in from Europe, Merrell remained icily sanguine about the drug. Despite Kelsey's objections, the company had distributed the drug to about twelve hundred US physicians as an "investigational agent." (Smith, Kline & French, another company, was also involved in the patient trial.) In February 1962, Merrell wrote a calmly worded letter to the doctors, casually advising them to keep prescribing the drug: "There is still no positive proof of a causal relationship between the use of thalidomide during pregnancy and malformations in the newborn."

By July, as the wave of cases in Europe crested, the US Food and Drug Administration relayed an urgent message to its officers: "In view of the great public interest in this situation, it is one of the most important [assignments] we've had in a long time. Every effort must be made to contact the physicians within the prescribed period . . . no later than Thursday morning, August 2nd [1962]." Later that month, all prescriptions were stopped. Thalidomide was dead.

In the fall, the FDA began to examine whether Merrell had broken the law in prescribing thalidomide as part of its "investigational trial" and whether it had prevaricated by concealing information in the safety documents filed with the government agency. FDA lawyers listed twenty-four independent counts of legal violation. And yet, in 1962, Herbert J. Miller, the assistant attorney general at the US Department of Justice, chose not to prosecute the company, arguing, with tragicomical absurdity, that it had distributed the drug to "physicians of the highest professional standing" and that only "one malformed baby" had been definitively proven to be harmed. Both claims were untrue. It concluded that "criminal prosecution is neither warranted nor desirable." The case was closed. Merrell, meanwhile, had quietly withdrawn its application from the FDA and tabled the drug for good. Thalidomide had been responsible for a crime of unfathomable proportions — but not a criminal was to be found.

How does thalidomide cause birth defects? As the zygote develops, its cells need to determine their identities and positions by integrating extrinsic factors (proteins and chemicals coming from neighboring cells that signal a cell telling it where to go and what to become) and intrinsic factors (proteins in the cell, encoded by genes, that are turned on and off in response to these signals).

Thalidomide, we now know, binds to one (or several) of the proteins in the cell that break down other specific proteins; it acts as a protein-specific degrader. An intracellular protein eraser. As we saw with the cyclin genes, the regulated breakdown of a particular protein in a cell is critical to the cell's capacity to integrate signals — signals to divide, differentiate, integrate extrinsic and extrinsic cues, and determine its fate. In cell biology, the *absence* of a protein, as much as the presence of a protein, can be just as important to regulate the growth, identity, and position of a cell.

In particular, cartilage cells, certain kinds of immune cells, and heart cells are likely affected by the regulated destruction of proteins that thalidomide alters, although some of these are still hypothetical targets. Unable to integrate the signals that they receive, the cells likely die or become dysfunctional. A multitude of cells are affected, resulting in the dozens of diffuse congenital malformations that were caused by thalidomide. The effect is extraordinarily potent: a single 20-milligram tablet was found to be sufficient to cause birth defects. Tens of thousands of women across the world do not know if their child was miscarried, stillborn, or maimed by an irreversible congenital defect because of thalidomide.

Frances Kelsey likely saved tens of thousands of lives by standing, like the final regulatory bulwark, against the relentless onslaught of a pharmaceutical giant. In 1962, she was awarded the Presidential Medal of Honor. This chapter serves to memorialize her service and tenacity.

If this book is about the birth of cellular

medicine, it must also mark the birth of its demonic opposite: the birth, and death, of a cellular poison.

I titled part two "The One and the Many" not just to mark the transition in our story from the single cell to multicellular organisms, but also to capture an essential tension in science. Biologists often work singly or sometimes in pairs, but, like cells themselves, they also coalesce into scientific communities. And those communities, in turn, belong to, and must respond to, the community of all humans. There is the one and the many, and also the "many many."

We've broached the fundamental properties of the cell in this part: autonomy, organization, cell division, reproduction, and development. What, then, are the permissible limits, and perils, of tampering with these first, fundamental properties, and how are our perceptions of "tampering" changing as new technologies advance? With in vitro fertilization, for instance, "medically assisted" reproduction — once considered radical, prohibited, and even abhorrent to some — has been transformed into the norm. And as Denis Rebrikov, the Russian biologist, readies his lab to gene-edit embryos with hearing disorders, we are confronting yet new ways of manipulating reproduction that disrupt our sense of norms. The saga of thalidomide is obviously a cautionary study in tampering (unwittingly) with the developing fetus. But in recent years, surgery to correct birth defects in the fetus in utero has advanced radically, and drug delivery systems specifically intended to target the fetus are being developed in animal models. Is it that "natural" processes that have evolved untouched since the birth of humans are part of the past, while "tampering" with the developing cell is our inevitable future?

This much is undeniably true: we've thrown open the black box of the cell. To snap the lid shut now might be to foreclose the possibility of a magnificent future. To keep it jammed open without guidelines and rules would be to assume that we've reached some tacit global agreement about what is permissible and impermissible in the manipulation of human reproduction and development — which, assuredly, we have not. We used to think about the fundamental features of our cells as our destiny, manifest. We are now beginning to treat these properties as legitimate arenas of scientific annexation — manifest destiny.

These debates — the manipulation of reproduction and development, or of embryos to change their genes — are ricocheting around the globe as I write this (and I wrote extensively about the promises as well as the perils of these technologies in *The Gene*). The arguments won't be resolved easily, for they impinge not just on the fundamental features of cells, but also on the fundamental features of humans. The only way to find a reasonable answer, or even a compromise, lies in a continuous engagement with evolving debate about the limits of scientific intervention, and the advancing front of cellular technologies. Every human is a stakeholder in this debate. It involves the one, the many, and the "many many."

PART THREE
BLOOD

PART THREE

BLOOD

Multicellularity, the evolutionary transition that made single-celled organisms organize themselves into many-celled beings, may have been inevitable, but it wasn't easy. Multicellular organisms needed to evolve specialized and separate organs to serve their many functions. Each such being had to evolve functional units — separated but connected — to handle its multifarious requirements: self-defense, self-recognition, the movement of signals through the body, digestion, metabolism, storage, waste disposal.

Every organ in the body exemplifies these features: cooperativity between cells and cellular specialization to achieve the function of the organ. But perhaps more than any cellular system, blood represents a model to describe how an entire system of cells achieves these functions. The constant circulation of blood works as the body's central highway to deliver oxygen and nutrients to all tissues. It ensures a coordinated response to injury: platelets and clotting factors use the circulatory system to survey and navigate the body to respond to acute injury. And it enables a response to infection: white blood cells traffic through the same system of vessels to provide layer upon layer of defense against pathogens.

Deciphering the biology of each of these systems has led, in turn, to the creation of new cellular medicines — blood transplantation, immune activation, and platelet modulation, among others. And so, from single cells, we now move to systems of multiple cells: to cooperation, defense, tolerance, and self-recognition, the hallmarks embodying the benefits and liabilities of multicellularity.

THE RESTLESS CELL
CIRCLES OF BLOOD

---

*The cell . . . is a nexus: a connection
point between disciplines, methods,
technologies, concepts, structures,
and processes. Its importance to life,
and to the life sciences and beyond,
is because of this remarkable position
as a nexus, and because of the cell's
apparently inexhaustible potential to be
found in such connective relationships.*
— Maureen A. O'Malley, philosopher
of microbiology, and Staffan Müller-
Wille, science historian, 2010

*There is much that is irresolute
and restless about me.*
— Rudolf Virchow, in a letter
to his father, 1842

Consider where we are in our story. We began with the discovery of cells: their structure, their physiology, their metabolism, their respiration, and their inner anatomy. We journeyed, if briefly, into the world of single-celled microbes and the transformative effect of that discovery on medicine: antisepsis, and the eventual discovery of antibiotics. We next encountered cell division: the production of new cells from existing cells (mitosis) and the genesis of cells for sexual reproduction (meiosis). We witnessed the identification of the four phases of cell division (G1, S, G2, M), the characterization of its crucial regulators — Cyclins and CDK proteins — and the co-ordinated, yin-yang dance of their functions. We discovered how our understanding of cell division is transforming cancer medicine and in vitro fertilization (IVF), and how reproductive technologies, coupled with cell biology, have forced us to enter the ethically unfamiliar landscape of interventions on human embryos.

But thus far we have dealt with cells in isolation: the single-celled microbe, invading the body and precipitating infections. The dividing zygote, floating alone in the petri dish like a lone planet. The egg and sperm, in separate vials, spirited between hospitals in a taxi across Manhattan. The retinal ganglion cell, rescued from degeneration by gene therapy.

The purpose of a cell in a multicellular organism, though, is not to be alone, or to live alone; it is to serve the needs of the organism. It has to function as part of an ecosystem; it must be an integral part of the sum. "The cell is . . . a nexus," Maureen O'Malley and Staffan Müller-Wille wrote in 2010. Every cell lives, and functions, in "an apparently inexhaustible potential to be found in such connective relationships."

It is to these *connective relationships* — between cells and cells, between cells and organs, and between cells and organisms — that we shall now turn.

I spend most of my Mondays with blood. I am a hematologist by training. I study blood and treat blood diseases, including cancers and precancers of white blood cells. On Monday, I arrive much earlier than my patients, when the morning light is still aslant across the black slate of the lab benches. I close the shutters and peer through the microscope at blood smears. A droplet of blood has been spread across a glass slide, to make a film of single cells, each stained with special dyes. The slides are like previews of books, or movie trailers. The cells will begin to reveal the stories of the patients even before I see them in person.

I sit by the microscope in the darkened room, a notepad by my side, and whisper to myself as I go through the slides. It's an old habit; a passerby might well consider me unhinged. Each time I examine a slide, I mumble out the method that my hematology professor in medical school, a tall man with a perpetually leaking pen in his pocket, taught me: *"Divide the main cellular components of blood. Red cell. White cell. Platelet. Examine each cell type separately. Write what you observe about each type. Move methodically. Number, color, morphology, shape, size."*

It is, by far, the favorite time of my day at work. *Number, color, morphology, shape, size.* I move methodically. I love looking at cells, in the way that a gardener loves looking at plants — not just the whole but also the parts within the parts: the leaves, the fronds, the precise smell of loam around a fern, the way the woodpecker has bored into the high branches of a tree. Blood speaks to me — but only if I pay attention.

Greta B. was a middle-aged woman who had been diagnosed with anemia. Her doctors suspected that it stemmed from menstrual bleeding and prescribed iron supplements. But the anemia would not remit. Merely walking a few steps left her short of breath. When she vacationed in the Sierra Nevada mountains, six thousand feet above sea level, she could hardly breathe. Her doctors increased Greta's iron pills, but to no avail.

Greta's illness, it turned out, was more mysterious than her doctors had initially suspected. If you looked at the blood counts, this wasn't a simple anemia. Yes, her red blood count was lower than normal, as would be expected. But so, too, were her white cells — just a smidgeon below the normal limits for her age. And the platelets also fell short of the normal range, although barely so.

Under the scope, Greta's blood smear revealed a more complicated story. I ran my eyes through the smear like a wild animal sensing a new landscape — pausing, sniffing, sending shivers of thought through my brain. The red cells appeared almost normal. *Almost.* I underlined the word. Scanning the smear, I found some strange-looking ones that carried a distinct blue dot in the middle — the remnant of a nucleus that most mature red cells do not have, as they usually expel their nuclei in the bone marrow. *"That nuclear remnant shouldn't be there,"* I whispered aloud, and noted it in my book.

The white cells were the most odd-looking. Normal white cells come in two main forms: lymphocytes and leukocytes. (We will return to the distinction later.) In Greta's case, one type of leukocyte, termed a neutrophil, was the strangest-looking of all. The nuclei of normal neutrophils have three to five lobes, like an archipelago of three or five islands connected by narrow isthmuses. But some of Greta's neutrophils had only two nuclear lobes, perfectly round and connected by a narrow line of blue in between. They resembled a pair of eighteenth-century spectacles. "Pince-nez cells," I wrote. Gandhi's spectacles. And at least a couple of neutrophils had large, dilated nuclei with disorganized-looking chromatin. Immature blood cells, or blasts. The first signs of malignant white cells.

I read through my notes. The red cells and white cells — two of the main cellular components of blood — were abnormal. A bone marrow biopsy confirmed that she had myelodysplastic syndrome, a clinical syndrome in which the bone marrow does not generate normal blood. About one in three patients diagnosed with MDS progresses to leukemia — cancer of white blood cells.

Greta's iron pills were discontinued, and she was started on an experimental drug. Her blood counts normalized for about six months, but then the anemia returned, and the percentage of blasts in her bone marrow began to rise again. Under normal circumstances, blast cells make up, at most, 5 percent of the marrow; her blast count was several times that, indicating that the MDS was in the process of transforming to frank leukemia. At that point, her treatment options would be limited to chemotherapy to kill the leukemia or the possibility of trying another experimental drug to keep the disease at bay.

In medical school, my professors taught me how to speak the language of blood; now, at long last, the tissue speaks back to me. In fact, blood speaks to everyone and everything: it is the central mechanism of long-distance communication, of transmission, in humans. Be it hormones, nutrients, oxygen, or waste products, blood delivers and connects — *talks* — to every organ and from one organ to the next. It even speaks to itself: its three cellular components, red cells, white cells, and platelets, in particular, engage in an elaborate system of signaling and cross talk. Platelets band together to form a clot. A single platelet, in isolation, cannot congeal into a clot, but millions of platelets, in conjunction with proteins in the blood, collaborate to seal a bleeding site. White blood cells have the most complex system of all: they signal one another to coordinate an immune response, the healing of wounds, fighting microbes and surveying the body for invaders as a *system* of cells. Blood is a network. As with M.K., the young pneumonia patient with immunodeficiency, the collapse of one piece of the net can lead to the collapse of the whole net.

The idea of blood as an organ of communication, or transmission, between organs has a long history. Around AD 150, Galen of Pergamon — the Greek surgeon to the Roman gladiators, and, eventually, physician to Emperor Lucius Aurelius Commodus — had proposed that normal bodies were composed of some "balance" of four humors: blood, phlegm, yellow bile, and black bile. This humoral theory of disease predated Galen:

295

Aristotle had written about it, and Vedic doctors often referred to the interplay of internal fluids. But Galen was among its most vociferous champions. Illness, he contended, occurred when one of the humors was off-balance in a body. Pneumonia was an excess of phlegm. Jaundice (or rather hepatitis) came from yellow bile. Cancer was a disease of the accumulation of black bile, a fluid also associated with melancholia or depression (*melancholia* literally means "black bile") — a magnificent theory that was as metaphorically seductive as it was mechanistically flawed.

Of the four fluids, blood was the most familiar. It poured out of gladiatorial wounds; it was acquired easily from slaughtered animals for experimental manipulations; it was, indeed, built into the very vocabulary of common human language. It was warm, active, and red to start, Galen noticed, and then, like the victims who spilled blood, it grew blue, sluggish, and cold. Galen associated its normal function with heat, energy, and nutrition. Its redness, or rubor, was a sign of its warmth and vitality. Blood existed to distribute nutrition and heat to the organs, Galen posited. The heart, he imagined, was the body's furnace — a heat-generating, smelting machine cooled by the bellows-like lungs. It was a restatement of Aristotle's idea of blood as the internal "cooking oil" of the body. Blood picked up the heated-up food from the heart, and, like a food delivery vehicle, kept the nutrients warm until they had reached the brain, kidney, and other organs.

In 1628, the English physiologist William Harvey upended this theory in his book *Exercitatio Anatomica de Motu Cordis et Sanguinis in Animalibus*. Early anatomists had proposed that the flow of blood was unidirectional, traveling from one's heart to the intestines, say, where it reached a dead end. Harvey argued that blood moved around in continuous circles: entering the heart, exiting it, and then returning to the heart again after having completed its delivery route. There were no separate conduits for heating and cooling. "I began privately to think that it might rather have a certain movement, as it were, in a circle," he wrote. "[Blood] flows through the lungs and heart and is pumped to the whole body. There it passes through pores in the flesh into the veins, through which it returns from the periphery everywhere to the center, from the smaller veins into the larger ones, finally coming to [the heart again]." The heart was not a furnace or a factory, or even a cooling fan for a furnace or factory. It was a pump — or, rather two pumps, conjoined with each other — that powered these two circuits. (We will return to Harvey's work on the heart in a few chapters.)

But what was the purpose of the circular movement of blood? What substance was blood carrying — in these restless, continuous circles — around the body?

Cells, of course, among other things. Red blood cells. Leeuwenhoek had seen them floating in blood. On August 14, 1675, he wrote: "[T]hose sanguineous globules [red blood cells] in a healthy body must be very flexible and pliant, if they are to pass through the small capillary veins and arteries, and that, in their passage, they change into an oval figure, reassuming their roundness when they come into a larger room." It was a prescient idea: that blood cells, as they moved through thin-bored capillaries, would deform their structures, and then resume their disk-like roundness. Marcello Malpighi, the seventeenth-century Italian anatomist, had also seen these red globules. So had the Dutch physician-scientist Jan Swammerdam, who had pulled a drop of freshly ingested human blood out of a louse's stomach in 1658. In the 1770s, a British anatomist and physiologist named William Hewson studied the shape of red cells more carefully. They weren't round globules, he concluded, but disk shaped, with an indentation in the middle, like a circular pillow that has just been punched.

The cells were so abundant that they must have a function, Hewson surmised. But the mystery of what red blood cells carried — why the circles circulated so relentlessly, and why they squeezed their way so purposefully through tiny capillaries, distorting their very shape — remained unsolved. In 1840, Friedrich Hünefeld, a German physiologist, found a protein in the red blood cells of earthworms. Hünefeld was surprised by the protein's abundance — more than 90 percent of a red cell's dry weight is composed of just one protein — but he did not understand its function. The name given to the protein — hemoglobin — was just a bland restatement of its cellular location. A glob in blood.

By the late 1880s, however, physiologists had begun to understand the importance of the "glob." They noted that hemoglobin carried iron, and the iron, in turn, bound oxygen, the molecule responsible for cellular respiration. The observations made by Harvey, Swammerdam, Hünefeld, and Leeuwenhoek began to crystallize into a theory. The principal purpose of the red blood cell was to ferry oxygen, bound to hemoglobin, to tissues in all the body's organs. Red cells pick up oxygen in the lungs, then are routed to the heart, which propels them on their voyage through the arteries to the rest of the body.*

*But why does one need a *cell* to transport oxygen? Why not float hemoglobin as a free protein in the plasma and let it move in the body? It's a conundrum that still remains unsolved and has to do with the structure of hemoglobin — a fascinating subject we will return to in this book's final pages.

In addition to cells, plasma, the fluid component of blood, carries other materials crucial to human physiology: carbon dioxide, hormones, metabolites, waste products, nutrients, clotting factors, and chemical signals.

One astonishing feature of the body's circulation is that, like all circles, it is recursive. Red cells carry oxygen to all body parts — and, in due course, to the muscles of the heart, the very organ that is responsible for propelling blood throughout the body. The heart draws oxygen from red cells to pump, thereby sending the red cells in another roundabout mission to bring them more oxygen to pump and so forth, in endless circles. In short, circulation depends on the heart, whose essential function depends on . . . well, circulation. The transmission of every substance in the body, and, by extension, the operation of *every* organ, then, depends on the most restless of all our cells.

But there is yet another kind of transmission that blood is capable of: it can be transferred from one human to another. Blood transfusion, the first modern form of cellular therapy, would lay the basis for surgery, for treating anemia, for cancer chemotherapy, for trauma medicine, for bone marrow transplantation, for the safety of childbirth, and the future of immunology.

Blood transfusion didn't have a particularly

propitious origin: early experiments to transfuse blood into humans ranged from the macabre to the mad. In 1667, Jean-Baptiste Denys, a personal physician to King Louis XIV of France, bled a young boy multiple times with leeches and then attempted to transfuse sheep's blood into him. Miraculously, the boy survived — likely because the amount of blood transfused was minimal, and there was no allergic response. Later that year, Denys tried to transfuse animal blood into Antoine Mauroy, a man with a psychiatric disorder. The blood of a calf, an animal known for its sober disposition, was chosen in the belief that it might calm the overheated madness in Mauroy — yet again reinforcing Galen's notion of blood as one of the carriers of the psyche. Three transfusions later, unfortunately, Mauroy was not just preternaturally calm; he was dead, his body and face bloated by an allergic reaction. His wife tried to sue Denys for murder, and the doctor narrowly escaped imprisonment. He stopped practicing medicine. The episode set off a minor furor in France, and experiments with animal-to-human transfusions were banned.

Other research into transfusing blood continued throughout the seventeenth and eighteenth centuries. Scientists noted that transfusion between identical twin animals was accepted, while transfusions between siblings, including fraternal (nonidentical) twins, were rejected — suggesting that some genetic compatibility was required in order for transfusion to succeed. But the nature of that compatibility remained a mystery.

In 1900, an Austrian scientist named Karl Landsteiner began to tackle the challenge of human blood transfusion more systematically. Where there had been madness before him — sheep and calf blood forced into leached boys or psychically disturbed men — Landsteiner was all about method. Blood was a liquid organ. It moved freely around the body. Why couldn't it be moved, just as freely, from one human body to another?

Landsteiner mixed blood from one individual (call him A) and serum from another (B) and watched the two react in test tubes and on glass slides. Serum differs from plasma: it is the fluid that's left over after blood has been clotted. It contains proteins, including antibodies, but no cells. Serum from A mixed with blood from A obviously yielded no reaction — a sign of compatibility. "[T]he result was exactly the same as if the blood cells had been mixed with their own serum," Landsteiner noted. The mixture melded together and remained liquid. But on other occasions, when blood from patient A was mixed with serum from patient B, the combination formed tiny, semisolid clumps. (My hematology professor described them as "seeds in strawberry juice.") The incompatibility could not lie in A's *cells* rejecting B's cells; serum, remember, has no cells. Rather, it must be a protein — later found to be an antibody — present or absent in A's blood that was attacking B's cells, a sign of immunologic incompatibility.*

By mixing and matching blood from various donors, Landsteiner eventually found that he could classify the blood from humans into four groups: A, B, AB, and O.† The groups indicated transfusion compatibility. Humans with the A blood group could accept blood only from others with the A blood group (and the O group). Those with B could accept blood only from others with B (and the O group). The O group was the strangest: O blood did not react with either A or B. Humans in this group could *donate* blood to someone with either type, but they could not *accept* blood from anyone but fellow Os. A fourth and final major blood group, AB, was discovered soon after. These individuals

---

*The antibody was later found to be reacting to a unique set of *sugars* found on the surface of red blood cells.

†Landsteiner initially found only three blood groups that he denoted A, B, and C. But in his papers, published in 1936, he had already distinguished four independent blood groups, now denoted A, B, AB, and O.

could *receive* blood from all donors but could donate only to other AB group humans. In common parlance, the four groups came to be known as A, B, O (universal donors), and AB (universal acceptors). In a single table (reproduced in his collected papers and later published in 1936) Landsteiner delineated the four basic blood groups and laid the basis for blood transfusion. It was an advance of such medical and biological significance that that one table alone would be sufficient to award Landsteiner the 1930 Nobel Prize in Physiology or Medicine.

In time, the blood group system would undergo further refinement. Other factors were added, such as Rh-positive (denoting the presence of an inherited protein called Rhesus-factor on the surface of red blood cells) and Rh-negative (indicating a lack of Rh-factor), for determining compatibility within each group: A+, B-, AB-, and so forth.

The discovery of blood compatibility transformed the field of blood transfusion. In 1907, at Mount Sinai Hospital in New York, Dr. Reuben Ottenberg began to use Landsteiner's reaction of compatibility to perform the first safe transfusions between humans. By matching blood between donors and recipients *before* the transfusion, Ottenberg demonstrated that blood could be safely transferred between mutually compatible humans. Transfusion slowly turned into a systematic and safe science. In 1913, after more than a half decade's experience with matching blood, Ottenberg wrote:

"Accidents following transfusion have been sufficiently frequent to make many medical men hesitate to advise transfusion, except in desperate cases [but] since we began making observations on this question in 1908, such accidents could be prevented by careful preliminary tests. . . . Our observations on over 125 cases have confirmed this view and we believe that untoward symptoms can be prevented with absolute certainty."

But even so, early blood transfusions were still extraordinarily cumbersome. Timing was crucial; it was like a hectic relay race, with a blood-filled syringe as the moving baton. One technician would draw pints of blood repeatedly through a needle inserted in the donor's arm, another would rush the crimson liquid across the room as fast as she could, and a third would inject the blood into the recipient's arm. Or a surgeon might create a *physical* link between the donor's artery and the recipient's vein — binding them, literally, by blood — so that the fluid could flow directly from the donor's circulation to the recipient's without touching free air. But without such interventions, the liquid form of blood was fleetingly evanescent outside the body. Left alone, even for a few extra minutes, it would coagulate, turning it from a life-saving fluid to an unusable clog of gel.

A few final technological advances were necessary to make blood transfusions amenable to use in the field setting. The addition of a simple salt found in lime juice, sodium citrate, kept blood from coagulating, prolonging its storage. In 1914, the year the Great War ignited, an Argentinean doctor, Luis Agote, transfused citrated blood from one person to another — a glorious instance of technology preempting its need. "This great stride forward in the technique of blood transfusion coincided so nearly with the beginning of the war," British surgeon Geoffrey Keynes wrote in 1922, "that it seemed almost as if *foreknowledge* of the necessity for it in treating war wounds had stimulated research." Another advance, refrigeration, added to the longevity of stored blood. Further innovations involved the use of paraffin-coated storage bags, and the addition of simple sugar (dextrose) to keep the blood from spoiling. The number of transfusions soared in hospitals across the world. In 1923, there were 123 transfusions at Mount Sinai Hospital. By 1953, there were more than 3,000 annually.

The real trial for blood transfusion — its field test, as it were — was on the blood-soaked battlefields of World Wars I and II. Shelling tore off limbs; internal wounds hemorrhaged;

arteries severed by bullets could bleed out in a matter of minutes. By 1917, when the United States joined the Allies in the fight against Germany and the other Central Powers, two military medical specialists, Major Bruce Robertson and Captain Oswald Robertson, had pioneered transfusions for acute blood loss and shock. Plasma, too, was used extensively to revive seriously wounded soldiers. Though a short-term solution for blood loss, it could be stored more easily and required no typing and matching.

The two Robertsons were unrelated. Oswald, serving in the French front of the US Medical Corps, began to think of blood as a mobile organ — restless, not just inside humans or between humans, but between national borders and battlefields. He collected O group blood from convalescing soldiers at one site, then packed sterile two-liter glass bottles containing citrated, dextrose-supplemented blood in ammunition boxes filled with sawdust and ice, and shipped them to the battlefield for use. In effect, Captain Oswald had established one of the first blood banks. (A more formal bank would be set up in Leningrad in 1932.)

Gratitude poured in. "On the 13th June, you took my leg off above the knee," one soldier wrote to Major Bruce Robertson in 1917, "and until I received blood from someone else, you considered the betting about 3 to 1

307

on my pegging out. . . . Can you find time to let me know the name and address of the man who gave me blood? I should much like to write to him."

By the time World War II broke out just two decades later, banking, matching, and transfusion had become common practices in the field. Compared to the First World War, the mortality rate of wounded soldiers who reached a field hospital nearly halved — due partly to blood transfusions. In the early 1940s, the United States, aided by the American Red Cross, launched a nationwide program for blood donation and banking. By the end of the war, the Red Cross had collected thirteen million units of blood, and within a matter of years, the US blood system had fifteen hundred hospital-based blood banks. There were forty-six community donation centers, and thirty-one regional centers to donate blood.

As one writer put it in a 1965 issue of the journal *Annals of Internal Medicine,* "War has never lavished gifts on humanity; an exception may be made for the impetus and popularization of the use of blood and plasma . . . attributable to the Spanish Civil War, World War II, and the Korean conflict." Perhaps more than any other intervention, transfusion and banking — cellular therapy — stands as the most significant medical legacy of the war.

It is virtually impossible to envision the development of modern surgery, safe childbirth, or cancer chemotherapy without the invention of blood transfusion. In the late 1990s, I resuscitated a man with liver failure who had one of the most severe forms of hemorrhage that I have ever witnessed. He was a sixtysomething man from South Boston with liver cirrhosis caused by something that the consultant hepatologists had never quite been able to pinpoint. A former restaurateur, he drank alcohol but insisted that the amount was well below the levels that would make his liver fail. There was no chronic viral infection. Some genetic predisposition must have compounded the alcohol use and caused the chronic cellular inflammation, eventually resulting in his burned-out, shrunken liver. His eyes were yellow from jaundice, and the level of albumin, a protein synthesized in his blood, was dangerously low. His blood failed to clot normally — again a sign of a diseased liver, for the organ produces some of the factors required for blood to clot. Now he was in the hospital, awaiting a liver transplant. But overall, he was generally fine and had been placed on a routine monitor.

At first, the evening was uneventful. But then the patient felt a wave of nausea, and his blood pressure dropped. A little monitor beeped. The blood pressure cuff read and reread the numbers: something was off. Within minutes, it was as if a spigot had opened in his guts, with blood pouring out everywhere. Liver failure often causes the blood vessels in the stomach and esophagus to dilate and become brittle; once they burst, the gush of blood can be unstoppable. Add to this the impaired clotting associated with cirrhosis, and the bleeding can spiral into a medical disaster. The nurses and doctors in the ICU tried to stop the bleeding and then urgently issued an emergency "code" call. I was the senior medical resident on call that night.

By the time I walked into the room, it was already humming with frantic activity. The IV lines inserted into his veins were too thin. "I need a line," I commanded, surprised by the volume and confidence of my own voice. We inserted two new ones, but the bags of saline, dripping sluggishly, could hardly be expected to keep up with the loss of blood.

By now, the man had begun to flail and lose consciousness. He spoke madly — cusswords, sitcom characters, childhood memories — and then, ominously, stopped talking altogether. I touched him. His feet felt ice cold: the blood vessels in his skin had compressed to conserve blood in his vital organs. The floor, meanwhile, was covered in white towels that had turned crimson red; there were clots of blood drying on my clogs. My scrubs were crusty and had turned a purplish-red color. One nurse replaced the blood-soaked towels with new ones, but within a few minutes, they were bright red, too.

A surgical resident managed to insert a large-bore line into his neck vein, while I looked frantically for an IV site in the groin.

Pulse, pulse, pulse, I said to myself. The man's blood pressure, meanwhile, kept dipping, and his pulse became thready. The code team continued working in a choreographed dance that reminded me of the early days of blood transfusion: this, too, was a relay race, with blood as the central baton.

It seemed like hours before the bags of blood were brought up, but, in fact, the whole process took less than ten minutes. We hung two bags. "Squeeze gently," I said, and the nurse managed to send in a bagful over a few minutes. "Squeeze hard," I said, changing my mind, as if I could speed up time. It took eleven, perhaps twelve, bags of blood to stabilize him. I lost count. We added a bag or two of clotting factors and platelets to help his blood clot. Two hours later, we had managed to restore the pulse, and the bleeding had slowed down. By late evening, the bleeding had stopped. His skin warmed up, and he began to respond to commands. "Move your left hand." He did. "Shake your toes." He did. I felt a joy that I cannot describe. He awoke the next day and managed to hold a cup of ice in his hand.

The lasting image I have from that night is walking down the lonely corridor of the sixth floor and ducking into the bathroom to disinfect my soaked clogs with a spray bottle and wash the dried blood off them. The leather was so deeply encrusted that it made me nauseous. It was a Macbethian moment: I could not wash off the stains. I threw the clogs in the trash and bought new ones from the hospital store the next morning.

Ever since that evening, I never use the word *bloodbath* casually. I happen to be among the few people that has actually been bathed in blood.

THE HEALING CELL
PLATELETS, CLOTS, AND A
"MODERN EPIDEMIC"

Imperious Caesar, dead and turned to clay,
Might stop a hole to keep the cold away
Oh, that that earth, which
kept the world in awe
Should patch a wall to expel
the winter's flaw.
— William Shakespeare,
Hamlet, Act V, Scene 1

It would be too glib to say that the surgeons, or the nurses, or I, for that matter, had stopped the man's bleeding that night in Boston. We were accessories. There was a cell — or, rather, a fragment of a cell — that played the central role in controlling the bleed.

In 1881, the Italian pathologist and microscopist Giulio Bizzozero found that human blood carried minuscule fragments of cells — tiny, shorn-up pieces, barely visible but always present. For decades, hematologists had wondered about these fragmentary pieces that floated in blood; in 1865, a German microscopic anatomist named Max Schultze had described them as "granular fragments." Schultze thought of them as pieces of shredded blood cells, found them in clots, and recommended that for "those who are concerned with the in-depth study of the blood of humans, the study of these granules in human blood is enthusiastically recommended."

Bizzozero recognized them as an independent component of blood. "The existence of a constant blood particle, differing from red and white blood cells, has been suspected by several authors for some time," he wrote. "It is astonishing that none of the previous investigators made use of the observation of circulating blood in living animals." He assigned the shards a name: *piastrine,* in Italian, for their flat, round, platelike appearance. In English, they were called platelets — small plates.

Bizzozero was more than a microscopist; he was a physiologist in full. Having observed these cell fragments in the blood, he began to wonder about their function. Were they just debris — flotsam in the red ocean of blood? When he pricked the artery of a rabbit with a needle, he observed platelets accumulating at the injury site: "Blood platelets, swept along by the bloodstream, are held up at the damaged spot as soon as they arrive at it," he wrote. "At first, one sees only 2 to 4 to 6 (platelets); very soon the number climbs to hundreds. Usually, some white blood cells are held up amongst them. Little by little, the volume increases, and soon the *thrombus* [blood clot] fills the lumen of the blood vessels, and impedes the bloodstream more and more."

Platelets are unusual in their biology right from their birth. In the early 1900s, the Boston hematologist James Wright developed a new stain to visualize cells in the bone marrow. Nestled within the various cell types — maturing neutrophils, slowly unfurling their early, egg-shaped nuclei into many-lobed ones; red cells developing in tight clusters — he found a massive cell that seemed to have defied the conventions of cell biology. Rather

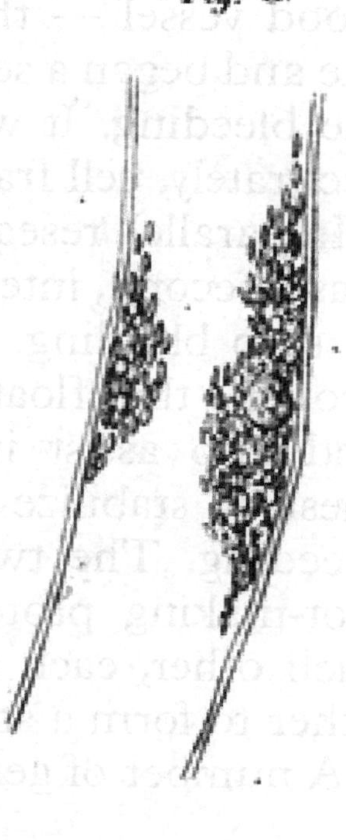

Fig. 8.

Bizzozero's illustration from his paper on clotting showing the growth of a clot around a site of vascular injury. Note the central large cell, likely a neutrophil, drawn in by inflammation, surrounded by platelets.

315

than possessing a single nucleus, it was a cell with more than a dozen nuclear lobes. It had been born, presumably, from a mother cell that had replicated its nuclear contents but halted the division or birth of daughter cells — preferring, instead, to mature and then splinter into a thousand fragments. Indeed, as Wright followed the fate of these megakaryocytes (massive, multi-nuclear-lobe-carrying cells), he found that they broke up, like fireworks, into thousands of little shards — platelets.

This early anatomical work led to a period of intense investigation into the function and physiology of these cells. As Bizzozero had observed, platelets were found to be the central component of a clot. Activated by signals from an injury — a wound, say, or a broken blood vessel — they swarmed to the wound site and began a self-perpetuating loop to plug the bleeding. It was a healing cell (or, more accurately, cell fragment).

In parallel, researchers discovered that there was a second, intersecting system in the blood to stop bleeding. This involved a cascade of proteins that float in the blood, sense injury, and also assist in congealing into a dense mesh to stabilize the platelet clot and stanch bleeding. The two systems — platelets and clot-making proteins — communicate with each other, each amplifying the effect of the other to form a stable clot.

A number of genetic disorders involving the

failure of platelet function — and resulting in abnormalities in clotting — further elucidated how a platelet senses an injury. In 1924, a Finnish hematologist, Erik von Willebrand, described the case of a five-year-old girl from the Åland Islands in the Baltic Sea whose blood did not clot properly. By analyzing the blood of family members, several of whom had similar clotting disorders, von Willebrand discovered that they all possessed an inherited abnormality that impairs platelet function. In 1971, researchers finally apprehended the culprit: people with this disease, named after von Willebrand, were either missing or deficient in a key clotting protein called, appropriately enough, von Willebrand factor (vWf).

Von Willebrand's factor circulates in blood and is also strategically located right under the cells that line blood vessels. Injury to the blood vessel exposes vWf. Platelets carry receptors that bind to vWf and thus have the capacity to "sense" when a wound has exposed the vessel, and they begin to gather around the site of injury.

But the formation of a clot is a much more complex process. Proteins secreted by the injured cells send out further signals to summon platelets to the site of injury, amplifying their activation. And clotting factors floating in the blood use yet other sensors to detect the injury. A cascade of changes is launched. Ultimately, the cascade leads to the conversion of a protein called fibrinogen into a mesh-forming protein called fibrin. Platelets, trapped in the fibrin mesh, like sardines in a net, ultimately form a mature clot.

If the vagaries of ancient human life involved plugging wounds to maintain homeostasis, the vagaries of modern life have unleashed the opposite problem: *too much* platelet activation. The process meant to heal wounds has become pathological; as Rudolf Virchow might have put it, cellular physiology has undergone an about-face and become cellular pathology. In 1886, William Osler, one of the founders of modern medicine, described platelet-rich clots that had formed in the valves of the heart, and in the aorta, the large arch-shaped blood vessel that courses through the body. Nearly three decades later, in 1912, a cardiologist in Chicago described the mysterious case of a fifty-five-year-old banker who "fell like a dud." As doctors investigated the case, they found that the artery that brought blood to the patient's heart had become occluded by a clot. The condition became commonly known as a "heart attack" — the word *attack* signifying the speed and suddenness of the crisis.

And so, as much as ancient humans may have desired a drug to activate platelets to heal their wounds, modern humans are in search of drugs that *dampen* platelet activity. Our lifestyles, lifespans, habits, and environments

— fat-rich diets, lack of exercise, diabetes, obesity, hypertension, and smoking, in particular — have led, in turn, to the accumulation of plaques: inflamed, calcified, cholesterol-rich blobs that hang on the walls of the arteries, like precarious mounds of debris alongside highways, accidents waiting to happen.*

*The unraveling of the mechanisms of cholesterol metabolism, its link to heart disease, and the creation of novel medicines to manipulate cholesterol is an exemplary story of how astute clinical observations, cell biology, genetics, and biochemistry can synergize to solve a mysterious clinical problem. The story begins with clinical observations on several families that had unusual symptoms from extraordinarily high levels of cholesterol in their blood. In 1964, for instance, a three-year-old child named John Despota was brought to see his primary care doctors in Chicago. His skin had erupted with yellow-brown cholesterol-filled bumps. His blood cholesterol was six times normal. By age twelve, he had signs of cholesterol plaques in his arteries, and he was experiencing regular bouts of chest pain. Clearly, John had a genetic predisposition for abnormal cholesterol accumulation — he was having heart attacks at age *twelve* — and so his doctors sent a biopsy of his skin to two researchers who were investigating cholesterol biology. Over the next decade, analyzing cases such as John's, researchers Michael Brown and Joe Goldstein discovered that normal cells carry receptors on their surface for a certain kind of cholesterol-rich particle

When a plaque ruptures or breaks, it is sensed as a wound. And the ancient cascade to heal wounds is activated and released. Platelets rush in to plug that "wound" — except the plug, rather than sealing an injury, blocks the vital flow of blood into the heart muscle. The healing platelet now turns into a deadly platelet.

"The modern epidemic of heart disease," the physician-historian James Le Fanu writes, "started quite suddenly in the 1930s. Doctors had no difficulty in recognizing its gravity because so many of their colleagues were among its early victims, apparently healthy, middle-aged physicians who, for no obvious reasons, suddenly collapsed and died. . . . This new disease needed a name. The cause, it seemed,

---

that circulates in the blood: low-density lipoprotein, or LDL. Under normal circumstances, the cells internalize the cholesterol and metabolize it, pulling it out of the blood and leading to low levels of circulating LDL. In patients such as John Despota, this process of internalization and metabolism is interrupted due to genetic mutations. High levels of LDL cholesterol circulate in the blood, ultimately leading to those porridge-like deposits in the arteries, including the arteries of the heart and leading to chest pain and heart attacks. Over the next years, Brown and Goldstein uncovered dozens of rare genetic mutations that disrupted cholesterol metabolism.

was a clot of blood in the arteries of the heart which had been narrowed by a porridge-like substance . . . made up of a fibrous material and a type of fat called cholesterol."

If you choose to read obituaries in local newspapers from the fifties and sixties — an admittedly morbid preoccupation — you can witness the modern epidemic being born. Newspaper obituaries brimmed with the names of men and women who had experienced "sudden chest pain," followed by collapse and death: Elmer Sweet, Mendocino, California, superintendent, fifty-three years old in 1950; John Adams, a Pine City, Minnesota, tinsmith, seventy-seven years old in 1952; Gordon Mitchell, supervisor of a

---

But in a grand synthesis of this work that followed, cardiologists began to realize that a high level of LDL was a culprit in cholesterol deposits not just in rare individuals with genetic mutations, but also in a vast swathe of the population at risk for heart attacks. This, in turn, led to the development of Lipitor and other cholesterol-lowering drugs that have had an enormously positive impact on cardiac disease. Brown and Goldstein were awarded the Nobel Prize in 1985; their work has saved millions of lives. In the 1980s, working in Brown and Goldstein's lab, Helen Hobbs and Jonathan Cohen found other genes that altered the internalization and metabolism of LDL cholesterol, leading to yet another generation of drugs that decrease LDL levels and prevent heart attacks.

321

spinning mill, forty years old in 1962; Lloyd Ray Luchsinger, sixty-one years old in 1963; and so forth, day upon day. As the death toll from heart attacks mounted, pharmacologists shifted their attention to finding drugs that blocked the cascade of clotting. The most prominent among these was aspirin. Its active ingredient, salicylic acid, originally found in willow extract, had been used by the ancient Greeks, Sumerians, Indians, and Egyptians to control inflammation, pain, and fever.

In 1897, a young chemist named Felix Hoffman, working for the German pharmaceutical company Bayer, found a way to synthesize a chemical variant of salicylic acid. The medicine was called aspirin, or ASA, short for acetyl salicylic acid. (The name was drawn from the *a* in *acetyl,* and *spir* from *Spiraea ulmaria,* the plant from which salicylic acid was extracted.)

Hoffman's synthesis of aspirin was a marvel of chemistry, but the pathway from molecule to medicine was tortuous. A senior executive at Bayer, Friedrich Dresser, suspicious of aspirin, almost stopped production, claiming that the drug had an "enfeebling" effect on the heart. He preferred concentrating on the development of another drug — heroin — as a cough syrup and pain reliever. But Hoffman persisted obstreperously with the production of aspirin, pushing the Bayer executives to the brink of firing him. Eventually the tablets were made and marketed to the public. Ironically, to satisfy Dresser's concerns, the medicine, initially sold to relieve aches, pains, and fevers, had to carry the label "Does Not Affect the Heart" on their packaging.

In the 1940s and 1950s, Lawrence Craven, a suburban general practitioner in California, began giving patients aspirin to prevent heart attacks. Craven experimented on himself, increasing the dose of aspirin to twelve tablets — far above the recommended dose — until his nose began to bleed spontaneously and profusely. Having tamped down his bleeding with napkins, and now convinced that aspirin was a potent anticlotting agent, Craven treated nearly eight thousand patients with the drug. He noted that their rates of heart attacks decreased markedly.

But Craven wasn't a traditional physician scientist; he had no control group of untreated patients to compare to those that he had treated with aspirin. His study was dismissed for decades until, in the seventies and eighties, massive randomized trials proved that aspirin was indeed among the most effective therapies to prevent and treat a heart attack in progress.

In the 1960s, deeper investigations into the biology of platelets revealed how aspirin works to prevent clots. Platelets, in concert with some other cells, produce chemicals to signal injury and get activated. Aspirin, at low doses, blocks the key enzyme that produces these injury-sensing chemicals, thereby decreasing platelet activation and subsequent clots. As a prevention mechanism for heart attacks, aspirin may well rank among the most important medicines of the past century.

A heart attack, or myocardial infarction, occurs when a plaque in one of the coronary arteries ruptures and incites a clot. In the 1990s, I was a trainee at an internal medicine clinic run by a balding octogenarian in polished wing tips and with a refined, patrician air. He told me of a time, during his own medical training, when the only treatment for a heart attack was bedrest, oxygen, and sedation with morphine delivered through a glass syringe. It was a long way from the current diagnostic tests and therapies: the mad rush to the hospital (every minute wasted is a minute's worth of heart muscle dying, causing irreparable damage); an electrocardiogram (ECG), to measure the heart's electrical activity, conducted in the ambulance and then transmitted digitally to the hospital; aspirin, oxygen, and the frantic race to get the patient to a cardiac catheterization lab, where the patient might be administered an intravenous drug known as a thrombolytic, which can dissolve a blood clot rapidly, or undergo a procedure to open the clotted artery using an inflatable, balloon-shaped device.

My preceptor claimed he could diagnose coronary artery disease by physical examination alone. First, he would make a mental list of the patient's risk factors, some of them avoidable, some not — obesity, high levels of a particular kind of cholesterol, chronic tobacco use, hypertension, and/or a family history of coronary disease — assigning each points in a calculus that he kept to himself. He would place his stethoscope against the person's neck and listen for bruits (pronounced *broo-eez*) — gurgling sounds — that might indicate plaque buildup in the carotid arteries that course through the neck and up to the brain; fatty gunk in one artery typically indicated gunk in another. And he would carefully note any history of chest pain, or even the slightest tingle, when the patient walked or ran. With the grandiosity of a magician, he would then declare that the patient had, or did not have, coronary disease, before sending him or her off for confirmatory tests. He was usually right. With a little bit of the same grandiosity, he called the coronary arteries that supply blood to the heart "the rivers of life."

Like a growing collection of trash and silt on the edge of a river, coronary plaque usually builds up over decades — bulging out toward the center of the hollow vessel and slowing down blood flow, although never obstructing it entirely. Plaque contains deposits of cholesterol, inflamed immune cells, and calcium, among other components. The artery's opening (lumen) is narrowed, and the backed-up traffic manifests in the intermittent, searing chest pain called angina pectoris, as the heart muscle strains to get enough oxygen-laden blood to meet its demands.

But angina can portend a much more acute crisis. One day the debris might rupture, spilling into the center of the river. Platelets, the body's injury detectives, rush into the blown-open injury site to plug it up. What was designed to be a physiological response to a wound becomes the pathological response to a plaque. The slowed-down traffic in the river now turns into a standstill jam — a heart attack.

Over the years, pharmacologists have discovered a gamut of drugs and procedures to prevent or treat heart attacks. There's aspirin, of course, which prevents platelets from forming clots. There are clot-dissolving drugs that break up an active clot, and platelet-inhibiting drugs that ensure that the platelets don't get activated. And, in the realm of prevention, there's Lipitor, one of the many drugs that reduce the level of a particular form of cholesterol, carried in the blood in blob-like particles, called LDL. Drugs like Lipitor decrease the blood levels of LDL, and that, in turn, prevents the buildup of cholesterol-rich lumps of trash to clog our arteries.

But these drugs need to be taken every day for life. Recently, a newly launched biotech company in Boston, Verve Therapeutics, has proposed an audacious strategy for reducing blood levels of LDL cholesterol. Its founder, geneticist and cardiologist Sek Kathiresan, trained at Massachusetts General Hospital just a few years ahead of me. MGH was an "each one, do one, teach one" hospital; the most experienced doctors taught the senior residents, who, in turn, taught their juniors and the interns. It was from Sek, a senior resident when I was an intern, that I learned how to snake an intravenous line into the jugular vein of a man writhing in the ICU, or to insert a catheter through the neck veins into the ventricle of a woman's heart to precisely measure the pressures in it. Years later, I would discover that Sek's interest in heart disease was deeply personal: his brother, then in his forties, had returned from a run, and collapsed and died from a heart attack. Over the next decades, Sek's pathbreaking work would identify scores of genes that, when inherited in an altered form, increase the risk of heart attacks.

Many of the crucial proteins that enable the formation, trafficking, and circulation of so-called bad cholesterol are synthesized in the liver. Recall the gene-editing technologies used by He Jiankui to alter genes in human embryos — in essence, rewriting the genetic script of human cells. Neither Sek nor Verve has any interest or desire to change genes in human embryos; rather, they hope to use gene-editing technologies to inactivate the genes that encode for these cholesterol-related proteins in human liver cells — and that, too, without removing the liver from the body. Scientists at Verve have devised ways to insert catheters into the arteries leading to the liver. (The dexterity that Sek learned from decades of practice in cardiology helped.) These catheters will deliver gene-editing enzymes, loaded inside tiny nanoparticles, to the organ. Once these particles off-load their cargos inside the liver cells, the gene-editing enzymes will change the scripts of genes that aid and abet cholesterol metabolism, thereby drastically decreasing the amount of circulating cholesterol in the blood — in essence, activating the LDL metabolizing pathways. It's a one-and-done infusion. Once the genes have been altered, they are altered for life. If successful, Verve's gene therapy would transform you into a human with permanently lowered cholesterol level, permanently protected from coronary artery disease, permanently safe from myocardial infarction. It would be the ultimate feat of cellular reengineering for heart disease. The river of life (to use my preceptor's favored phrase) would be cleansed forever.

THE GUARDIAN CELL
NEUTROPHILS AND THEIR
KAMPF AGAINST PATHOGENS

In 1736 I lost one of my sons, a fine boy of four years old, by the small-pox, taken in the common way. I long regretted bitterly, and still regret that I had not given it to him by inoculation [vaccination].

— Benjamin Franklin

Blood is so red — that color so dominantly ingrained in our image of what blood *is* — that white blood cells were not even noticed, or discovered, for centuries. In the 1840s, a French pathologist in Paris, Gabriel Andral, looked down a microscope and found what two generations of microscopists had seemingly missed: yet another type of cell in blood. Unlike red blood cells, these cells lacked hemoglobin, possessed nuclei, and were irregularly shaped, occasionally with pseudopods — fingerlike extensions and projections. They were termed "leukocytes," or white blood cells. (They are "white" only in the sense that they are not "red.")

In 1843, an English doctor named William Addison astutely proposed that these white cells — "colorless corpuscles," as he called them — played some crucial role in infection and inflammation. Addison had been compiling autopsy reports of tubercules: white, pus-filled nodules associated typically with tuberculosis but also with some other infections. In one case report, he noted: "A fine young man, aged 20, reported that he had cough and pain in the side . . . he had a little, [hacking] cough which plagued him." Soon the symptoms progressed to "an obscure, deep-seated mucous rale, and on coughing, a very characteristic *plash.*" The man died four months later, "with all the attendant symptoms of a deep and rapid decline." When Dr. Addison examined his lungs at autopsy, he found them studded full of "tubercles, in considerable numbers." Placed between glass slides, the tubercles often crumbled or melted into globs. Under the microscope, the globs were made of pus and thousands of white blood cells, as if these cells had been recruited especially to the inflamed sites. Some of them were "filled with granules," Addison noted. Perhaps, he reasoned, they were delivering this granular cargo to the infected areas of the body.

But what was the connection between white cells and inflammation? In 1882, a wandering professor of zoology, Elie (or Ilya)

Metchnikoff, quarreled with his colleagues at the University of Odessa and huffed off to Messina, Sicily, where he set up a private laboratory. He was a temperamental man with a depressive streak — he would attempt suicide twice in his lifetime, once by swallowing a strain of pathogenic bacteria — often at odds with the orthodoxy of science, but with an unerring eye for experimental truth.

In Messina, where the warm, shallow, windy beaches yielded a constant wealth of marine animals, Metchnikoff began to experiment with starfish. Alone one evening — his wife and children had gone to watch apes at the local circus — Metchnikoff began to devise an experiment that would define his career and change our understanding of immunity. The starfish were semitransparent; he had been watching cells move about in the bodies. He was particularly interested in the movement of the cells after injury. What if he stuck a thorn in one of the starfish's feet?

He spent a sleepless night and returned to the experiment the next morning. A group of motile cells — a "thick cushion layer" — had accumulated busily around the thorn. He had, in essence, observed the first steps in inflammation and immune response: the recruitment of immune cells to the site of injury, and their activation once they had detected a foreign substance (in this case, the thorn). Metchnikoff noted that the immune cells moved toward the site of inflammation autonomously as if impelled by a force or an attractant. (Later, these attractants would be identified as specific proteins, called chemokines and cytokines, released by cells upon injury.) "[T]he accumulation of mobile cells round the foreign body is done without any help from the blood vessels or the nervous system," he wrote, "for the simple reason that these animals do not have either the one or the other. It is thus thanks to a sort of spontaneous action that the cells group round the splinter."

Over the next few years, Metchnikoff took the seed of this idea — immune cells being summoned actively to inflammatory sites — and conducted a series of experiments. He extended his observations to other organisms and to other forms of injury. He introduced infectious spores that penetrated the guts of *Daphnia*, tiny crustaceans commonly referred to as water fleas. The immune cells, he found, did not merely travel to the sites of inflammation. They tried to ingest — *eat* — the infectious agent or irritant that had accumulated at the site. He called the phenomenon phagocytosis: the engulfment and consumption of an infectious agent by an immune cell.

In a series of papers published in the mid-1880s that would eventually win him the Nobel Prize, Metchnikoff used the German word *Kampf,* meaning "fight," "combat," or "tussle," to encapsulate the relationship between an organism and its invaders. He described a "drama unfolding within organisms" that was like a perpetual struggle. (It is tempting to speculate that his relationship with the scientific establishment was its own perpetual *Kampf*.) According to Metchnikoff: "A battle takes place between the two elements [the microbe and the phagocytic cells]. Sometimes the spores succeed in breeding. Microbes are generated that secrete a substance capable of dissolving the mobile cells. Such cases are rare on the whole. Far more often it happens that the mobile cells kill and digest the infectious spores and thus ensure immunity for the organism."

The human versions of the phagocytic cells that Metchnikoff discovered — macrophages, monocytes, and neutrophils — are among the very first cells to respond to injuries and infections. Neutrophils are produced in the bone marrow. Their name references the fact that they can be stained by neutral dyes but not by acidic or basic ones; hence "neutrophil" or "neutral loving."*

*This classification of white blood cells based on their dye staining was yet another one of Paul Ehrlich's seminal contributions to biology. Working with thousands of dyes, he found that some had a remarkable capacity to bind to a cell or one of its

Neutrophils live for just a few days after entering the circulation. But what dramatic days! Incited by an infection, the cells mature from the bone marrow and flood into blood vessels, hot for combat, their faces granulated, their nuclei dilated — a fleet of teenage soldiers deployed to battle. They have evolved special mechanisms to move quickly through tissues, squirming their way through blood vessels like contortionists. It is as if they are maniacally driven to reach sites of infection and inflammation — in part, because they so keenly perceive the gradient of cytokines and chemokines released by injury. They are lean, energetic, mobile machines built for immune attack. Professional killers — guardian cells — on a mission.

Their arrival at the site of an infection

---

substructures. Initially, Ehrlich used this binding characteristic to differentiate cells from one another — hence *neutrophils,* which were colored blue when they bound neutral dyes, and *basophils,* another cell type found in blood that bound a nonacidic dye. Ehrlich, calling this idea specific affinity, began to wonder if the specific affinity of a chemical for a particular cell could be used not just to stain a cell but also to kill it. This idea was the basis of his discovery of the antibiotic Salvarsan in 1910 and would form the underpinning of his desire to find a magic bullet for cancer: a chemical with a specific affinity, and toxicity, for a malignant cell.

inaugurates an elaborate, soldierly deployment. First, they marginate toward the edges of a blood vessel. Then they begin to roll along its walls, tumbling into action as they stick and unstick to specific proteins in the walls. Finally, they tether themselves more firmly to the edge of a vessel and then migrate, actively, into the tissue — the lung, or the skin — where they bombard the microbe with the toxic substances carried in their granules. They might begin phagocytosing the microbe or its bits, internalizing the pieces, and directing them to lysosomes — special compartments filled with toxic enzymes to break down the microbe.

An astonishing feature of this early immune response is that its cells, neutrophils and macrophages among them, are *intrinsically* armed with receptors that recognize proteins (and other chemicals) found on the surface or the interior of some bacterial cells and viruses. Pause for a moment to consider that fact. We — multicellular animals — have been at war with microbes for such a long time in evolutionary history that, like ancient, conjoined enemies, we've been defined by each other. We are in a lockstep dance. Our first-responder immune cells carry pattern-recognition receptors that are inherently designed to latch on to molecules found in microbial cells or injured cells that are not specific to a specific pathogen (*Streptococcus,* say) but are broadly present in all bacteria and viruses. Some receptors recognize a protein found in bacterial cell walls but not in animal cell membranes. Some bind to a protein found uniquely in the swimming tails of some bacteria. Yet others sense signals sent out by cells that have been infected by viruses. Speaking generally, these receptors come in two classes: those that recognize "damage-associated molecular patterns" (substances released upon cellular damage) and those that sense "pathogen-associated molecular patterns" (components of microbial cells). In short, they sniff around the body looking for *patterns* of injury and infection — substances that signal invasion and pathogenicity.

When a neutrophil or a macrophage meets a bacterial cell, it is already prepared for combat. Theirs is not a "learned," or adaptive, form of immunity; the response is intrinsic to the cell, and the sensors of the response exist in the neutrophil from its very inception. In short, we carry reverse images of some microbes, or memories of what they incite in our bodies, like photographic negatives, on the surfaces of our cells. Us and them: they are within us even when they are not within us. It is a symbol of our *Kampf*.

In the 1940s, this wing of the immune response — neutrophils, macrophages, among other cell types, with their attendant signals

336

and chemokines — began to be termed the "innate immune system."* *Innate,* in part, because it exists inherently in us, with no requirement to adapt to, or learn, any aspect of the microbe that caused the infection. (We will come to the adaptive wing of the immune response, with B cells, T cells, and antibodies, in the next chapter.) Innate, also, because it is the most ancient wing of the immune system and therefore innate to our ancestors. Starfish have it, as Metchnikoff first observed. So do water fleas, sharks, elephants, lorises, gorillas, and, of course, humans.

Some version of the innate response is found in virtually every multicellular creature. Flies

---

*The innate immune system has multiple other cells, including mast cells, natural killer (NK) cells, and dendritic cells. Each of these cell types performs a different function in the early immune response to pathogens. The one feature they share in common is that they do not have any learned or adaptive capacity to direct their attack on a specific pathogen. Nor do they retain any memory of a particular pathogen (although recent work has shown that subsets of natural killer cells might have a limited adaptive memory for certain pathogens). Rather, as first-responder cells, they are activated by general signals released upon infection, inflammation, and injury, and have mechanisms to attack, kill, and phagocytose cells while summoning and activating the B cell and T cell responses.

have only an innate system; if you mutate the genes of this system, flies — the very creatures associated with decomposition — become infested with microbes and begin to decompose. Among the most haunting images I have ever encountered in cell biology is that of a fly — its innate immune system destroyed — being eaten alive by bacteria.

The innate system is not just among the most ancient, but, being the first responder, the most crucial to our immunity. We associate immunity with B and T cells, or with antibodies, but without neutrophils and macrophages, we would meet the fate of the decomposing fly.

Despite the centrality of the innate immune response, or perhaps *because* of its centrality, innate immunity has proven difficult to manipulate medically. But, unknowingly, perhaps, we have been playing with innate immunity for longer than a century. This age-old instance of manipulating innate immunity is vaccination — although, of course, when vaccines were first invented, the vocabulary of innate immunity did not exist, nor was the mechanism of protection known. Even the word *vaccine* would be coined centuries after vaccination itself was being practiced widely across China, India, and the Arab world.

In April 2020, on a sweltering morning in Kolkata, India — the hawks outside my hotel room were circling upward, lifted by the warming air currents — I visited a shrine to the goddess Shitala, the deity that presides over the healing of smallpox. She shares the shrine with Manasa, the goddess of snakes, the healer of poisonous bites and the protectress against venom. Shitala's name means the "cool one": the myth runs that she arose from the cooled ashes of a sacrificial fire. But the heat that she is supposed to diffuse is not just the intractable wrath of summer that hits the city in mid-June but also the internal heat of inflammation. She is meant to protect against smallpox in children and to heal the pain of those that contract it. She is the anti-inflammatory goddess.

The shrine was a small, damp room on the edge of College Street, a few miles from Kolkata Medical College. In the inner sanctum, moistened with sprays of water, there was a figurine of the goddess sitting on a donkey and carrying a jar of cooling liquid — the way she has been depicted since Vedic times. The temple was 250 years old, the attendant informed me. That would place it, not by coincidence, perhaps, from around the time when a mysterious sect of Brahmans began to wander up and down the Gangetic Plain to popularize the practice of *tika:* taking a live pustule from a smallpox patient, mixing it with a paste of boiled rice and herbs, and inoculating a child by rubbing the mixture on a sharp nick on the skin. (The word *tika* comes from the Sanskrit word for "mark.")

"The place where the punctures were made commonly festures [*sic*] and comes to a small suppuration," a rather incredulous English physician wrote of the practice in 1731, "and . . . if the punctures do suppurate and no fever or eruption ensues, then they are no longer subject to the infection."

The Indian *tika* practitioners had likely learned it from Arabic physicians who, in turn, had learned it from the Chinese. As early as AD 900, medical healers in China had realized that people who survived smallpox did not catch the illness again, thus making them ideal caregivers for those suffering from the disease. A prior bout with an illness somehow protected the body from future instances of that illness, as if it retained a "memory" of the initial exposure. To harness this idea, Chinese doctors harvested a smallpox scab from a patient, ground it into a dry, fine powder, and used a long silver pipe to insufflate it through a child's nose. Vaccination was a tightrope walk: if the powder contained too much inoculum of live virus, the child would acquire not immunity but the disease — a devastating outcome that occurred about one in a hundred times. But if the child survived the inoculum and its "fester," he or she would develop merely an attenuated, local disease, with either no

340

symptoms or mild ones, and be immunized for life.

By the 1700s, the practice had spread throughout the Arab world. In the 1760s, traditional healers in Sudan were known to practice Tishteree el Jidderee — "buying the pox." The healer, typically a woman, would approach a sick child's mother to purchase the ripest pustules for inoculation, haggling over the price. It was an exquisitely measured art: the most astute healers recognized the lesions that were exactly right in their maturity to yield *just enough* viral material to confer protection, but not so much as to introduce the disease. The various sizes and shapes of pustules led to the European name for small-pox: variola, from the word *variation*. And the immunization against the pox was called variolation.

In the early eighteenth century, Lady Mary Wortley Montagu, the wife of the British ambassador to Turkey, was herself affected by the pox, her perfect skin left pitted with lesions. In Turkey, she witnessed variolation in practice and, on April 1, 1718, wrote to her lifelong friend Mrs. Sarah Chiswell in wonder:

> *There is a set of old women, who make it their business to perform the operation, every autumn, in the month of September, when the great heat is abated. . . . The old*

woman comes with a nut-shell full of the matter of the best sort of small-pox, and asks what vein you please to have opened. She immediately rips open that you offer to her, with a large needle (which gives you no more pain than a common scratch) and puts into the vein as much matter as can lie upon the head of her needle, and after that, binds up the little wound with a hollow bit of shell, and in this manner opens four or five veins. Then the fever begins to seize them, and they keep their beds two days, very seldom three. They have very rarely above twenty or thirty in their faces, which never mark, and in eight days time they are as well as before their illness. Where they are wounded, there remains running sores during the distemper, which I don't doubt is a great relief to it. Every year, thousands undergo this operation, and the French ambassador says pleasantly, that they take the small-pox here by way of diversion, as they take the waters in other countries. There is no example of any one that has died in it, and you may believe I am well satisfied of the safety of this experiment, since I intend to try it on my dear little son.

Her son never got the pox.

Variolation had one further legacy: it gave rise to perhaps the first time the word *immunity* was used. In 1775, a Dutch diplomat who dabbled in medicine, Gerard van Swieten, used the word *immunitas* to describe the fever and smallpox resistance induced by variolization. The history of immunity, and of smallpox, were therefore to be forever intertwined.

Set in 1762, the story, perhaps apocryphal, has it that an apothecary's apprentice named Edward Jenner heard a dairy girl say, "I shall never have smallpox, for I have had cowpox. I shall never have an ugly, pockmarked face." Perhaps he had overheard it from local folklore, for a "milkmaid's milky skin" was a recurrent English meme. In May 1796, Jenner proposed a safer approach to smallpox vaccination. Cowpox, a virus related to smallpox, caused a far less severe form of the disease, with no deep pustules and no risk of death.

Jenner harvested pustules from a young dairymaid, Sarah Nelmes, and inoculated his gardener's eight-year-old son, James Phipps, with it. In July, he inoculated the boy again, but this time with material from a smallpox lesion. Although Jenner had breached virtually every boundary of ethical human experimentation (for instance, there is no record of informed consent, and the subsequent "challenge" with live virus might well have been lethal to the child), it apparently worked: Phipps did not develop smallpox. After facing initial resistance from the medical community, Jenner increased his vaccination efforts and became broadly celebrated as the father of vaccination. In fact, the word *vaccine* carries the memory of Jenner's experiment: it is derived from *vacca,* Latin for "cow."

Yet this story, retold and recycled in textbooks, is potentially riddled with misattributions. The virus carried in Sarah Nelmes's pox lesions was likely horsepox, not cowpox. In a book he self-published in 1798, Jenner acknowledged the fact: "Thus the Disease makes its progress from the Horse [as I conceive] to the nipple of the Cow, and from the Cow to the Human Subject." Furthermore, Jenner might not have been the first vaccinator in the Western world: in 1774, Benjamin Jesty, a hefty, prosperous farmer from Yetminster village in the county of Dorset, also convinced by the stories of dairymaids who contracted cowpox and then appeared to be immune to smallpox, supposedly harvested lesions from the udder of an infected cow, and inoculated his wife and two sons. Jesty became an object of ridicule among physicians and scientists — yet his wife and children survived the smallpox epidemic without catching the disease.

But how did inoculation generate immunity, particularly long-term immunity? Some factor produced in the body must be able to counter the infection and also retain a memory of the infection over multiple years. Vaccination, as we will soon learn, generally works by inciting specific antibodies against a microbe. The antibodies come from B cells, and they are retained in the cellular memory of the host because some of these cells live for decades — long after the initial inoculum was introduced. We will turn to how B cells manage to achieve memory, and how T cells help, in the next chapter.

But the underappreciated fact of vaccination is that it is, at first, a manipulation of the innate immune system. Long before B and T cells appear on the scene, the first step in vaccination is the activation of the first-responder cells: macrophages, neutrophils, monocytes, and dendritic cells. It is these cells that pick up the inoculum, especially if it is mixed with an irritant; the paste of boiled rice and herbs that I referred to earlier might unwittingly have served this purpose. Then, through various signaling processes, including phagocytosis, they digest and process the inoculum to initiate the immune response.

And here is the central conundrum of immunology: if you disable the ancient, nonadaptive innate system — the system designed to attack microbes without discrimination — you also disable the adaptive B and T cells, the system that discriminatingly retains the memory of a specific microbe. In mice, genetic inactivation of innate immunity makes the animals respond poorly to vaccines. Humans that lack a functional innate system — typically children with rare genetic syndromes — are severely immunocompromised, and their response to vaccines is also severely diminished. They die of bacterial and fungal infections just as the flies lacking innate immunity die with tragic immune failure: infested, overtaken, overwhelmed by microbes.

Vaccination, more than any other form of medical intervention — more than antibiotics, or heart surgery, or any new drug — changed the face of human health. (A close contender might be safe childbirth.) Today there are vaccines against the deadliest of human pathogens: diphtheria, tetanus, mumps, measles, rubella. Vaccines have been devised to prevent infection by human papillomavirus (HPV), by far the major cause of cervical cancer. And we will soon encounter the triumphal discovery of not just one but several independent vaccines against SARS-COV2, the virus that released the Covid pandemic.

But the story of vaccination is not the story of progressive scientific rationalism. Its hero is not Addison, who first found white blood cells. Nor is it Metchnikoff, whose discovery of phagocytes might have opened a door on protective immunity. Not even the scientists who discovered the innate response to bacterial cells merit being lauded as the heroes behind this medical milestone.* Rather, its history is one of veiled hearsay, gossip, and myth. Its heroes are nameless: the Chinese doctors who air-dried the first pox pustules; the mysterious sect of worshippers of Shitala who ground viral matter with boiled rice and inoculated it into children; the Sudanese healers who came to discern the ripest lesions.

On an April morning in 2020, I switched on a microscope in my New York lab. The tissue culture flask was teeming with mobile monocytes that one of my postdoctoral researchers had been growing.

Here you are, I said to myself. It was one of those mornings when there's no one in the lab, and I can have an internal conversation out of human earshot. These monocytes, cells of the innate immune system that can "eat" pathogens and their debris, had been genetically modified to become super-phagocytes, their hunger multiplied tenfold. We've inserted a gene that makes them want to eat ten times more cellular material than normal phagocytes consume, and eat them ten times faster. The project, a collaboration with the

*Much of our knowledge of innate immunity, and the genes that activate this wing of the immune response, arises from experiments performed in the 1990s by Charles Janeway, Ruslan Medzhitov, Bruce Beutler, and Jules Hoffman.

scientist Ron Vale, involves engineering a new kind of immunity. Remember that monocytes, along with macrophages and neutrophils, are agnostic when it comes to specific stimuli; instead, they carry receptors that bind to factors common to many bacteria and viruses, and they migrate toward cells that send out general SOS signals of injury or inflammation.

But what if we could redirect the monocyte to eat and kill a specific cell? What if we armed these cells with genes that, rather than detecting general patterns of infection, were attuned to a specific protein present only on the surface of, say, a cancer cell? The soldier, usually deployed to a battalion, now becomes a directed assassin sent to hunt down a specific target. That's what we were trying: we had created a new class of receptors that would be expressed on monocytes, bind to a protein on cancer cells, and elicit a hyperactive form of phagocytosis — resulting, hopefully, in the monocyte consuming the cancer cell with unprecedented, insatiable gusto. In essence, we had attempted to make an intermediate cell that lived somewhere between a monocyte, with its indiscriminate cell-eating propensities, and a T cell, with its ability to go after a particular target. It's a type of cell that has never existed in biology — a chimera. Such a cell, we hoped, would fuse the toxic, indiscriminate fury of innate immunity and the more discerning killing capacity of adaptive immunity — thereby directing a potent punch on cancer, but without stirring up the inflammatory response in general.

In early animal experiments, we had implanted tumors into mice and infused them with millions of these super-phagocytes. These cells had eaten tumors alive. We are now growing these cells in massive numbers and testing all kinds of mechanisms by which they might be redirected against breast cancers, melanomas, and lymphomas.

It's been nearly two years since that April morning when I first saw the super-phagocytes eating cancer cells in my lab. And it is an eerie coincidence that as I am writing this sentence — on the morning of March 9, 2022 — the very first patient, a young woman in Colorado with a deadly cancer of T cells, is being infused with this experimental therapy (the protocol has undergone all the necessary approvals from the FDA and review boards).

It will be months before we know whether the treatment has worked. The only thing I am told about the outcome is that the woman survived the treatment without complications. But as the infusion drips in her body, it is as if I can feel every droplet enter her veins. *What is she thinking about? What is she looking at? Is she alone?*

When I finally fell asleep that night at about four a.m., I dreamed of my childhood. In my dream, I am a ten-year-old boy in Delhi, thinking about — what else? — droplets. The monsoons would hit the city in July and August, and I would play a game: as the rains began, I would install myself by the window, open my mouth, and try to catch drops of water. In my dream last night, I caught the droplets in my mouth at first, but a sudden splash of water landed on my eye. And then there was the clap of distant thunder, and the rain stopped.

It's hard to describe the heady mixture of terror, anticipation, and exhilaration that you experience when a discovery from your lab makes the transition into a human medicine. Thomas Edison, the inventor, used to define genius as 90 percent perspiration and 10 percent inspiration. I have no claim to genius; I only feel the perspiration. I cannot get the image of the woman in the trial out of my mind. The only moment I have felt anything similar was during the first minutes after the birth of my two children.

But this, too, is a moment of birth. Perhaps, a new therapy is being born. And, with it, a new human.

I switched off the microscope and thought about the strange temple of Shitala — and of how long and how hard it has been to cool or heat innate immunity to make it an agent of our medical needs. Shitala, the cool goddess, is also known to have a tetchy side: anger her, and she might wreak havoc on the body with inflammation from poxes, fevers, plagues. Sometime in the near future, we will learn to pitch the innate immune system's wrath against cancer cells; to calm it in the case of autoimmune diseases; to augment it to create a new generation of vaccines against pathogens. Once we teach our innate immune cells to attack malignant cells in humans, we will have invented an entirely new mode of cell therapy that harnesses inflammation. Perhaps we might describe it, metaphorically, as a pox on cancer.

THE DEFENDING CELL
WHEN A BODY MEETS A BODY

Gin a body meet a body
Comin thro' the rye,
Gin a body kiss a body —
Need a body cry.

— Robert Burns, "Comin
Thro' the Rye," 1782

It is hardly a coincidence that the shrine to the goddess Shitala in Kolkata also pays devotion to a second deity: Manasa, the goddess of snakes and the protectress against venoms and snakebites. She is usually depicted as a stately, if steely, being, often standing on a cobra and haloed by a canopy of cobras rearing their heads. Snakes descend from her coils of matted, Medusa-like hair. Portrayals of Manasa in tribal Bengal are much more fearsome: she bears a snake's body and is often encoiled entirely by snakes.

The conjoining of the two ancient pestilences carries an old memory: snakebites and smallpox haunted seventeenth-century India like twin demons, and goddesses to protect against each might well share a shrine. (India still reports eighty thousand snakebites a year, the largest number in the world.)

It is only fitting, then, that if the story of the innate immune system begins with Shitala, then the story of the second wing of the adaptive immune system — the one composed of antibodies, B cells, and T cells — might begin with a snakebite.

The legend comes in so many variations that it is difficult at times to unbraid fact from myth. In the summer of 1888, Dr. Paul Ehrlich, working in Robert Koch's laboratory in Berlin, became infected with the very tuberculosis strain he was using for his experiments. Ehrlich diagnosed himself, in fact, using a test he'd devised, acid-fast staining, to detect the bacterium in his sputum. He was sent to convalesce in Egypt, where the warm air along the Nile River was thought to be salubrious.

One morning during his sojourn in Egypt, Ehrlich was summoned urgently to help with a medical case. A man's son had been bitten by a snake, and the locals were aware that Ehrlich was a visiting doctor. It isn't known whether the boy survived, but his father told Ehrlich an extraordinary account of his own experience: he, too, had been bitten as a child, and several times as an adult. He survived the first snake attack, and with each subsequent bite, the symptoms became milder and milder. Over multiple exposures to the venom from this particular species of snake, the man had become virtually resistant to it. Variations of this story are common among the snake catchers in India. The legend runs that they make tiny nicks in their skin and expose themselves to minuscule and progressively larger doses of venom starting in childhood and into adolescence. After several exposures, they, too, are resistant to the bite.

The father's story stuck in Ehlrich's mind. Evidently, the man had developed some response to the venom — an antivenin — and then retained immunological memory. But what was the mechanism that armed a human body to generate protective immunity? Why, we might wonder, did a *single* exposure to a dried smallpox pustule render lifelong immunity to the disease?

In the early 1890s, soon after his return from Egypt, Ehrlich met biologist Emil von Behring, who had just joined the newly founded Royal Prussian Institute for Infectious Diseases, in Berlin. At the institute, von Behring and a visiting Japanese scientist, Shibasaburo Kitasato, soon launched a series of experiments on specific immunity. Among the most dramatic of these was an experiment that implicitly reminded Ehrlich of the Egyptian man's protective immunity: Kitasato and von Behring demonstrated that the serum of an animal exposed to the bacterium that caused either tetanus or diphtheria could be transferred to another animal and confer immunity to the disease. In a rather desultory footnote to the diphtheria paper, von Behring first used the word *antitoxisch,* or antitoxin, to describe the activity of the serum.

The question remained: What was this *antitoxisch,* and how was it generated? Von Behring had imagined it as a property of the serum — an abstraction. Or was it perhaps a *material* substance made in the body? In a wide-ranging, speculative 1891 paper entitled "Experimental Studies on Immunity," Ehrlich pushed his fellow scientists to think not just of the potential but also about the *material* nature of the substance. He boldly coined the word *Anti-Körper* (antibody). *Körper,* from *corpus,* or *body,* signaled his growing conviction that an antibody was an actual chemical substance: a "body" produced to defend the body.

How were such antibodies made? And how could they be specific to one toxin and not to another? By the 1890s, Ehrlich had begun to build a magnificent theory. Every cell in the body, he argued, displayed an immense set of unique proteins — side chains, he called them — attached to its surface. A chemist at heart, Ehrlich had returned to the language of dye making. He knew that you could change the color of a dye by attaching a different chemical side chain to it. And so, perhaps it was with antibodies: by changing the side chain of a chemical substance, you could change the binding properties, or specific affinity, of an antibody. When a toxin or pathogenic substance bound to one such side chain in a cell, the cell increased the production of that antibody. With repeated

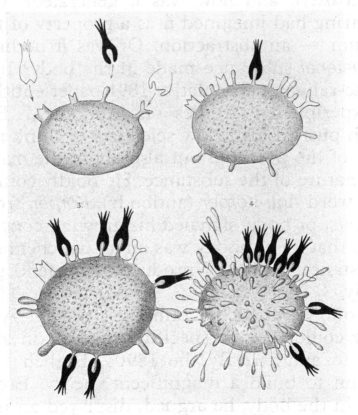

Ehrlich's illustration of how antibodies are generated. The German scientist imagined that B cells (shown in 1) had many side chains on their cell surface. When an antigen (black molecule) bounds one such side chain (2), the B cell makes more and more of that particular side chain (3) to the exclusion of others, until it finally begins to secrete that antibody (4).

exposures, Ehrlich speculated, the cell turned out so much cell-bound antibody that it was ultimately secreted into the blood. And the presence of the antibody in blood resulted in immunological memory. The substance bound by the antibody — the toxin or the foreign protein — was soon termed an anti*gen:* a substance that *generates* an antibody.

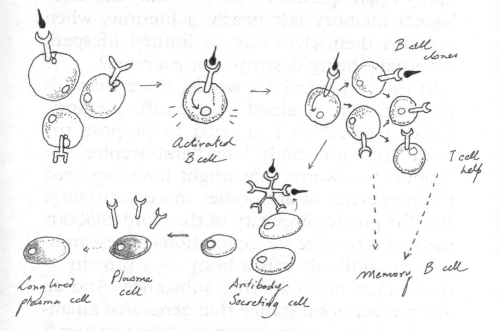

Author's illustration of the actual process of antibody genesis through clonal selection, using similar graphic motifs as Ehrlich. Each B cell expresses a unique receptor on its cell surface. When an antigen is bound, that particular B cell expands and gives rise to a short-lived antibody secreting cell (the initial antibody is usually a complex of five antibodies, a pentamer). Eventually, an antibody-secreting plasma cell is formed. Some of these plasma cells become long-lived plasma cells. The activated B cells, with T cell help, also become memory B cells.

357

Ehrlich's theory was a wrong constructed out of many rights. He had guessed correctly that an antibody bound physically to its cognate antigen like a key binding to a lock. He was also right in surmising that antibodies were eventually secreted into the blood and were the source of one kind of immunological memory. But Ehrlich's side chain theory left many open questions. How could immunological memory last nearly a lifetime, when proteins themselves have a limited lifespan, eventually being destroyed or excreted?

In the end, Ehrlich's words, more than his theories, were retained in scientific memory. Other researchers had tried to propose the term "immune body," or "amboceptor" or "copula" — words that might have captured the properties of antibodies more accurately. But the poetic simplicity of the word *antibody* made it attractive for generations of researchers. An antibody was a body — a protein — that locked on to another substance. And an antigen was a substance that generated an antibody. As one scientist wrote: "the two words were destined to form one of those inseparable pairs like Romeo and Juliet or Laurel and Hardy." The names, like the chemicals, were locked together, like inseparable pairs. They stuck.

By the early 1940s, experiments on birds had shown that antibodies were made by cells in a strange organ near their anuses (cloacae), named the bursa of Fabricius after its saclike structure (bursa) and its discoverer, the sixteenth-century anatomist Hieronymus Fabricius of Aquapendente. The antibody-making cells were called B cells, after the word *bursa*. Mammals, including humans, don't have a cloacal bursa. Our bodies produce B cells primarily in the bone marrow (thankfully, another *B*), which then mature in the lymph nodes.

Thus far, Ehrlich's side chain theory — that antibodies are made by cells with attached side chain receptors for antigens — had been largely left intact. The true molecular "shape" of an antibody would be discovered years later: between 1959 and 1962, Gerald Edelman and Rodney Porter, working at Oxford University and the Rockefeller Institute in New York, respectively, would discover that antibodies are Y-shaped molecules with two sharp heads. The heads, or tines, of the Y bind to the antigen, each acting like a prong: most antibodies thus have two binding prongs. The shaft, or stem, of the Y serves many purposes. Macrophages — the eating cells — use the antibody's stem to engulf and then gulp down antibody-bound microbes, viruses, and peptide fragments, much like the shaft of a fork is used to deliver food into the mouth; specific receptors on macrophages grab the shaft, just as a fist might grab a fork. This, indeed, is one mechanism of phagocytosis, the phenomenon that Elie Metchnikoff had observed.

The shaft or stem of the Y has yet other purposes: once bound to a cell, it also attracts a cascade of toxic immune proteins from the blood to attack microbial cells. An antibody, in short, can be conceived as a molecule with multiple parts — the binding prongs that attach themselves to the antigen, and a shaft that enables it to liaise with the immune system to become a potent molecular killer. These two distinct functions of the antibody — antigen binder and immune activator, are combined in one molecule, with a form — an immunological pitchfork — that is consummately linked to its function.

But let's move back a decade: by the 1940s, long before the pitchfork shape of antibodies was known, the philosophical and mathematical questions that Ehrlich's idea had raised were profound and troubling. The crucial linchpin of his theory was that cells were capable of displaying hundreds, or even thousands, of *premade* receptors to an antigen on their surface, like some mythical hedgehog capable of displaying a million different shaped quills. The immune response involved just a ramping up of the production of these antibodies — the active shedding of one quill — when one of these receptors happened to bind an antigen.

But the numbers did not make sense. How many premade antibodies could possibly exist on the surface of a cell? How many quills could a hedgehog hold? Was the whole universe of antigens "mirror imaged" in receptors on cells — an infinitely quilled hedgehog? How could there even be enough *genes* in a B cell to make such a counter-universe of antibodies? If Ehrlich was correct, then each of our B cells must perpetually carry an inverted cosmos of all things capable of immune reactions. To every conceivable antigen? There is an Indian legend of Yashodhara, mother of Krishna, one of the major Hindu deities, opening his infant mouth because he has swallowed a clod of dirt. She pries his teeth apart and witnesses the whole universe inside him: the stars, the planets, the million suns, the whirling galaxies, the black holes. Was each of our B cells carrying a reflected cosmos — the cognate reverse of every antigen in the universe?

In 1940, the fabled chemist at the California Institute of Technology, Linus Pauling, proposed an answer — an answer so wrong that it would eventually point to the truth. Pauling's scientific achievements were legendary. He had solved an essential feature of protein structure, and described the thermodynamics of the chemical bond — but he could also be spectacularly off-track. There's a story that quantum physicist Wolfgang Pauli, as notoriously cantankerous as he was brilliant, supposedly read a student's paper and remarked that it was "so bad that it was not even wrong." Pauling, with his daring, off-the-wall theories, often lobbed casually during scientific meetings, achieved the converse provocation: his hypotheses or models were sometimes so wrong that they weren't even bad. Pauling's colleagues had gotten used to his madcap theories; they even cherished them. By analyzing the internal contradictions of Pauling's models — in other words, by reasoning through *what* was wrong with the proposal, and *why* it couldn't be correct — they often found that they could arrive at the real mechanism, the truth.

Pauling imagined that when antibodies confronted their antigens, they were actively twisted and turned into shape by the antigen. In short, the antigen (part of a bacterial protein, say) "instructed" — as he put it — the shape of the antibody, acting as a template on which the antibody was built or molded, like melted wax poured to create a death mask.

But researchers had a difficult time reconciling Pauling's instructional theory of antibodies with the fundamentals of genetics and evolution. Proteins are, after all, encoded by genes, and if genes are fixed in their code, then the protein, built from that code, is fixed in its structure. An antibody — a protein — is a biological chemical with a predetermined physical form, not some kind of shape-shifting funereal linen that can drape itself, in perfect form, around a mummified antigen.

There was only one possible answer: if the structure of antibodies was malleable, then the genes that encoded them must also be malleable — by mutation. At Stanford, the geneticist Joshua Lederberg challenged Pauling's ideas and proposed an alternative: "Do antigens bear instructions for antibody specificity, or do they select cell lines that arise by mutation?" To Lederberg, theoretically, at least, the answer was obvious. In cell biology and genetics — in fact, in most of the biological world — learning and memory typically happen by mutation, not instruction or aspiration. A giraffe's long neck isn't the product of generations of its ancestors aspiring to stretch their necks to reach tall trees. It is the consequence of mutations, followed by natural selection, that produces a mammal with an extended vertebral structure that, in turn, creates a long neck. How on earth would antibodies "learn" to become twisted to conform to the shape of an antigen? Why would an antibody behave out of character, like some kind of malleable, medieval drapery that could spontaneously change its shape to fit an antigen?

Lederberg was right, of course. The correct answer to the conundrum of antibody genesis was eventually to be found buried in an obscure paper published in 1957 in the *Australian Journal of Science* by an Australian immunologist. (Even today, professors of immunology confess to having never read it.) In the 1950s, Frank Macfarlane Burnet, drawing on earlier work by Niels Jerne and David Talmage, realized that neither Pauling nor Ehrlich had clinched the answer to the puzzle. An antibody wasn't created by instruction or aspiration. Nor could a single B cell display the universe of all potential antibodies to bind every potential antigen.

Burnet turned Ehrlich down under. Ehrlich's idea, recall, was that every cell — an infinitely-quilled hedgehog — displayed a vast array of antibodies and the antibodies were selected when they bound the antigen. But what if, Burnet reasoned, every B cell displayed only *one* receptor for an antigen, and it was the *cell* — not the antibody — that was selected, and grew, when it bound the antigen? Proteins don't grow on command, but cells do. A B cell bearing a single antigen-binding receptor on the cell's surface protein, given an appropriate signal, could do precisely that.

The pointed comparison, Burnet argued, could be drawn from neo-Darwinian logic. Imagine an island of finches, each carrying a mutation that gives a finch a unique and slightly different beak: some large and flat, some thin and pointed. Then imagine that natural resources become suddenly limited: the fruit trees are demolished in a storm, and all the soft fruit vanishes; the only food that remains are hard-shelled seeds. A gross-beaked finch, capable of cracking the fallen seeds, might be naturally selected and survive, while a thin-beaked finch, meant to dine on fruit nectar, would die.

In short, individual finches, like individual cells, don't have an infinite repertoire or cosmos of beaks and happen to choose or adapt the one that is best suited to its circumstances. *Rather, natural selection chooses the individual finch that happens to have an ideal beak for the natural disaster.* The population of such selected finches grows. And the memory of the previous disaster persists.

Burnet extended the analogy to B cells. Imagine an enormous cohort of B cells in a body, each of which carries a unique receptor bound to its surface — each cell a finch with a unique beak, if you will. Imagine each receptor as an antibody — except it is bound to the surface of a B cell (and is connected to a network of signaling molecules to activate the cell). When an antigen binds to one such B cell (a clone), it is stimulated and begins to outgrow all the others. The finch (or B cell) that happens to carry the right beak (or antibody) is selected. This is not natural selection but *clonal* selection: the selection of an individual cell capable of binding an antigen.

A wondrous process occurs when a B lymphocyte, displaying the right receptor, meets a foreign antigen. As Lewis Thomas wrote in his book *The Lives of a Cell: Notes of a Biology Watcher* (1974): "When the connection is made, and a particular lymphocyte with a particular receptor is brought into the presence of the particular antigen, one of the greatest small spectacles in nature occurs. The cell enlarges, begins making new DNA at a great rate and turns into what is termed, appropriately, a blast. It then begins dividing, replicating itself into a new colony of identical cells all labeled with the same receptor." In the end, the dominant B cell clones, displaying the "right" receptor (the one that best binds the antigen) blast away, outgrowing all others. It is a Darwinian process, much like the finch with the right beak is "chosen" by natural selection.

As Ehrlich had imagined back in 1891, these blasts now begin to secrete the receptor into the blood. Freed from the B cell's membrane and now floating in the blood, the receptor "becomes" the antibody.* And

*I have simplified the process somewhat, but the basic details of antibody genesis are captured here. The activation of a B cell receptor by an antigen, the secretion of that receptor into blood, the refinement of the antibody over time, the sustained secretion of

when the antibody is bound to its target, it can summon a cascade of proteins to poison the microbe and can recruit macrophages to devour, or phagocytose, it. Decades later, researchers demonstrated that some of these activated B cells don't simply peter out. They persist in the body in the form of memory cells. In Thomas's words, "The new cluster [of cells stimulated by the antigen] is a memory, no less." Once the fulminant infection has ceased and the microbe cleared, some of these B cells become more quiescent, but they persist — finches huddled in the cave. When the body encounters the antigen again, the memory B cell is reactivated. It arises out of dormancy into active division to mature into an antibody-making plasma cell, thereby encoding an immunological memory. The locus of immunological memory, in summary, is not a protein that persists, as Ehrlich may have imagined. It is a B *cell,* previously stimulated, that bears the memory of the prior exposure.

---

the antibody by plasma cells and the transformation of some activated B cells into memory B cells, essentially captures the process. As we shall soon see, some antibody secreting cells — plasma cells — also become long lived. Both appear to contribute to the memory of the prior infection. Helper T cells are essential for this process, and we will turn to these cells in the next chapters.

How does each B cell acquire its unique antibody? Darwin's finches had developed their individual beaks through mutations in sperm and egg cells, which changed the morphology of each beak. These mutations are germ line: they are present in the DNA of every cell of the finch, and are carried, intact, from one generation to the next; therefore, a gross-beaked finch will give rise to a gross-beaked finch, and so on.

In the 1980s, a series of illuminating experiments performed by the Japanese immunologist Susumu Tonegawa showed that B cells also acquire their unique antibodies through mutations, albeit a precisely regulated form of mutation that occurs in these cells, not in sperm and eggs. B cells rearrange a set of antibody-making genes, mixing and matching genetic modules, like articles of clothing. The analogy oversimplifies the process, but it is important. To give you an example, one antibody might be composed of three mixed-gene modules: a vintage jacket paired with yellow pants and a black beret, while a second might use a different arrangement of modules — perhaps a dark coat matched with blue pants and wing-tip shoes. There is a large wardrobe of genetic modules that every B cell can try on; imagine fifty shirts, thirty hats, twelve shoes, and so forth. To become a mature B cell, it merely has to open its closet and select some unique permutation of gene modules and re-arrange the modules to produce an antibody.

Each such gene rearrangement is also a mutation, although a highly regulated, deliberate kind of mutation in a B cell. A special apparatus fashions the gene rearrangements in an individual B cell, giving each antibody a unique conformational identity, and therefore a unique affinity to bind and hold a particular antigen. The distinctive genetic arrangement in every mature B cell allows it to display a particular receptor on its surface. When an antigen binds to it, the B cell gets activated. It switches from displaying the receptor on its surface to secreting it, in the form of an anti-body, into the blood. Even further mutations accumulate in the B cell, refining the anti-body's binding to the antigen.* Ultimately, the B cell matures into a cell so single-mind-edly dedicated to antibody production that its structure and metabolism are altered to facili-tate the process. It is now a cell dedicated to making antibodies — a plasma cell. Some of these plasma cells also become long lived and retain the memory of the infection.

The new knowledge of B cells, plasma cells, and antibodies burst upon medicine in

*This process is called affinity maturation, and it continues until the antibody reaches an incredibly high binding affinity for an antibody.

unexpected ways. We've already broached the role of the innate system — macrophages and monocytes among them — in the effect of a vaccine. But the ultimate activity of a vaccine depends on the adaptive system: it's the B cell that makes antibodies, and these antibodies are typically responsible for long-term immunity. (As we know, T cells also contribute to this.) A macrophage or monocyte might present digested bits of a microbe or summon B cells to the site of an infection, but it's the antibody-secreting B cell that binds some part of the microbe. The cell that carries a receptor which binds the microbe is activated to clonally expand and begins to secrete the antibody into the blood. Finally, that B cell changes its internal landscape and becomes part of the memory B cell compartment, thereby retaining the memory of the original inoculum.

But beyond vaccines, the discovery of antibodies reignited Paul Ehrlich's fantasy of a magic bullet: if an antibody could somehow be persuaded to attack a cancer cell or a microbial pathogen, it would work as a natural drug against the cell. It would be a medicine like none other: a drug tailor-made to attack and kill its target.

The challenge of making such drug-like antibodies was solved by an Argentinian scientist, César Milstein, at Cambridge University.

Milstein had originally arrived there as a visiting student to work on protein chemistry in bacterial cells. The lab was a one-room chamber. He needed a pH meter to measure the acidity of his chemical solutions, and Fred Sanger, the legendary protein chemist next door, had only one such instrument in a corner room in the Department of Biochemistry. Over casual conversations, and pH measurements, the two became close friends. In 1958, Sanger was awarded the Nobel Prize for solving the structure of a protein — a monumental achievement in molecular biology. And in 1980, he would win a second Nobel for learning how to sequence DNA.

In 1961, Milstein returned to the Instituto Malbrán in Argentina to head the Department of Molecular Biology. But the move, prompted by his dreamy fervor to be back in his homeland, soon devolved into a nightmare. Argentina was permeated by sectarian, divisive nationalism. On March 29, 1962, barely a year after Milstein had settled in the capital of Buenos Aires, the country was torn apart by yet another bloody political coup — Argentina's fourth, with two more to follow.

Chaos reigned. Jews were expelled from universities, Milstein's department was partially disbanded, Communists were killed at gunpoint, and civilians, especially Jews, were thrown in jail. Milstein, with his Jewish name and background, and liberal sympathies, lived in fear of arrest and of being accused as a dissenter or a Communist. Sanger, through his elaborate connections, arranged for Milstein to be smuggled out of Argentina and back to Cambridge. The shared pH meter, tucked away on the top floor of a lab, turned out to be a talisman — the unwitting ticket to Milstein's return to England.

Back in Cambridge, Milstein changed his interest from bacterial proteins to antibodies. Fascinated by their specificity, he began to imagine making magic bullets out of B cells. Could you take a single plasma cell, capable of secreting a single, chosen antibody, and turn it into an antibody factory? Could that antibody become a new drug?

The problem was that single plasma cells were not immortal. They would grow for a few days, then struggle to stay alive, and finally shrivel and die. Milstein, working with the German cell biologist Georges Köhler, came up with a solution that was as brilliant as it was unorthodox: using a virus that could glue cells together, they fused the B cell with a cancer cell. I am still awestruck by the idea. How did they even *think* of using the undead to resuscitate the dying? The result was one of the strangest cells in biology. The plasma cell retained its antibody-secreting property, while the cancer cell conferred its immortality. They called their peculiar cell a hybridoma — a, well, hybrid of *hybrid* and *oma,* the suffix of *carcinoma*. The immortal plasma cell was now capable of perpetually secreting only one kind of antibody. We call this antibody of a single type (in other words, a clone), a monoclonal antibody.

Milstein and Köhler's paper was published in *Nature* in 1975. Weeks before publication, the government-run National Research Development Corporation (NRDC) in the United Kingdom was alerted to the wide-ranging commercial applications of such antibodies; they could be the basis for new highly specific drugs. But the NRDC chose not to patent the method or any materials. "It is certainly difficult to identify any immediate practical applications," the NRDC said in a written statement. In the decades since, that cursory judgment about the applicability of monoclonal antibodies has likely cost the NRDC and Cambridge University several billion dollars in revenue.

The practical implications were immediate. Monoclonal antibodies, abbreviated MoAb, could now be used as detection agents or as markers for cells. But their most important, lucrative, and best-known application was medical: they could form a cosmic array of new drugs.

A drug typically works by binding its target — as Paul Ehrlich had pointed out, like a key to a lock — and inactivating or, occasionally, activating its function. Aspirin, for instance, jams itself into the lock cyclooxygenase, an enzyme involved in blood clotting and inflammation. By that same logic, antibodies, designed to bind other proteins, could also be made into drugs: What if an antibody could bind a protein on the surface of a cancer cell and summon the cascade to kill it? Or recognize a protein of a hyperactive immune cell that was causing rheumatoid arthritis and harpoon it to death?

In August 1975, N.B., a fifty-three-year-old man from Boston, noticed that the lymph nodes in his armpits and neck had become swollen and painful. There were drenching night sweats and relentless fatigue. Yet a full year passed before he finally went to see doctors at the Sidney Farber Cancer Institute in Boston.* Upon examining him, oncologists noted that in addition to the swollen glands, N.B.'s spleen was massively enlarged, to the extent that they could feel its outer edge when palpating his abdomen.

Next, they checked some laboratory values. The patient's white cell count hovered just a little above normal. However, it was the *pattern* of the white cells in the blood that was striking: not only was the number of lymphocytes elevated, but they also appeared to

*The medical facility is now known as the Dana-Farber Cancer Institute.

be malignant. A thin, long biopsy needle was inserted into one of the swollen lymph nodes to withdraw a sample of tissue, which was then sent to a pathologist for analysis. N.B. was diagnosed with a lymphoma — a diffuse, poorly differentiated, lymphocytic lymphoma (or DPDL).

Advanced DPDL — with a swollen spleen, lymph nodes, and circulating lymph cells — is a disease with a dim prognosis. The man's spleen, chock-full of malignant cells, was surgically resected, and he was started on chemotherapy. Drug after cell-killing drug was infused intravenously. None of them worked. The counts kept rising.

Lee Nadler, an oncologist at the institute, drafted a new plan. Lymphoma cells have a number of proteins of their surface. Injected into mice, the animals make antibodies against the malignant cells. Nadler, following a modification of Milstein and Köhler's method, used N.B's cancer cells to create antibodies against his tumor cells and then injected him with serum containing one of the antibodies, hoping for a response. This was an extreme example of personalized cancer therapy — or, more accurately, personalized cancer *immuno*therapy.

The first dose of serum, at 25 milligrams, was seemingly shrugged off by the lymphoma. The second dose, at 75 milligrams, effected a marked dip in his white cell count. The cancer responded but bounded right back. A third dose, at 150 milligrams, again elicited a response: the lymphoma cells in the blood dropped by almost half. But then N.B's tumor cells grew resistant and stopped responding. The serotherapy, as Nadler described it, was discontinued, and N.B. died.

But Dr. Nadler persisted in looking for proteins on the membranes of lymphoma cells that might be targets of antibodies. Eventually he found an ideal candidate called CD20. But could an antibody against CD20 be harnessed as an anti-lymphoma drug?

Three thousand miles away, at Stanford University, the immunologist Ron Levy was also on a hunt for an antibody that would attack lymphoma cells. In the early 1970s, Levy had returned from a sabbatical at the Weizmann Institute of Science in Israel. An investigator there, Norman Kleinman, had developed a method to isolate single plasma cells that could produce antibodies — antibodies against cancer, possibly — but the cells were so short lived that it seemed an impossible effort. "We'd isolate single plasma cells that could produce a single kind of antibody, but they would inevitably die," Levy told me.

"And then," Levy continued, "in 1975, suddenly, Milstein and Köhler came up with this method of fusing a plasma cell with a cancer cell. The fusion enabled the anti-body-making cell to live forever." Levy's face grew animated; his hands began to drum his desk. "It was a revelation. A bon-aaaaaa-nza. Ironically, we could use the immortality of a cancer cell [fused with a plasma cell] to make an immortal cell to produce an anti-body against cancer. We could fight fire with fire."

Levy set off to find antibodies against B cell lymphomas — cancers of B cells. At first, he focused on personalized antibody therapy, in which a unique antibody would be custom-built, so to speak, for every patient. He found a company called IDEC that would make the antibodies. But although some patients responded to the antibodies that were made, IDEC and Levy soon realized that the approach was absolutely unscalable: how many antibodies, against how many individual antigens, could a company possibly produce?

The second set of immunizations produced a MoAb against CD20, the molecule that Nadler had discovered sitting on the surface of both normal and malignant B cells. Levy admits that he wasn't impressed: he believed that the experimental intervention "was going to destroy the immune system and not be safe," he told me. "But they [IDEC] convinced us to do the clinical trial anyway."

Levy was both wrong and incredibly lucky. Fortuitously, humans can live without B cells that express CD20, in part, because once B cells have matured into antibody-secreting cells, or plasma cells, they don't have CD20 on their surface and are therefore resistant to the antibody. Attacking CD20-expressing lymphoma cells *would* inevitably precipitate a concomitant attack on normal B cells, rendering patients partly immuno-compromised, but it wouldn't kill them; they would still retain plasma cells to make antibodies. "There was an off chance that it would work," Levy said. In 1993, he recruited two fellows, David Maloney and Richard Miller, to run the study.

One of the first patients to receive the antibody was a voluble, articulate internist named W.H. She had follicular lymphoma, a slow-progressing, or indolent, cancer that is marked by CD20. "She responded to the first dose," Dr. Levy recalled. However, she relapsed just a year later and had to go back on the experimental MoAb. This time W.H. had a complete response, the tumors melting away. The pattern persisted, though: a third recurrence in '95, for which she was administered the monoclonal antibody in combination with chemo. Another response.

In 1997, the FDA approved the therapy, rituximab, sold under the brand name Rituxan. That year, W.H.'s lymphoma returned. Rituxan delivered a knockout blow, yet the disease came back for rematches in 1998 and in 2005 and in 2007. Twenty-five years after her original diagnosis, W.H. is still alive. Since then, Rituxan has found its place in treating a variety of cancers as well as noncancerous diseases. It has been used in conjunction with chemotherapy to treat and even cure aggressive, lethal lymphomas that express CD20, as well as for rare lymphatic cancers. In the early 2000s, I met a young man with a very unusual splenic cancer that involved CD20-expressing cells. He spiked fevers daily and found it impossible to walk. We surgically removed his swollen spleen — so bulky that it would not fit in the standard surgical tray and had to be placed in a cart for the ride to the Pathology Department — then put him on a course of Rituxan. The nodular tumors dissolved slowly, and the fevers defervesced. He remains in a remission twenty years later.

Rituxan was among the first monoclonal antibodies against cancer. A host of such MoAbs have populated the pharmacopeia, including Herceptin (used to treat certain forms of breast cancer), Adcetris (Hodgkin's lymphoma), and Remicade (immune-mediated diseases such as Crohn's disease and psoriatic arthritis). I reminded Levy of how the NRDC in England had doubted the "practical applicability" of antibody therapy. He laughed, saying, "I'm not even sure that *we* knew its potential."

"Using cells to fight cells," he marveled. "We never really thought about all that we could do when we raised that first antibody."

THE DISCERNING CELL
THE SUBTLE INTELLIGENCE
OF THE T CELL

---

*For centuries, the thymus has been
an organ in search of a function.*
— Jacques Miller, 2014

In 1961, a thirty-year-old PhD student in London, Jacques Miller, discovered the function of a human organ that most scientists had long forgotten. The thymus, so-named because it vaguely resembles the lobe-shaped leaves of the thyme plant, was, as Galen described it, "a bulky and soft gland" that sits above the heart. Even Galen, practicing medicine in the second century AD, noted that it slowly involuted as humans grew older. And when the organ was removed from adult animals, nothing significant happened. A dwindling, dispensable, involuting organ; how could it possibly be essential for human lives? Doctors and scientists began to think of the thymus as a vestigial detritus left behind by evolution, not unlike an appendix or a tailbone.

But might it have a function during fetal

development? Using minute forceps and the thinnest silk sutures, Miller removed the thymus from neonatal mice about sixteen hours after birth. The effect was as unexpected as it was dramatic: the blood level of lymphocytes — the white cells in the circulation that are not macrophages or monocytes — dropped precipitously, and the animals became increasingly susceptible to common infections. B cells dropped in number, but some other white cell — a previously unknown type — was diminished even more dramatically. Many of the mice died of the mouse hepatitis virus; many had bacterial pathogens colonize their spleens. Stranger still, when Miller placed a piece of foreign skin on the animal's flank, the graft was not rejected. Rather, it remained alive and intact, growing "luxuriant hair." It was as if the mouse had no mechanism to distinguish its own tissues from foreign tissues. It had lost its sense of "self."

By the mid-1960s, Miller and other researchers had realized that the thymus was far from vestigial. In newborns, it was the site of maturation for a different kind of immune cell: not a B cell, but a T cell (*T* for *thymus*).

But if B cells generated antibodies to kill microbes, what did T cells do? Why were the mice that lacked T cells colonized by infections, and why did they so meekly accept foreign skin grafts that should have been rejected immediately? How and why did they lose their sense of self? And what is the "self" anyway?

It is a testament to the infancy of cell biology as a science that the physiology of one of the most essential cells of the human body remained a mystery as late as the 1970s. T cells were discovered only about fifty years ago. And it was barely two decades after Miller's experiment — in 1981 — that these cells would become the epicenter of one of the defining epidemics in human history.

Alain Townsend's lab sat atop a steep hill at the Institute for Molecular Medicine* at the edge of Oxford University. In the fall of 1993, when I arrived at Oxford to study with Alain as a graduate student in immunology, the mysteries of T cell function were still being deciphered. The Institute was a modernist steel and glass building. The security guard at the front desk, a woman with a thick Welsh accent, checked IDs before letting you in. Without the right card, she would refuse to let you enter. Two years passed with me fumbling through my pocket for the card, until I finally got the courage to confront her. I had been there, every day, for twenty-four straight months. Didn't she recognize me by my face?

She looked at me stonily. "I'm just doing

*Now called the Weatherall Institute of Molecular Medicine.

my job." Her job, I suppose, was to detect intruders — as if I might well have been James Bond, who'd driven up the hill in an Aston Martin and a Mukherjee mask on a top-secret mission to feed my T cells cultures at night. In retrospect, I've grown to appreciate her diligence. She had internalized immunity.

In Alain's lab, I was assigned a problem that continues to fascinate and frustrate scientists: How does a chronic virus, such as herpes simplex virus (HSV), cytomegalovirus (CMV), or Epstein-Barr virus (EBV), remain hidden persistently inside the human body, while other viruses, such as influenza, are completely eliminated after infection? Why aren't the chronic viruses vanquished by the immune system — especially by T cells?*

*Each of these viruses, we now know, has evolved a specific method of avoiding immune detection — a phenomenon called viral immune-evasion. In the case of EBV, the immunologist Maria Masucci's studies and my own graduate work converged on the same answer. The genome of the Epstein-Barr virus encodes many genes. But once it enters B cells, EBV can turn off most of these genes, except for two: EBNA1 and LMP2. The protein EBNA1 would be an ideal candidate for T cells to detect — but, surprisingly, it is invisible to them. Part of the reason is that EBNA-1 resists being chopped up into pieces inside the cell. As we shall soon learn, Alain Townsend discovered that T cells can recognize only pieces of viral proteins

384

The lab was a buzzing intellectual paradise, full of a frenetic energy that I had never encountered. At four in the afternoon, an old brass bell clanked, and the whole institute descended en masse to the cafeteria to consume weak, lukewarm, nearly undrinkable tea and tough, nearly inedible biscuits. Ita Askonas, one of the pioneers of immunology, used to occasionally hold court in a corner; Sydney Brenner, the Nobel-winning geneticist from Cambridge, might drop in for a casual chat, his fabulously bushy eyebrows, like twin caterpillars, lifting and waving with joy every time we told him of a new experimental result.

An Italian postdoc, Vincenzo Cerundolo, was my direct mentor. Cerundolo, known as Enzo, was short, garrulous, and effervescent. Yet during my first weeks in the lab, he ignored me entirely; he rushed around the lab,

---

— peptides — loaded on a molecule known as major histocompatibility complex, abbreviated MHC. And EBNA-1, it turns out, doesn't produce any peptides. LMP2 may have other means of immunoevasion, but these aren't yet known. Herpes simplex virus relies on a different approach to immune-evasion, disabling the mechanism by which peptides are transported to be loaded on to the MHC molecules. Likewise, cytomegalovirus has yet another evasive maneuver: it makes a protein that can destroy MHC — the very molecule that enables T cells to spot a CMV-infected cell.

navigating his way past me as if I were some irritating piece of equipment that someone had off-loaded in the wrong place. He was trying to finish a research paper, and teaching a freshly arrived graduate student the rococo details of immunology seemed hardly worth his time or energy.

One aspect of Enzo's project involved manufacturing viruses to infect mouse and human cells. The virus was designed to deliver genes into human cells so that Enzo could test the genes' functions. To expand the virus — that is, to make more viral particles — you had to infect a layer of cells. Then you extracted the virus by placing the whole culture in a tube, and freezing and thawing it exactly three times. It was a procedure that required precision and patience. Without freeze thawing, you could not release the viral particles; but if you overdid the process, you might kill the virus entirely. One morning, soon after I had arrived in the lab, I found Enzo fussing over one such tube. A research technician, also Italian, had made a viral prep for him, but she had left for a vacation, and Enzo had no idea whether the virus had been extracted or whether the tube had been left without the viral extraction. It was a tense moment. A low viral count, and the whole experiment, critical to his paper, would go down the drain. He cussed in Italian under his breath: *Cavolo*.

I asked him whether I could look at the tube, and he handed it to me.

On its bottom, in barely visible ink, I saw that the technician had scribbled the letters *C, S, C, S, C, S.*

"What's the word for *freeze* in Italian?" I asked him.

"*Congelare,*" Enzo replied.

"And for *thaw*?"

"*Scongelare.*"

So *that's* what the technician had written: Freeze. Thaw. Freeze. Thaw. Freeze. Thaw, except in a kind of Italian Morse code: C, S, C, S, C, S. Three times each.

Enzo regarded me sharply. Perhaps I wasn't a waste of time after all. He finished his experiment and then asked me if I wanted a coffee. He made two cups. Something had thawed between us.

We became friends. He taught me virology, cell culture, T cell biology, Italian slang, and the secret to making a good Bolognese. I would bike uphill every morning through the incessant rain to work with him, and bike down every evening in the rain again. I came and went as I pleased — sometimes up and down the hill at midnight, while my experiments cooked in the laboratory incubators.

My inner world was suffused by thoughts about T cells and their interactions with chronic viruses. I would rethink my experiments as I biked downhill, running through the data in my head, imagining the life of the virus inside the cell. *"To understand T cell virology, learn to think like a virus,"* Enzo told me. And so I did. I would "become" EBV one afternoon, and herpes the next. (The latter involved having some sense of humor.)

Even after I left Oxford, Enzo and I continued to collaborate, publishing papers together. He sent me vials of cells for my experiments in the lab. I sent him recipes from my mother for his experiments in the kitchen. We met at seminars across the globe, each time resuming our conversation as if we had never left off from the last one. Our interests had shifted, almost simultaneously, from immunology to cancer, and, finally, to the immunology *of* cancer. Over the decades, I matured from a mentee to a colleague and friend. But I could never make Enzo an espresso that he found satisfactory. I tried once, and he spat it out. It was, as Wolfgang Pauli might have put it, so bad that it was not even wrong.

In early 2019, I learned that Enzo had been diagnosed with advanced lung cancer. The news of his illness numbed me with shock; I couldn't bring myself to call him for days. A week passed, perhaps two, until I finally dialed his number from New York. He picked up immediately. He was matter-of-fact about his condition. Perhaps those T cells whose inner mysteries he had spent his life uncovering would find a way to fight his cancer. As Alain Townsend wrote about Enzo in the journal *Nature Immunology:* "One hears the phrase 'fight with cancer,' but this description is a pale shadow of the intense, personal, grinding immunological battle he waged with the rebellious cells that challenged him. He grappled and wrestled with every resource he could muster at home and from around the world, employing every bit of his deep knowledge and experience. He did this . . . unperturbed, never missing a seminar, always available to his students and colleagues. It was a demonstration of supreme courage."

In 2020, a few weeks before I was about to visit Oxford to deliver a talk, I found out that Enzo had died. I cancelled my trip. That evening, I sat in the lab silently, recalling my mentor, my Bolognese instructor, my friend, holding back the memories until they had hardened. I felt dazed, desiccated, congealed in gloom. It was only then, hours later, that the grief burst inside me in wave upon liquid wave.

Congelare; scongelare.

Inner worlds, outer worlds, separated by membranes. What do T cells do during an infection? Imagine, as a human immune system might see it, that there are two pathological worlds of microbes. There is an "outer" world of a bacterium or a virus floating outside the cell, in lymph fluid or blood, or in tissues. And there is an "inner" world of a virus that is embedded and living within a cell.

It is the latter world that presents a metaphysical, or rather a physical, problem. A cell, we said before, is a bounded, autonomous entity with a membrane that seals it from the outside. Its inside — the cytoplasm, the nucleus — is a closed sanctum, largely inscrutable from the outside, except for the signals, or receptors, that the cell chooses to send to its surface.

But what if a virus has taken residence within a cell? What about, say, a flu virus that has infiltrated the cell and hijacked its protein-making apparatus in order to churn out viral proteins that are indistinguishable from the cell's own? This is what viruses do: they "go native." A flu virus turns its hostage into a veritable flu factory, producing thousands of virions per hour. And because antibodies cannot enter cells, how are they to identify one of these rogue cells masquerading as a normal cell? What, then, prevents any virus from using every cell in our body as a perfect microbial sanctuary?

The answers to all these questions, I would soon find out, lay in the cell whose seductive song had tugged me all the way from California to Alain Townsend's lab at Oxford; the cell that could, with nearly miraculous sensitivity, discern a virus-infected cell from an uninfected one, and the cell that can discriminate the self from the nonself. The subtle, wise, discerning T cell.

In the 1970s, Rolf Zinkernagel and Peter Doherty, immunologists working in Australia, found the first clue to decipher T cell recognition. They began with so-called killer T cells: T lymphocytes that recognize virus-infected cells and souse them with toxins until they shrivel and die, thereby purging the microbe taking refuge there. These cytotoxic (cell-killing) T cells, brandished a particular marker on their surfaces: CD8, a type of protein.

The peculiar thing about these CD8-positive T cells, Zinkernagel and Doherty discovered, was that they had a capacity to recognize viral infections *only in the context of the self.* Consider that thought: your T cells can recognize virally infected cells only if they come from *your* body, not someone else's.*

A second feature of killer Ts was just as

*If the T cell and the target cell are "unmatched" — meaning that they come from different bodies and carry different protein markers on their surfaces — then the immune system will kill them anyway, regardless of whether they are infected. This is the basis for graft rejection: if you implant a stranger's cells into your body, those cells will be rejected. We will return to this "nonself" recognition in future pages.

puzzling. Although a CD8 T cell could recognize a cell from the same body, it killed only *infected* cells from the same body. No viral infection, no kill. It was as if the T cell was capable of asking two independent questions. First: *Does the cell that I am surveying belong to my body?* In other words, is it self? And second: *Is it infected with a virus or a bacterium?* Has the self been changed? Only if both were true — the self *and* infection — would a T cell kill its target.

In short, T cells had evolved to recognize self, but an *altered* self that happens to harbor an infection. But how? Using genetic techniques, Zinkernagel and Doherty tracked the detection of the self to a set of molecules called MHC class I.★

It is as if the MHC protein is a frame. Without the right frame, or context ("yourself"), the T cell cannot even see the picture, even if it is a distorted version of the "self." And without the picture in the frame (presumably, some part of the virus — an *infected* self), again, the T cell cannot recognize the

---

★The MHC class I protein comes in thousands of variants. Each of us carries a unique combination of MHC class I genes. It is this self MHC that the T cell first detects. If the infected cell and the CD8 T cell come from a person (with the same MHC class I), then recognition happens, and the infected cell is killed.

infected cell. It needs both the pathogen *and* the self — the picture *and* the frame.*

Zinkernagel and Doherty had solved one piece of the puzzle: that T cells recognize the infected "self." But the second piece was just as thorny a problem. This molecule — MHC class I — is involved, yes, but how does a cell signal an *altered* self — in other words, a self with an infection? How does a CD8 cell find a self-cell with a flu virus embedded within it?

Alain Townsend, my former advisor, who has, over the years, become a close friend, took up this question in the 1990s, first at Mill Hill in London and then at Oxford. Alain is among the most brilliant and prescient scientists I have ever met. He was, at times, the very caricature of the Oxford academic: he despised traveling to scientific meetings in exotic destinations. The word *tropical* inspired terror. He ate doughy, meat-filled pasties for lunch virtually every day, and he had perfected the English habit of deadly

---

*There's presumably a deep evolutionary logic behind this. A peptide fragment displayed by a macrophage or a monocyte indicates a bona fide infection. A free-floating one — lacking the frame provided by a phagocytic cell, and not presented appropriately — might be incidental debris, or worse, a fragment from a human cell. Mounting an immune response to the "self" fragment would unleash autoimmunity — a devastating consequence of T cell immunity.

euphemism. If he thought an idea was stupid, or unscientific, he would look vaguely into the distance, pause, and say. "Oh! That notion seems . . . um . . . er . . . rather *subtle*." At lab meetings, I must confess, I was often rather subtle.

In the late 1980s and early 1990s, Townsend, among others, began to uncover how a killer T cell detects a virus-infected cell. Townsend began his experiments with CD8 killer T cells. His particular interest was cells infected with the influenza virus. How are these infected cells recognized and cleared? Just as Zinkernagel and Doherty had demonstrated previously, Townsend found CD8 T cells would kill flu-infected cells that came from the same body — in other words, they depended on the recognition of the self. But as mentioned above, the self cell had to carry an infection — and the expression of a viral protein — to be killed. What viral protein was being recognized? Some of these killer T cells, researchers had found, were detecting the presence of the influenza protein, called nucleoprotein (NP), inside a flu-infected cell.*

*Nucleoprotein is an influenza protein that is made inside the cell. It is then packaged into the influenza virion. The protein has no signals to enable it to reach the surface of the cell — hence Alain Townsend's puzzlement about how a T cell could possibly detect it.

394

But that's where the mystery began. It was an inside-outside problem. "That protein, NP, never makes it to the cell surface," Alain told me. We were sitting in a London taxi cab, returning from a lecture. It was dusk, London dusk, with its sudden shards of oblique English light, and the streets, as we moved through them — Regent Street, Bury Street — were full of endless rows of houses with partially lit windows and impregnable doors. How could a detective, moving door to door, find a resident inside one of these houses, unless the person happened to poke his head outside?

T cells can't get *inside* cells — there are membranes that separate them — but then how can a T cell assess the components of the inside of an infected cell?

"NP is *always* inside the cell," Alain continued. His eyes were shining — glowering — now as he remembered the experiments. He performed the most sensitive tests — assay upon assay, week upon week — to find even the faintest trace of the NP protein on the flu-infected cell's surface, where a T cell might detect it. But it wasn't there. It never pokes its head out of the cell's membrane. "As far as cell surface proteins are concerned, there is nothing for a NP-detecting T cell to see," he said. "It's invisible on the cell surface — *it isn't even there* — and yet it's perfectly visible to the T cell." The taxi halted at a flickering light, as if waiting for an answer.

How, then, was the T cell detecting NP? The crucial discoveries came in late 1980s. The CD8 killer cells, Alain found, were not recognizing the intact NP poking its face outside the cell. Rather, the cells were detecting viral *peptides* — small pieces, or fragments, of the viral protein, NP. And crucially, these peptides had to be "presented" to the T cells in the right "frame" — in this case, carried, or loaded, by the MHC class I protein, and brought to the surface of the cell. The self, yes, but the altered self.

The MHC class I protein — the very protein that Zinkernagel and Doherty had implicated in the killer T cell response — was actually a carrier, a peptide bearer, and a "frame." *The MHC was turning the inside to the outside, constantly sending out a sampling of a cell's innards.*

Imagine it as a spy — our man in Havana — sending out semaphores about a cell's interior for the immune system to recognize. The T cell needs the right spy — therefore the requirement for self-recognition. And it needs the right semaphore — therefore the requirement for a foreign pathogen inside a cell. It is yet another example of biology's codes. Combine the right inside man and the right semaphore — the self MHC carrying a piece of a viral peptide — and the T cell moves in for its kill.

In biology, there is rarely a more poignant moment than when a structure of a molecule melds with its function: what a molecule *looks* like, and what is *does*, fall into perfect union. Take DNA, the iconic double helix. It *looks* like an information carrier — a string of four chemicals, A, C, T, and G, with a unique sequence (ACTGGCCTGC) just like a four-letter Morse code. The double helix also allows us to understand how replication occurs. The strands are complementary, yin and yang: the A on one strand is matched with T on the other, and C matched with G. When a cell divides to make two copies of DNA, each strand serves as a template to make the other. The yin dictates the formation of the yang; the yang shapes the yin — and two new yin-yang double helices of DNA are formed.

The sperm's tail that whips around to make the sperm squirm toward the egg *looks* like a tail, except that it is built out of an assemblage of proteins. The motor that makes the tail spin *resembles* a motor, with a set of moving parts arranged in a circle. And the hook that connects the motor to the tail, transforming circular motion to the propeller-like, swimming motion of the sperm, *looks* like a hook engineered precisely to achieve this transformation.

And so it was with the MHC class I. When its structure was finally solved by the crystallographer Pam Bjorkman at Caltech University, it seemed to fall into perfect step with its function. The molecule looks, well, as you might expect: like a hand holding up two open halves of a hot dog bun. The two sides of a bun — two protein helices of the MHC molecule — leave a perfect groove in the middle. The viral peptide ready to be presented is the sausage stuck in the groove between the two halves of the bun, waiting to be served to a T cell.

"Everything came together in that image. Everything clicked," Alain said. The foreign element (the viral peptide in its groove) and the self element (the spiral MHC edges of the molecule) are *both* visible to the T cell. Alain was infinitely moved by looking at that structure; he could actually visualize the presentation of a viral peptide to a T cell. "[E]very immunologist's pulse will race as he or she sees the three-dimensional structure of the binding site of an MHC molecule displayed for the first time," he wrote in the pages of *Nature,* because it will explain the "structural basis" of antigen recognition. That image of the class I molecule answered a thousand questions from immunologists and raised a thousand more. Alain titled his 1987 article with a fragment of a poem borrowed from William Butler Yeats:

Those images that yet
Fresh images beget.

And, indeed, the image of the MHC, with its bound peptide, begot fresh images. What about the MHC class I allows T cell recognition? And since the MHC class I — the carrier protein — is a molecular platter that displays both self and foreign elements, what about the structure of the cognate recognition molecule on the T cell's surface? What does the protein that detects the carrier MHC peptide complex look like?

At around the same time that the molecular structure of the MHC class I was solved, several groups, including Mark Davis's at Stanford, Tak Mak's in Toronto, and Jim Allison's in Houston, homed in on the gene that encodes the T cell receptor — the molecule on the T cell that recognizes the MHC bound to the peptide. And when its structure was finally solved, there was, yet again, a profound matching of structure and function.

The T cell receptor looks like two outstretched fingers. Parts of the two fingers touch the self — that is, the raised hinges of the MHC molecule around the sides of the peptide. And parts of it touch the foreign peptide carried in its groove. Both self and foreign are recognized *simultaneously:* the two requirements for the detection of an infected cell are contained in the structure. One part of a finger touches the self, one part makes contact with the foreign. When both are touched, recognition is achieved.

The matching of form and function is one of biology's most beautiful ideas, first articulated centuries ago by thinkers such as Aristotle. In the structures of the two molecules — MHC and T cell receptor — one can identify the fundamental themes of immunology and cell biology. Our immune system is built on both the recognition of self and of its distortion. It is designed, by evolution, to detect the altered self. As Alain concluded in his seminal article: "T-cell recognition can now be explored in a rational way."

Let's leave aside the matching of structures and functions for a moment. The solution to the problem of T cell recognition, Townsend knew, had created another problem. It had begotten yet another fresh image: How did a viral protein — let's say NP — synthesized *inside* a cell make its way outside to where a T cell might find it?

As their molecular studies deepened, Townsend and others began to uncover an elaborate internal apparatus that would precisely accomplish this task of turning the cell's innards inside out to display it to the outside world. The process begins, we now know, as soon as the viral protein is made inside the cell. The cell doesn't know whether the protein is part of its normal repertoire or whether it's foreign; there's no special feature of a viral protein that identifies it as "viral."

And so, like all proteins, NP is eventually sent into the inherent waste-disposal mechanism of the cell, the cell's meat grinder — the proteasome — that then chews them into smaller pieces (peptides), and ejects the peptides into the cell. And then, using special channels, these peptides are transported into a compartment where they can be loaded on to MHC class I. The loaded class I proteins carry the viral peptides to the cell surface and present them to the T cell. The class I molecules, as their structures indicated, are like molecular platters, constantly presenting teasers — *amuse-bouche* — of the cell's insides for T cell surveillance.

It is one of the cleverest ways of repurposing a cell's intrinsic molecular apparatus: it takes the natural waste-disposal factory of the body, treats the viral protein as if it were any other protein meant for disposal, mounts it on a protein carrier, and pushes it out of a hatch and onto the cell's surface.

The inside is now outside. The cell has sent a sampling of its inner life, bound in the correct frame, to be surveyed by the immune system. When a CD8 cell comes by, sniffing the cell surface, it will find a large selection of peptides from the interior of a cell loaded on its surface — including, of course, the peptide from the virus. And only if that foreign peptide is presented by the self MHC (the altered self) will it trigger an immune response, killing the infected cell.

So far, we've focused on the "inner" world of the cell — that is, on pathogens lodged inside a cell. But the "outer" world — when pathogens are free-floating in the body — presents its own questions: How do viruses and bacteria present *outside* the cell activate a T cell response?

In principle, activating a T cell response *before* a virus has infected its target cell — while it's still coursing through the blood, say, or moving through the lymph system — would carry many advantages for an organism: it could prepare the various wings of the immune response for the impending infection. It could trigger alarms in the body — fevers, inflammation, and the production of antibodies, all in an effort to thwart the infection at an early stage.

As we've discussed already, the cells of the innate immune system — macrophages, neutrophils, and monocytes — are constantly surveying the body to find signs of injuries and infections. Once such an infection is detected, these cells swarm to the infected site to ingest, or phagocytose, bacterial cells or the viral particles. They devour the invaders, internalize them, and route them to special compartments. These compartments — lysosomes, among them — are chock-full of enzymes to degrade the virus into smaller fragments, including those bits and pieces of proteins known as peptides.

This, too, is a form of "internalization" — although not an internalization that causes infection. Here the virus is noticeably a foreigner, destined for destruction. It is yet to enter a cell, make new virions, and "go native." Alain Townsend's work, discussed earlier, had focused on the CD8 T cell response that occurs *after* a virus has sheltered itself inside a cell. But what about readying a T cell response as soon as the surveillance system in the body has detected a pathogen?

In the 1990s, Emil Unanue, now a professor at the Washington University School of Medicine, began to explore the T cell response to microbes outside the cell. This form of immune detection, he discovered, follows nearly analogous principles to the ones that Townsend had found.

Once phagocytosed, targeted to the lysosome and degraded, bacteria and viruses are chopped into peptides.* And just as the class

*A word of caution here: a small fraction of peptides from the interior of the cell — typically waste products — are also sent to the lysosome for destruction and are presented on class II MHCs.

I MHC molecule frames and presents a cell's *internal* peptides to T cells, a related class of proteins — called class II MHCs — presents mostly *external* peptides to T cells. Its structure, too, is similar: a hand holding up two halves of a bun, with a groove for the peptide in the middle.

In other words, broadly speaking:

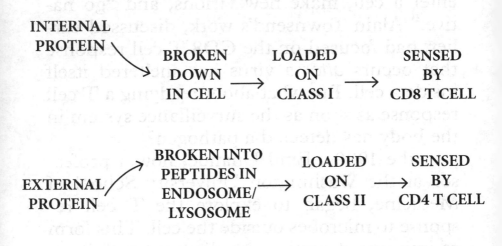

But it's here that the immune response diversifies, incorporating a second wing of attack. The internal peptides, presented by class I MHCs, as Zinkernagel and Doherty had found, are detected by a set of T cells called CD8 killer T cells. The CD8 cells, you'll remember, kill the infected cell, purging the virus in the process.

In contrast, a majority of peptides derived from pathogens outside the cell (and a few

404

from the cell's interior that end up in the lyso-some) are presented by class II MHCs. These are detected by a second class of T cells, called CD4 T cells.

The CD4 T cell isn't a killer (again, there's a logic to this. The virus is already dead and minced to pieces; why kill the cell that's alert-ing the T cell to a dead virus?). Rather, this T cell is an *orchestrator*. Having sensed the MHC II peptide complex, the CD4 cell be-gins to coordinate an immune response. It incites B cells to start synthesizing antibodies. It secretes substances that amplify the macro-phage's capacity to phagocytose. It causes an upsurge of local blood flow and summons yet other immune cells, including B cells, to chal-lenge the infection.

In the absence of the CD4 cell, the transi-tion between innate immunity and adaptive immunity — i.e., between the detection of a pathogen and antibody production by B cells — would fall apart. For all these properties, and especially for supporting the B cell anti-body response, this type of cell is called the "helper" T cell. Its job is to bridge the innate and adaptive immune system — macrophages and monocytes on one end, and B and T cells on the other.*

*So desperate, and so constant, is our battle with pathogens that even the helper needs helpers. Many different kinds of cells — monocytes, macrophages,

Antigen processing and presentation to CD4 and CD8 cells — the mainstays of T cell recognition — are slow but painstakingly

---

and neutrophils that we've encountered earlier — can present peptide/MHC complexes, those molecular platters loaded with their inner contents, to engage helper and killer T cells; this is after all, a *general* surveillance system for virus-infected cells. But there's a specialized cell so acutely attuned to engaging a T cell — so inherently specialized for antigen presentation — that its main, and only, function may be to detect pathogens and spark an immune response. This cell, discovered by the scientist Ralph Steinman, lives mainly in the spleen and sends out dozens of branches — almost beckoning the T cell to come and look. Steinman had found them by looking through a microscope in the 1970s, and spent nearly four decades deciphering their function. This cell possesses one of the most effective mechanisms to trap viruses and bacteria, one of the most efficient processing systems to present peptide/MHC complexes, the densest collections of surface molecules to activate T cells, and among the most potent mechanisms to secrete molecular alarm bells that activate both the adaptive and innate immune responses. It is called the "dendritic cell," from the Greek word for *branch,* for the many branches and fingerlings that extend from its body (one might almost imagine these branches evolved to create separate docking sites for T cells). But metaphorically, too, it is many-branched — capable of coordinating every aspect of methodical processes. Unlike an antibody, a gunslinging sheriff itching for a showdown with a gang of molecular criminals in the center of town, a T cell is the gumshoe detective going door to door to look for perpetrators hiding inside. In *The Lives of a Cell,* Lewis Thomas wrote: "Lymphocytes, like wasps, are genetically programmed for exploration,

---

the multipronged immune system and prime it to respond to an infection. The dendritic cell, perhaps, is the first among the first responders to jump-start immunity to a pathogen. Ralph Steinman died in New York on September 30, 2017, a few days before the Nobel Prize committee awarded him the prize for his discovery (for a while, tragically, there was a prize but no prizewinner. The Nobel is not awarded posthumously, but the decision to award Steinman the prize had been made long before his death, and so the honor was still conferred on him). Obituaries and tributes to Steinman poured out from scientists, physicians, and Steinman's trainees. But the one tribute that I find the most evocative, written by the Seattle immunologist Phil Greenberg, carries a title that brings us back to the very roots of cell biology — to Van Leeuwenhoek, Hooke, and Virchow, looking down their scopes and unveiling a new cosmos of biology. The article is called "Ralph M. Steinman: A Man, a Microscope, a Cell, and So Much More." It is the story of virtually every researcher who inhabits this book captured in three words: A scientist. A scope. A cell.

but each of them seems to be permitted a different solitary idea. They roam through the tissues, sensing and monitoring." Unlike the B cell, though, the T cell isn't looking for a culprit to come bursting out of the saloon, guns a-blazing. It is, like some omniscient Sherlock Holmes with a pipe and an umbrella, seeking the *portent* of a person. The debris left behind by an inner presence. A torn-up letter, with the fragment of a name, discarded in the trash can outside. (You might think of that crumpled piece of stationery, mounted within a trash can, as a peptide presented on an MHC molecule.)

There is a duality in the immune system: one recognition system needs no cellular context (B cells and antibodies), while the other is triggered only when the foreign protein is presented in the context of a cell (T cells). It's this duality that ensures that viruses and bacteria are not just cleared from the blood by antibodies but are also cleared from infected cells where they could otherwise be harbored safely, by T cells.

It is, contrary to Alain's use of the word, truly rather subtle.

The first patients began to arrive at hospitals and clinics in 1979 and 1980. It was the winter of '79, and a Los Angeles doctor, Joel Weisman, noted a spike in the number of young men, typically in their twenties and thirties, presenting to his clinic with a bizarre illness: a "mononucleosis-like syndrome, marked by a hectic fever, weight loss, and swollen lymph nodes." On the other side of the country, similar clusters of unusual illnesses had also begun to emerge suddenly. In March 1980, in New York, a patient named Nick came down with an odd, wasting disease: "lassitude, weight loss, and a slow consumption of the whole body."

By early 1980, there were more patients — again, primarily young men in New York and Los Angeles, many of whom had acquired a form of pneumonia seen previously only in severely immunocompromised patients, caused by a pathogen that was virtually unheard of outside textbooks: pneumocystis. The disease was so rare that the only drug to treat it, pentamidine, was dispensed through a federal pharmacy. In April 1981, a pharmacist at the US Centers for Disease Control (CDC) noted that demand for the antifungal had almost tripled, and all the requests seemed to come from various hospitals in New York and LA.

On June 5, 1981, a landmark date, the *Morbidity and Mortality Weekly Report (MMWR)*, the CDC's weekly record of the nation's illnesses, published five cases of young men with pneumocystis pneumonia (PCP), noting the highly unusual fact that they had all occurred in Los Angeles, within a few miles of each other. Often, they later learned, the patients were men who'd engaged in sexual contact with other men. "The occurrence of pneumocystis in these 5 previously healthy individuals without a clinically apparent underlying immunodeficiency is unusual," the report stated. "[A]ll 3 patients tested had abnormal cellular-immune function, two of the 4 reported recent homosexual contacts. All the above observations suggest the possibility of a *cellular-immune dysfunction* [my italics] related to a common exposure that predisposes individuals to opportunistic infections."

And then, on both coasts, men began showing up at doctors' offices with a rare cancer of the skin and the body's mucous membranes. Kaposi's sarcoma, rarely seen in the United States, was an indolent malignancy later associated with a viral infection. It typically presented as purplish-blue skin lesions and had occasionally turned up among aged Mediterranean men and patients in an endemic belt in subequatorial Africa. But in New York and Los Angeles, the sarcomas were aggressive, invasive cancers, covering arms and legs with erosive, violet welts that invaded the skin. In March 1981, the *Lancet* published a case report of eight such cases — another peculiar cluster. By then, Nick, the man with the wasting disease, was already dead of a cavitating brain lesion caused by *Toxoplasma gondii,* a common, typically noninvasive pathogen found in, of all creatures, harmless household cats.

410

By the late summer of 1981, peculiar illnesses seen previously only in severely immunocompromised patients seemed to arise out of nowhere. Week after week, the *MMWR* reported the grim record of a plague with a thousand faces, composed of seemingly unconnected diseases: more cases of pneumocystis pneumonia, cryptococcal meningitis, toxoplasmosis, violet sarcomas arising in young men, strange, indolent viruses suddenly turned active and furious, unusual lymphomas surging out of nowhere.

The only common epidemiological connection was that these diseases had a powerful predilection for men who had sex with other men — although by 1982, it was also clear that recipients of frequent blood transfusions, such as patients with the blood-clotting disorder hemophilia, were also at risk. And in virtually every case, there were signs of catastrophic immune collapse, particularly of cellular immunity. In a 1981 issue of the *Lancet*, a letter to the editor suggested the name "gay compromise syndrome." Some called it "gay-related immune deficiency," or, more perversely (and with evident discriminatory intent), "gay cancer." In July 1982, as doctors still searched frantically for a cause, the name of the illness was changed to acquired immunodeficiency syndrome, acronymed AIDS.

But what was the cause of this immunological collapse? As early as 1981, three independent groups in New York and Los Angeles had studied the patients and discovered that their cellular immune systems had been decimated. (Even the June '81 report in the *MMWR* had noted a collapse in "cellular immunity.") By sifting through each type of immune cell, the crucial defect was soon identified as a dysfunctional CD4 helper T cell that was also low in number. A normal CD4 count is around 500 to 1,500 cells per cubic millimeter of blood. Patients with full-blown AIDS had only 50, or even 10. AIDS, as one group of researchers described it, was "the first human disease to be characterized by the selective loss of a specific T cell subset, namely, CD4+ T helper/inducer cells." The threshold for AIDS was set at 200 CD4 helper T cells per cubic millimeter of blood.

That an infectious agent, likely a virus, was involved was soon apparent. It could be transmitted sexually, both through homosexual and heterosexual sex, through blood transfusions, and by infected needles introduced into the bloodstream, typically for illicit intravenous drug use. Routine tests revealed no known virus or bacteria. This was an infection with an unknown virus from an unknown source that happened to attack cellular immunity. It was also a perfect storm, for such a virus is, biologically and metaphorically, a consummate pathogen, killing the very system designed to kill it.

The identity of the AIDS-causing virus was finally revealed on March 20, 1983, when the French researcher Luc Montagnier, working with Françoise Barré-Sinoussi, published a paper in *Science* magazine describing the isolation of a novel virus from the lymph nodes of several patients with AIDS. Over the next year, as the disease swept through Europe and America, killing thousands, virologists debated whether this virus was, indeed, the cause of AIDS. In 1984, biomedical researcher Robert Gallo's lab at the National Cancer Institute settled the debate for good: the team published four papers in the journal *Science* providing unequivocal evidence that the new virus caused AIDS. It was christened human immunodeficiency virus, or HIV. Gallo's lab described a method to cultivate the virus and developed antibodies to it that would form the basis for the first tests for infection.

We typically think of AIDS as a viral disease. But it is also equally a cellular disease. The CD4-positive T cell sits at the crossroads of cellular immunity. To call it a "helper" cell is to call Thomas Cromwell a mid-level bureaucrat; the CD4 cell is not so much a helper as it is the master machinator of the entire immune system, the coordinator, the central nexus through which virtually all immune information flows. Its functions are diverse. Its work begins, as we read before, when it detects peptides from pathogens, loaded on class II MHC molecules, and presented by cells. Then it jump-starts the immune response, activating it, sending alarms, enabling B cell maturation, and recruiting CD8 T cells to sites of viral infection. It secretes factors that enable cross talk between the various wings of the immune response. It is the central bridge between innate immunity and adaptive immunity — between all the cells of the immune system. The collapse of CD4 cells thus cascades rapidly into a collapse of the immune system in total.

The tall, thin man who came to see me on a Friday afternoon had only one complaint: he was losing weight. No fevers, no chills, no night sweats. Yet his weight kept falling precipitously. Every day, he stood on a scale at home and found that another pound had gone. He stood up to show me: he had tightened his belt buckle, hole after hole, in the last six months until he had reached the last of the holes. And still his pants were falling off his waist.

I probed a little deeper. He was a retired real estate agent from Rhode Island. He had been married, but now lived alone. The man had an unusual affect: while totally frank about his medical symptoms and risks, he was reserved about his personal life, offering only the vaguest of details.

414

"IV drug use?" I asked.

"No," he replied emphatically. Never.

"A history of cancer in the family?"

Yes. His father had died of colon cancer. His mother had breast cancer.

"Unprotected sex?"

He looked at me as if I were mad.

"No." He claimed to have been celibate for years.

I performed a physical examination. Nothing stood out. "Let's order the basic tests," I said. Asymptomatic weight loss is a tough medical conundrum. We would check for occult bleeding or any signs of cancer. Tuberculosis seemed like an unlikely scenario. His risk for HIV was low, but we could circle back to that.

Our time having finished, he got up to leave. He was wearing sneakers with no socks on. And it was as he turned around that I saw it out of the corner of my eye: a blue-purple lesion on one ankle, just above the shoe.

"Come back for a minute," I told him. "Take your sneakers off."

I examined the lesion carefully. It was a small mound on his skin, about the size of a kidney bean, with a deep eggplant color. It looked like Kaposi's sarcoma. "Let's add a CD4 count," I said, then added delicately, "and an HIV test?" He seemed unfazed.

A week later, the numbers were back: he had full-blown AIDS. His CD4 count was one-tenth of what we consider normal, and a biopsy of the blue-purple lesion tested positive, as I'd suspected it would, for Kaposi's sarcoma, one of the AIDS-defining diseases.

I sent the man to an HIV specialist. The next time he came to see me, he again denied vehemently any involvement in the risky behaviors associated with HIV/AIDS: no unprotected sex with men or women, no IV drug use, no blood transfusions. It was as if the virus had appeared out of thin air. There was no point probing further. An impenetrable sheet of privacy stood between us. In his 1981 novel *Midnight's Children,* Salman Rushdie writes of a doctor who is allowed to examine his patient, a young woman, only through a hole in a white sheet of cloth. It seemed, at times, that I could visualize my patient only through a hole in a sheet of cloth — of what? Homophobia? Denial? Sexual shame? Addiction? We started him on antiretroviral therapy. His CD4 counts began to climb, slower than we had expected, but edging up, day by day. His weight plateaued for a while.

Then it began to drop again. In an unusual plot twist, two new lesions — purple-blue — suddenly turned up on his arm. A new bruise? More Kaposi's? But the timing made no sense. By then, he had begun to spike spectral fevers and chills. Lumps had appeared under his armpits, and those two new blue-black lesions enlarged. A few afternoons later, he was back in the emergency room.

From there, things spun quickly out of control. His blood pressure plummeted, and his toes turned blue. His blood cultures grew out the bartonella bacterium often found in patients with AIDS. And the tongue twister of the case took another turn: the blue-black lesions that had grown afresh on his skin were not Kaposi's after all but were tumorlike protrusions caused by blood vessels inflamed by bartonella. What are the chances that two identical lesions in the same patient would have two radically different causes? Sometimes medical mysteries are deeper than we can possibly imagine.

We treated him with the antibiotics doxycycline and rifampin until his symptoms subsided. He stayed in the hospital for two weeks. I went to see him a week into his hospitalization, and he was back to his reticent self. Bartonella is almost always caused by a cat scratch; typically, fleas introduced into the skin by the scratch transmit the disease.

We sat for a while in silence, as if each of us was figuring out a strategy in this mutual battle of concealment.

"Cats?" I asked him. "You've never told me you had cats."

He looked at me with puzzlement. No cats.

No risk factors for HIV. No drugs. No unprotected sexual intercourse. No cats. No scratches. I shrugged and gave up.

Fortunately, the man recovered from his infections. The antiretroviral drugs are working, and his CD4 counts have returned to normal. But the black box of causes still remains perfectly sealed. Sometimes human mysteries are deeper than medical ones.

Combination antiviral drug therapy, with three to four drugs, has changed the landscape of HIV treatment. The pharmacopeia against the virus has increased year by year. There are drugs that prevent the virus from replicating efficiently, drugs that prevent the virus from duplicating its RNA or integrating into the host genome, drugs that prevent the virus from maturing into infective particles, drugs that prevent the virus from fusing to the vulnerable cells — five or six separate classes of drugs in all. Therapy with these medicines is so effective that patients with HIV can live for decades without any sign of the virus — *undetectable,* in the parlance of medicine. They are not cured, but so deeply controlled, with such vanishingly low viral loads, that they cannot infect others.

And laboratories around the globe are hunting for vaccines against HIV that might prevent infection altogether, thus eliminating the need for chronic, multidrug therapy. Indeed, some of the most far-reaching drug trials have moved from treatment to prevention. In one study, a two-dose regimen of the antiviral nevirapine — prescribed to an HIV-positive mother before delivery, and one dose given to the newborn within three days of birth — decreased the risk of transmission from 25 percent to about 12 percent. It costs around $4. More potent combinations of drugs to prevent transmission in pregnant mothers or in high-risk individuals after sexual contact are being put to the test virtually every month.

But while we await the HIV vaccine, at least one route to the cure of a cellular disease might involve cellular therapy. On February 7, 2007, Timothy Ray Brown, an HIV-positive man, underwent a bone marrow transplant. Brown, originally from Seattle, was diagnosed with HIV in 1995 while a university student in Berlin. He had been treated with antiviral drugs, including the then-new protease inhibitors, and had lived symptom free for a decade. His CD4 counts were barely below normal, and his viral load was undetectable.

However, in 2005, he began to feel suddenly exhausted and weak, unable to finish his usual bicycle rides. He was found to be moderately anemic, even though his HIV was under control. A bone marrow biopsy revealed that Timothy had acute myelogenous leukemia, a deadly cancer of white blood cells. (Brown happened to be extremely unlucky. This cancer and the HIV infection are thought to be unrelated; HIV-infected men have a higher risk of certain lymphomas, but not of AML.)

He was treated initially with standard chemotherapy, but the leukemia relapsed in 2006. For the next step of treatment, his oncologists suggested high doses of chemo intended to wipe out his malignant cells — and with them, his defenses against disease — followed by a bone marrow transplant from a matched donor. Such matched donors are usually hard to find, but, surprisingly, Brown turned out to have a glut of 267 matches in the international registry. And so, faced with a wealth of choices, his doctor, Gero Hütter, a Berlin hematologist with an experimental bent, suggested seeking a donor who also happened to have a natural mutation in CCR5, the coreceptor that HIV uses to enter CD4 cells. In some humans, all cells, including CD4 cells, have a natural mutation in the CCR5 gene called CCR5 delta 32 — the very same mutation that Chinese geneticist He Jiankui had tried to create in Lulu and Nana via gene editing. Humans who inherit two copies of this mutant CCR5 gene are resistant to HIV infection. Brown's transplant, then, would be not just an innovative medical treatment but also a once-in-a-lifetime experiment.

Hütter knew of an earlier patient, also from Berlin, who had been taken off HIV drugs because he was thought to have naturally inherited a gene that conferred resistance to HIV. The patient's viral load had not rebounded even after his HIV meds had been discontinued — suggestive evidence, but not solid proof that a patient's genetic background could alter his or her susceptibility to HIV.

Brown's case, Hütter knew, would represent a significant leap forward. For one thing, the stem cell *donor,* not the host, would supply the resistance gene. And although the primary goal of the transplantation was to cure Timothy Brown's leukemia, Hütter reasoned, why not try to defeat the HIV infection in the same cellular swoop?

Unfortunately, the leukemia recurred a little more than a year after the transplant, necessitating a second attempt using stem cells from the same donor. It was an intensely grueling ordeal. "I became delirious, nearly went blind, and was almost paralyzed," Brown wrote in a reflective article in 2015, the twentieth anniversary of his cancer diagnosis. Recovery took months, and then years. Gradually, he relearned to walk, and his vision was restored. But he stayed off his HIV medicines, as had been planned after the first transplant. And as the new stem cells, with the naturally resistant version of CCR5 delta 32 cells, engrafted, he remained HIV negative. He was cured of leukemia — and, perhaps more astonishingly, of HIV.

Brown's case is still widely discussed in the medical community. Initially referred to anonymously as the "Berlin patient," Brown decided to reveal his identity in the media and in scientific journals in early 2010, the year he returned to the United States. He remained HIV free for thirteen years and began to refer to himself as "cured." In 2020, Timothy Ray Brown died of relapsed leukemia at the age of fifty-four — but still with no sign of HIV in his blood.

Let's make one fact clear: the world's pandemic of HIV is not going to be solved by bone marrow transplantation with CCR5 delta 32 donor cells. The procedure is far too expensive, too toxic, and too labor intensive to be considered a practical option for a large swathe of the human population.

But there are potent lessons, and open questions, in Brown's story that are relevant for vaccine and antiviral drug development. First, altering the cellular reservoir of HIV in the blood can potentially cure the disease, or at least achieve deep permanent control on viremia. In the wake of Timothy Brown's HIV cure, a second patient, in London, was also cured of HIV with a bone marrow transplant. Unless these two cases are anomalies, it is unlikely that there is a "secret" reservoir, beyond blood, where HIV might hide and get reactivated when the drugs are discontinued — a potential problem that has concerned researchers for decades. (Note that I specified blood, not just CD4 T cells. Macrophages, also derived from blood, for instance, are known to act as reservoirs for HIV.)

It is impossible to know whether a dormant reservoir of HIV remained in Brown's body after his supposed cure, but the fact is that he had lived virus free for more than a dozen years. And if there *were* such stores in his remnant macrophages, then perhaps the virus, unable to infect his CD4 positive T cells, was permanently trapped, like a man under a locked cellar door.

What factors had contributed to the potential cure? The particular strain of HIV? The low viral load before the transplants? The "engineering" of Brown's immune system after the transplant? Answers to these questions will guide the next wave of HIV therapies. We will learn about where the virus hides, how to attack its reservoirs, how cells might resist infection. Most importantly, we'll learn about how the immune system can be taught to recognize this most devious of pathogens.

THE TOLERANT CELL
THE SELF, HORROR AUTOTOXICUS, AND IMMUNOTHERAPY

And what I assume you shall assume,
For every atom belonging to me
as good belongs to you.
— Walt Whitman, "Song
of Myself," 1892

It is time to return to the question: What is the self? An organism, as I suggested before, is a cooperative union of units; a parliament of cells. But where does the union begin and end? What if a foreign cell tried to join the union? What passport must it carry that enables it to pass? As the Caterpillar in *Alice in Wonderland* asked Alice, "Who are *you*?"

Sponges on the ocean floor extend their branches out toward one another, but the branches stop growing once a sponge reaches close to a neighbor. As one spongiologist described it: "A clear nonconfluent margin separates different species or [even] different specimens belonging to the same species." What stops cells from transiting between one

424

sponge and another — or from one human to another? How does one sponge know *itself*?

There is a related question implicit in the last chapter that must be answered: T cells, I wrote, recognize the altered self. But if you parse that phrase carefully, it turns into a clown car of questions; all sorts of conundrums come tumbling out the doors. Break the phrase into two parts. First, how does a T cell know its altered self? In other words, how does it know to kill its target when a viral or bacterial peptide is presented, but not when a *self* peptide is presented to it? The cell doesn't maintain a ledger of every overlapping self peptide — the number of all possible peptides in a cell would extend to more than hundreds of millions — and so what mechanism exists to ensure that a T cell does not attack its own body? And second, what about the self? How does a T cell know that the frame carrying the peptide — the MHC molecule — came from its own body, and not from another?

Take the question of self first. At first glance, it seems like a rather artificial problem. Humans hardly need to worry about cells from other humans invading and colonizing our bodies and trying to pass themselves off as selves (even though that fantasy continues to inspire horror films and books). But for more primitive multicellular organisms — sponges, say — for which competitive existence is an everyday battle, every morsel of food precious, constancy a looming threat, and territory a limited resource, the potential invasion by another self isn't a trifling matter. Such an organism must query: Where do *I* end and *you* begin? Its self can exist only if its borders are strictly enforced. Such an organism must constantly ask each of its cells, "Who are *you*?"

Long before the birth of cell biology, Aristotle imagined the self as the core of being; a unity of the body and the soul. The *physical* boundary of the self, he proposed, was defined by the body and its anatomy. But the totality of the self was a unity of that physical vessel with a metaphysical entity that occupied it — the body filled by the soul. In principle, Aristotle, too, might have fretted about the possible invasion of the physical vessel by a foreign soul — indeed, "possession" was frequently used by psychics to explain mental and behavioral breakdowns — but he didn't seem to sweat the question: once the physical vessel had been occupied by one soul, its potential invasion by — or fusion with — another soul didn't bother him.

In quite the opposite sense, some Vedic philosophers in India, writing between the fifth and second century BC, welcomed the erasure of the individual self and its fusion with the universal. They rejected the Greek dualism between the body and the soul — and, indeed, between an individual body and the cosmic soul. They termed the self *atman.* (There are many other Sanskrit words for the self besides *atman,* but it is the one that holds the most meaning.) The universal, multitudinous self, in contrast, was the Brahman. For these philosophers, the self was an ideal fusion of *atman* and Brahman, or perhaps more accurately, the seamless flow of the universal self through the individual self. However, this fusion/flow was reserved as a goal for spiritual attainment. There was a cosmic ecology that bound the individual and the spiritual collective into one Being. The phrase *"Tat Twam Asi"* — *"That* you are" — permeates the Upanishads and is an expression of the boundless self that permeates not just a single physical body but also the cosmos. *You,* the self, the Upanishads proclaim, is permeated and penetrated by *That,* the universal. In an ideal body, the universal flows through the individual. (The word *invades,* with its negative connotations, is obviously avoided.)

In science, this boundlessness of the individual body and the cosmic body has more recently found its echo in ecology. The whole ecosystem of living beings, we might say, is connected through a network of relationships and, to some extent, the erasure of the boundaried self. A human body and a tree, and the bird that dwells in that tree, say, are linked through such networks — networks that ecologists are just beginning to decipher. The bird eats the fruit from a tree and disseminates the seeds through its droppings; the tree, reciprocally, provides a perch for that bird. It's not invasion, the ecologists insist. It's interconnectedness.

But ecological interconnectedness is not physical or competitive; it is relational and symbiotic — a topic to which we will return. For cell biologists, though, it is physical fusion that continues to raise a fundamental conundrum. The notion of chimerism — the fusion of physical selves — is not a New Age fantasy but an age-old threat. Cellular selves don't particularly like to mix with other cellular selves. Why else would a sponge go to such lengths to limit its fusion with another sponge to form a blissfully boundless cosmic Brahman sponge?

Extend the same challenge to a T cell. A T cell, remember, is activated when a foreign peptide is presented on an MHC protein mounted on a cell's surface — but *only* if it is presented by an MHC that is from the same body. It is as if the T cell is activated only if the frame, or context, is right — "right" in the sense that the frame comes from the self and what is carried is foreign. But how does a T cell know the self?

Even early physiologists noted that the rejection of the nonself — and the strict definition of boundaries — was a feature of human tissues. Surgeons in India, particularly Sushruta, who lived sometime between 800 and 600 BC, had performed grafts of skin from the forehead to the nose. (This was a not-uncommon procedure in ancient India, for criminals and dissenters often had their noses chopped off as punishment, leaving medical practitioners to devise ways to reconstruct them.) But when early surgeons tried allografts — skin grafts from one body to another — they found that the recipient's immune system set upon and rejected the transplanted skin, causing it to turn blue and gangrenous and eventually degenerate and die.

During World War II, there was renewed interest in understanding the underlying science of grafting. Skin grafting, in particular, was of great need, as both soldiers and civilians were often wounded, burned, or scalded by bombs and fires. The British government appointed a War Wounds Committee within the Medical Research Council to encourage research on wounds and healing.

In 1942, a twenty-two-year-old woman was admitted to the Glasgow Royal Infirmary with "extensive burns on her chest, right flank, and right arm." The surgeon, Thomas Gibson, worked with a zoologist from Oxford, Peter Medawar, to graft small pieces of her brother's skin onto her wounds. Unfortunately, the transplanted tissue was swiftly rejected, leaving trails of charred and mottled tissue on the woman's wounds. When they tried again, the rejection was even more immediate. By studying serial biopsies of the grafts and examining the infiltrating cells, Medawar and Gibson began to decipher that it was the immune system — the T cell in particular — that was rejecting the graft. The nonself, Medawar argued, was recognized by the self's immunity mediated by T cells.

Medawar knew of the work of a British immunologist named Peter Gorer and of American geneticist Clarence Cook Little, who, working independently, had transplanted tissues from one mouse to another. If the donor and the recipient mice came from the same strain, the transplanted tissues — typically tumors — would be "accepted" and would grow; but when the tumors were moved from one strain to a different strain, they would be immunologically rejected. (Little's interest in "genetic purity" seemed, at times, all consuming and obsessive. He produced strains of inbred mice for transplantation experiments — key to the field of T cell tolerance. He tried to breed dogs for experiments and maintained a personal collection of inbred dachshunds as pets. But the same instincts, perhaps, made him an ardent advocate of American eugenics, which sullied his reputation as a scientist.*)

---

*Albeit a giant in the field of transplantation, Little would also be criticized for his collusion with cigarette

But what factors were responsible for this compatibility or tolerance — the recognition of the self versus the nonself? In 1929, searching for a contemplative place, far from the dissonance of university departments where arguments about compatibility and tumor transplantation broke out weekly, Clarence Little set up the Jackson Laboratory on a forty-acre campus by the Atlantic Ocean in Bar Harbor, Maine. There he could breed thousands of mice in peace. The view from the windows was spectacular; the long summers suffused the campus with a preternaturally lucid North Atlantic light. The field of transplantation, in contrast, remained a mess — an impenetrable biological puzzle with hundreds of accreted, entangled observations. It made very little sense to Little.

By serially transplanting tumors across strains, Little realized that not one, but multiple genes, were involved in the immune rejection of the transplant. By the early 1930s, the Jackson Lab had become a natural haven for transplantation researchers hunting for the mysterious compatibility genes that defined self versus nonself. A young scientist, George Snell, was drawn to the lab to deepen Little's transplantation studies. A graduate of

---

manufacturers in the 1950s, when he became involved in the Tobacco Research Institute, which insisted on the safety of cigarettes.

Dartmouth College and Harvard University, Snell bred mice, generation upon generation, to produce animals that would accept or reject grafts from each other. He was a man of few words, reclusive, as cool as ocean water, and doggedly persistent: once, when an entire colony of mice, bred for at least fourteen generations, died in a laboratory fire, Snell dusted off his overalls and started breeding them again.

The selective breeding, while monitoring self versus nonself tolerance, paid off. Immunologically speaking, Snell eventually created multiple twinned selves: mice whose tissues were perfectly compatible with each other. You could put the skin, or another tissue, from one such mouse into its compatible sibling, and it would be "accepted" — tolerated — as if it were from the self. Most crucially, the inbreeding experiment produced two strains of mice that were almost exactly genetically identical — except that they rejected grafts from each other.

Snell used these animals to dissect the genetics of self versus nonself. By the late 1930s, building on Gorer's work, he slowly narrowed in on a set of genes that determined tolerance. He termed them H genes, for histocompatibility genes — *histo* from *tissue,* and *compatibility* because of their capacity to render foreign tissue to be accepted as self. It was some version of these H genes, Snell realized,

432

that defines the boundary of the immunological self. If organisms shared the H genes, you could transplant tissue from one organism to another. If they didn't, the transplant would be rejected.

Over the next decades, more histocompatibility genes were identified in mice — all located, cheek-by-jowl, on chromosome 17. (In humans, they are mostly found on chromosome 6.) Perhaps the most profound advance in the field occurred when the identity of these H genes was finally revealed. Most of them turned out to encode functional MHC molecules — the very molecules, recall, that had been implicated in how a T cell recognizes its target.

Step back for a minute. In immunology, as with any science, there are moments of grand synthesis, when seemingly disparate observations and seemingly inexplicable phenomenon converge on a single mechanistic answer. How does the self know itself? Because every cell in your body expresses a set of histocompatibility (H2) proteins that are different from the proteins expressed by a stranger's cells. When a stranger's skin, or bone marrow, is implanted into your body, your T cells recognize these MHC proteins as foreign — nonself — and reject the invading cells.

What are these self-versus-nonself genes that encode the proteins? Well, they are the very genes that Snell and Gorer had found and called H2. Humans have multiple "classical" major histocompatibility genes, and potentially many others, of which at least three, and possibly more, are strongly related to graft compatibility versus rejection. One gene, called HLA-A, has more than a thousand variants, some common and some very rare. You inherit one such variant from your mother and one from your father. A second such gene, HLA-B, also has thousands of variants. You might have guessed already that the number of permutations between just two such highly variable genes is mind-boggling. The chances that you'd share such a barcode with a random stranger you met in a bar are vanishingly small (and all the more reason not to fuse with him or her).

And what do these proteins *do* when they aren't rejecting grafts and cells from strangers — obviously an artificial phenomenon, at least in humans (but not, perhaps, in sponges or other organisms)? As Alain Townsend and others had shown, their main job is to enable the immune response to survey cells for their inner components and thereby detect viral infections.

In short, H2 (or HLA) molecules serve two linked purposes. They present peptides to a T cell so that a T cell can detect infections and other invaders and mount an immune response. And they are also the determinants by which one person's cells are distinguished from another person's cells, thereby defining the boundaries of an organism. Graft rejection (likely important for primitive organisms) and invader recognition (important for complex, multicellular organisms) are thus combined into a single system. Both functions repose in the T cell's capacity to recognize the MHC peptide complex, or the altered self.

Let's turn now to the other half of the conundrum: the question of the "slightly altered" self. The T cell, as I mentioned above, uses the MHC molecules to recognize the self and reject the nonself. But how does it discern whether the peptide being presented by the self MHC comes from a normal cell (in other words, it's part of the cell's normal roster of peptides) or whether it comes from a foreign invader, such as an internally dwelling virus that has entered a cell and "gone native"? I have written so much about war: *Kampf,* the toxic attacks on pathogens; the rejection of grafts. What about *peace,* then? Why don't immune cells, loaded with toxins and straining for vengeance, turn against ourselves?

This phenomenon of self-tolerance equally puzzled immunologists. In the early 1940s, in Madison, Wisconsin, Ray Owen, a geneticist and the son of a dairy farmer, performed an experiment that was, in a sense, the conceptual opposite of Peter Medawar's.

435

In Medawar's test, he had tried to understand the phenomenon of rejection, or the *intolerance for the nonself:* Why does a sister's immune system reject her brother's skin? Owen inverted the question: Why doesn't a T cell turn against its own body? How does it acquire *tolerance for the self*?

Cows, Owen knew from his farmhand days, were known to occasionally give birth to twins sired by two different bulls: a Guernsey cow might bear a twin fathered by a Guernsey bull *and* a Hereford bull because they happened to have fertilized the cow during the same fecund period. The twins born from a Guernsey-Hereford fertilization share a placenta. But they had different red blood cells, carrying different antigens. Typically, a non-twinned Guernsey cow would reject blood from a Hereford. But in the rare placenta-sharing twins, Owen found that there was no such rejection. It was as if something in the placenta had educated the immune system of one animal to become "tolerized" to the cells of the other animal.

Owen's idea was largely ignored. But in the 1960s, as immunologists began to take tolerization seriously, they returned to his result. Something about embryonic *exposure* to an antigen must tolerize the immune system to recognize it as self and not attack a cell presenting it. In a 1969 book titled *Self and Not-Self,* Macfarlane Burnet (by then already awarded the Nobel Prize for his clonal theory of antibodies) advanced Owen's observations with a radical theory: "Recognition that an antigenic determinant is foreign requires that it shall not have been present in the body *during embryonic life*" (italics my own), Burnet wrote, acknowledging Owen's earlier experiments.

The basis for this tolerance was that T cells that reacted against "self" cells — immune cells that attacked our own (i.e., pieces of proteins derived from our own cells and presented on our own MHC molecules) — were somehow deleted or removed from the immune system during infanthood or prenatal development. Immunologists called the self-reactive cells "forbidden clones" — *forbidden* because they had dared to react to some aspect of a self peptide and were therefore deleted from existence before they could be allowed to mature and attack the self. Burnet likened them to "holes" in immune reactivity. It is one of the philosophical enigmas of immunity that the self exists largely in the negative — as holes in the recognition of the foreign. The self is defined, in part, by what is forbidden to attack it. Biologically speaking, the self is demarcated not by what is asserted but by what is invisible: it is what the immune system cannot see. *"Tat Twam Asi."* "That [is] what you are."

But where were these forbidden holes

generated? How do immune cells, such as T cells, make a hole in their repertoire of recognition that does not attack a self protein — say, the antigens on the surface of a red blood cell or a kidney cell — as foreign?

A series of experiments provided an answer. As Jacques Miller had shown, T cells are born in the bone marrow as immature cells and migrate to the thymus to mature. Philippa Marrack and John Kappler, an immunologist couple in Colorado, forcibly expressed a foreign protein in mouse cells, including the cells of the thymus. Normally, such a protein should be recognized and rejected by T cells. But, much as Burnet had predicted, they found that the immature T cells that recognized pieces of that protein — the ones that attacked the self — were deleted in the thymus by way of a process called negative selection. The deleted T cells never matured. They left the "holes" that Burnet had proposed in self-reactive T cells.

But T cell deletion in the thymus — a mechanism called central tolerance because it affects all T cells during their central maturation — isn't enough to guarantee that immune cells don't end up attacking the self. Beyond central tolerance, there is a phenomenon called peripheral tolerance; here tolerance is induced once the T cells have left the thymus.

One of these mechanisms involves a strange and mysterious cell called the T regulatory cell (T reg). It looks almost identical to a T cell, except that rather than incite an immune response, the T reg suppresses it. T regulatory cells zero in on sites of inflammation and secrete soluble factors — anti-inflammatory messengers — that dampen the activity of T cells. The most profound evidence of their activity is the disease that occurs when they go missing. In humans, a rare mutation disrupts the formation of these cells and results in a terrifyingly progressive autoimmune disorder in which T cells attack the skin, pancreas, thyroid, and intestines. Children affected by immune dysregulation, polyendocrinopathy, enteropathy, X-linked (IPEX) syndrome suffer intractable diarrhea, diabetes, and psoriatic, friable, peeling skin. They are under attack by themselves, because T cells that control other T cells, the policemen who police the police, are missing in action.

It is an unsolved quirk of the immune system that the cell type that confers active immunity and incites inflammation (the T cell) and the cell type that dampens these processes (the regulatory T cell) arise from the same parent cells: T cell precursors in the bone marrow. Indeed, aside from very subtle distinctions in genetic markers, T cells and T reg cells are anatomically indistinguishable. And yet they are functionally complimentary. Immunity and its opposite are twinned: the Cain of inflammation conjoined with the

Abel of tolerance. Sometime in the future, we will understand why evolution chose to pair these cells. But the regulatory T cell remains a mystery — a cell that looks like it might activate immunity, but actually suppresses it.

"But there are mountains beyond mountains," the Haitian proverb runs. An out-of-control T cell can be so toxic to the body that there are backup systems beyond backup systems. What happens when the major regulatory forces no longer keep the immune system from attacking its own body? Around the turn of the twentieth century, Paul Ehrlich, the eminent biochemist, called it horror autotoxicus — the body poisoning itself. The condition befits the name. Autoimmunity ranges from mild to absolutely furious. In alopecia areata, an autoimmune disease, T cells are thought to attack hair follicle cells. One patient might notice only a single bald spot, while in someone else, the T cells might assail every hair follicle, resulting in complete baldness.

In 2004, when I was a fellow in medicine, I volunteered as an assistant in a graduate course in clinical immunology. My job was to scour the hospital for patients with autoimmune diseases and, with their consent, discuss the physical manifestations, cause, and treatment with the graduate students. My only quibble with Ehrlich's vivid phrase is the singular. Horror autotoxicus — autoimmunity

— comes in so many manifestations and forms that there is not a single horror but rather a multitude.

We met a thirtysomething woman with scleroderma, in which the immune system attacks the skin and connective tissue. In her case, the illness had begun, as it often does, with a phenomenon called Raynaud's disease, in which the fingers and toes turn blue when exposed to the cold. "And then," she told the students, "my fingers began to get blue when I got emotionally stressed or tired, even without the cold." My mind returned to an image from Shakespeare's poem about winter from his play *Love's Labour's Lost*: "Dick the shepherd blows his nail," as the wind howls around him. But this patient's coldness was internal, caused by spasming blood vessels in the hands and feet. It was as if autoimmunity had created an inner freeze.

Stranger attacks fell upon the woman's body: patches of skin began to tighten around her as the immune system turned against her connective tissue. The patches turned shiny, as if pulled by some invisible force, and stretched across her bones. Her lips tightened and scarred. She was treated with immuno-suppressants and, to reduce inflammation, corticosteroids, which made her manic. "It felt as if my own skin had started to bind me up, like a Saran Wrap folded around my body."

Next was a man with systemic lupus erythematosus (SLE), often referred to simply as lupus. The illness is named after a wolf, either because Roman physicians thought that the skin lesions of this form of horror autotoxicus reminded them of wolf bites or, more likely, because the rash that spreads across the face, crossing the bridge of the nose and under the eyes, was reminiscent of a wolf's markings. Add to this the fact that sunlight could exacerbate the rash, often forcing lupus sufferers to live in the dark, coming out only in moonlight, and the sinister-sounding name stuck. The hospital room's shades had been drawn, letting in only a single shaft of angled light. We gathered around him, as if in some sort of sepulchral chamber.

The man had a mild rash — he had taken to wearing sunglasses to hide it — but his kidneys were also under attack. Excruciating, wandering pains migrated through his joints, from his elbows to his knees. Lupus is a slippery, motile disease. It might affect just one organ system, such as the skin or the kidneys, or it might suddenly go after multiple systems at once. This patient had volunteered to participate in a clinical trial of a new immunosuppressant, and it seemed to have abated the disease somewhat. What, precisely, the immune system is reacting against in lupus remains a mystery, but it often involves antigens in the nucleus of the cell, antigens on cell

442

membranes, and antigens on proteins bound to DNA. And sometimes, the list of organs being assaulted keeps increasing: the illness moves from joints to kidneys to skin. It's like a fire feeding on itself: once the barrier to the self has been broken, everything that is self can be under attack.

The horror autotoxicus had a profound scientific lesson buried within it, although it would take decades for immunologists to come to terms with it. Autoimmunity, the attack on self cells, generated an obvious question: What if the immunological toxicity could be turned on cancer cells? After all, malignant cells occupy the disturbing boundary of the self and the nonself; they are derived from normal cells and share many features of normality, but they are also malignant invaders — rhinoceroses in one perception, and unicorns in another. In the 1890s, a New York surgeon, William Coley, had tried to treat cancer patients with a concoction made from bacterial cells that came to be known as Coley's toxins. His hope was to elicit a powerful immune response that might turn on the tumor. But the reactions were unpredictable. And with the development of cell-killing chemotherapies in the 1950s, the idea of an immune attack on cancer fell out of fashion.

But as cancer after cancer relapsed following standard chemotherapy, the notion of immunotherapy was brought back to life. Recall, for a moment, the mechanisms that allow the body *not* to be eaten alive by its own T cells. There are "forbidden" clones that would otherwise react against normal tissues that are forced to disappear during the maturation of T cells. And there are regulatory T cells that can dampen the immune response.

In the 1970s, scientists discovered yet other mechanisms by which T cells can be tolerized to the body so that it did not attack the self. To kill its target — for instance, a virus-infected cell or a cancer cell — merely engaging the T cell receptor with the MHC-peptide complex was not sufficient. Other proteins on the surface of T cells had to also be activated to stimulate an immune attack. There was not one switch, but a multitude of them. These backups beyond backups — mountains beyond mountains — are like trigger locks and safety switches in guns, and they had evolved to ensure that T cells did not accidentally direct their friendly fire on normal cells. The trigger locks would act as checkpoints against the indiscriminate killing of our own cells.

But before understanding and disabling such trigger locks, there was the looming uncertainty of specificity: Could a human T cell response be directed against cancer? At the National Cancer Institute, in Bethesda, Maryland, a surgical oncologist named Steven

Rosenberg had drawn native T cells out of malignant tumors, such as melanomas, reasoning that immune cells that had infiltrated a tumor must have the capacity to recognize and attack the tumor. Rosenberg's team had grown these tumor-infiltrating lymphocytes, expanding their numbers to several million, and then infused them back into patients.

There had been some potent responses: melanoma patients treated with Rosenberg's transferred T cells saw their tumors shrink, and some experienced a complete regression that was maintained over time. But the responses were also hit or miss. The T cells harvested from a patient's tumor may have trained themselves to fight it, but they might also be bystanders, passive witnesses lingering at a crime scene. They might have become exhausted or inured — tolerized to the tumor.

Cancers, varied as they are, share some common features — among them, their invisibility to the immune system. In principle, T cells can be potent immunological weapons against tumors. As Clarence Little and Peter Gorer had shown as early as the 1930s, when a tumor is implanted onto genetically mismatched mice, T cells from the recipient mouse reject the tumor as "foreign." But the tumor/recipient systems that Little and Gorer had chosen were grossly mismatched: the tumor brandished, on its surface, an MHC molecule that could instantly be recognized as "foreign" and thus swiftly rejected. More recently, in Emily Whitehead's case, her CAR-T cells had been specifically modified to recognize a protein on the surface of her leukemia cells.

Most human cancers, though, represent a vastly more subtle challenge for the immune system. Harold Varmus, the Nobel-winning cancer biologist, called cancer a "distorted version of our normal selves." And so they are: the proteins that cancer cells make are, with a few exceptions, the same ones made by normal cells, except cancer cells distort the function of these proteins and hijack the cells toward malignant growth. Cancer, in short, may be a rogue self — but it is, indubitably, a self.

And second: the cancer cells that ultimately form a clinically relevant illness in a human arise through an evolutionary process. The cells that are left after their cycles of selection may have *already* become capable of evading immunity — like Sam P.'s immune cells that had, for years, simply brushed past his tumor, ignored it, and moved on.

This double-headed problem — cancer's kinship to the self and its immunological invisibility — is the oncologist's nemesis. To attack a cancer immunologically, one has to first make it *re*-visible (to coin a word) to the immune system. And second, the immune system must find some distinct determinant in

446

the cancer that can enable an attack, without concomitantly destroying the normal cell.*

*There is a third, and increasingly important, arena of investigation that concerns cancer's ability to resist destruction by drugs and by natural mechanisms in the body. Cancer cells evolve to create unique cellular environments around themselves — typically by surrounding themselves with normal cells — that cannot be penetrated by drugs, or that can actively cause drug resistance. Similarly, these cellular environments can evade immunity by rendering T cells, NK cells, and other immune cells inactive by preventing them from even entering the vicinity of the cancer cell, or by drawing out blood vessels to supply the malignant cells with nutrients. Trials to choke cancer's blood supply, using a variety of drugs, have yielded only modest successes. And so have trials to force immune cells to remain active in cancer's "microenvironment." One of the most terrifying scientific images that I have seen in recent times is a tumor surrounded by a shell of normal cells that has excluded activated T cells. The T cells form a ring around the cellular carapace that the cancer has formed around itself, but are unable to penetrate it. The immunologist Ruslan Medzhitov has called this the "client cell" hypothesis: cancer cells pretend (or, more accurately) evolve to resemble "client" cells of the organ that they grow in, just like a burglar might pretend to be a client of the shop being robbed while the police — the immune system, in this case — look elsewhere.

Steve Rosenberg's experiment was an early, flickering sign that both these challenges could be overcome: in a few cases, tumors could become immunologically detectable, and they could be killed by T cells. But what, exactly, were cancer cells *doing* to achieve invisibility? Might the cells be using the same mechanisms that the normal body uses to prevent attacks on itself — that is, activating the trigger-lock systems that prevent autoimmunity?

In the winter of 1994, Jim Allison, working at the University of California, Berkeley, set up an experiment that would revive the field of immunotherapy — in part, by unlocking the mechanisms that keep T cells in check. An immunologist by training, Allison had been studying a protein called CTLA4 that sat on the surface of T cells. The protein had been known since the 1980s, but its function remained a puzzle.

Allison implanted mice with tumors that were well known to be resistant to an immune response. The tumors grew obdurately as expected, shrugging off any immune rejection. Experiments by the immunologists Tak Mak and Arlene Sharpe in the 1990s had hinted that CTLA4 might be one of the trigger locks used to keep activated T cells in check; when they had deleted the gene in mice, T cells had run amok and the animals had developed deadly autoimmune diseases. Allison reconceived the experiment but with a twist: rather than delete the CTLA4 gene entirely, might a drug-induced blockade of CTLA4 unleash T cells against cancer?

Allison injected some of the mice with antibodies to block CTLA4 — in essence, blocking the protein's function. Over the next few days, the immune-resistant tumors in the mice injected with the CTLA4 blockers disappeared. He repeated the experiment over Christmas. Again, the malignancies in the mice injected with the CTLA4 blocker dissolved — eaten alive, he would later discover, by an infiltrate of tetchy, activated T cells.

Intrigued by this activation of T cells against a tumor, Allison and several other researchers spent more than a decade trying to deepen their understanding of the protein's function. As all the earlier experiments had shown, they found that CTLA4 was a system to prevent horror autotoxicus; it was a T cell's trigger lock. Under normal circumstances, when CTLA4 on activated T cells meets its cognate binder, called B7,* that is present on the sur-

*I have tried to avoid an enormous amount of immunological jargon here. B7 is actually a complex of two molecules, CD80 and CD86. And there are yet other backup systems to prevent the T cell from becoming inappropriately activated. One such protein, CD28, initially discovered by immunologist Craig Thompson, is also the subject of intense investigation in my lab and others.

449

face of cells of the lymph nodes, where T cells mature, the safety switch is turned on. The maturing T cells are disabled from attacking the self but also unable to reject tumors. If you blocked this disabling pathway, however, the safety lock was switched off, and you overrode tolerance. CTLA4 stood as a barrier between inactivated and activated T cells. It was termed a *checkpoint,* based on the idea that the protein held T cell activation in check.[*]

I'm writing this story as if all of these crucial realizations happened in minutes, but it took decades of labor and love. I met Allison in New York a few years ago, and we spoke about the tortuous scientific path to uncovering the function of CTLA4. He laughed jovially, as if the ten grinding years that he had spent on the project were merely a distant memory. "No one believed me," he said. "No one thought that there was yet another way to keep T cells in check against cancer cells. But we kept at the problem until we had solved it."

While Allison was uncovering the function of CTLA4, a Japanese scientist named Tasuku Honjo, working in Kyoto, was preoccupied with the function of another mysterious protein named PD-1. As with Allison, a

[*]Over time, researchers have discovered that T cells have multiple checkpoints. Each acts as a safety switch to prevent trigger-happy T cells from attacking the self.

decade went by, yielding bizarre, often contradictory results. But Honjo's team slowly converged on the function of PD-1. This protein, they found, was similar to CTLA4 in that it, too, was a tolerizer. Like CTLA4, PD-1 is expressed on T cells. Its cognate binder — effectively its "off" switch — is called PD-L1. It is present, at low levels, on the surface of some normal cells all over the body. If you imagine CTLA4 on T cells as a safety switch on a gun, then PD-L1, on normal cells, is an orange jacket worn by an innocent bystander that says, "Don't shoot. I'm harmless!"*

In a matter of decades, two new systems of peripheral tolerance had been discovered and potentially inactivated. The binding of CTLA4 on T cells makes them impotent. The presence of PD-L1 on normal cells makes them invisible. Somewhere in the combination of impotence and invisibility lie the twin mechanisms that prevent the body from swallowing itself.

Cancers, we now know, can use both these mechanisms to cloak themselves against immune attack. Some express PD-L1, in essence, stitching their own orange jackets of invisibility: "Don't shoot. I'm harmless!" When PD-1 inhibitors were injected into mice, Honjo

*In fact, PD-L1 is more than an orange safety jacket. It even induces T cell *death,* thereby completely disabling a T cell attack.

451

found, T cells were incited to attack even orange-jacketed, immune-resistant tumors; the cancer's bluff was called. Both Honjo and Allison had independently converged on the same paradigm: turn off the safety locks on a T cell, or strip away the orange jackets of the cancer cells, and the immune response could actually turn against cancer. They had checkmated the checkpoints.

A new class of drugs grew out of this work: antibodies to inhibit CTLA4 and PD-1 among them. The first clinical trials with these new drugs revealed their potency. Melanomas resistant to chemotherapies regressed and disappeared. Metastatic bladder tumors were attacked* and rejected. A novel form of

---

*Why, how — *why, why, why, how, how* — can a cancer cell circumvent T cells designed to recognize and kill it? This question haunts immunotherapy today. Something about a solid tumor — perhaps the environment that it has created around itself — can circumvent and inhibit even the most potent reactivation of T cells. What is that "something"? The most solid evidence, and this is not a play on words, is that the immune attack on a cancer can occur only if a fully active lymphoid organ, containing neutrophils, macrophages, helper T cells, killer T cells, and an organized cellular structure, can be formed within a solid tumor. This secondary lymphoid organ (SLO) is like a lymph node that is usually formed when T cells attack a virus or a pathogen, except in this

immunotherapy for cancer, termed "checkpoint inhibition" — the removal of tolerizing checks on T cells — was born.

Yet these therapies had their limitations: take off the trigger locks, and the activated T cells, eager to attack, might turn on normal cells. It was this autoimmune attack on his own liver cells that ultimately limited my friend Sam P.'s response to treatment. Checkpoint inhibitors did unleash T cells against his melanoma, bringing its malignant growth under control. But they also unleashed an assault on his liver that we could never overcome. It was a medically induced form of horror autotoxicus. He was caught on the border between his cancer and his self. Eventually the tumor cells circumvented the border and survived. Sam was left behind.

I finished this part of the book, coincidentally, on a Monday morning, the day I reserve to look at blood. I left the office where

---

case, it is organized against a tumor. Tumors that do not enable the formation of such SLOs are resistant to immunotherapy, while those that do form them are generally sensitive. But that's a correlation. The cause-and-effect relationships, and the mechanisms that enable or disable the formation of such SLOs, have yet to be uncovered. Once we understand them, a new generation of immunotherapeutic drugs, or combinations, can be unleashed on cancer cells.

I usually write and walked down the corridor to the microscope room. Empty and silent, thankfully. The lights went down, the scope flickered on. A box of glass slides was waiting on the table. I mounted one on the scope and turned the focusing knob.

Blood. A cosmos of cells. The restless ones: red blood cells. The guardians: multilobed neutrophils that mount the first phases of the immune response. The healers: tiny platelets — once-dismissed as fragmentary nonsense — that redefined how we respond to breaches in the body. The defenders, the discerners: B cells that make antibody missiles; T cells, door-to-door wanderers that can detect even the whiff of an invader, including, possibly, cancer.

As my eyes darted from one cell to another, I thought about the trajectory of this book. Our story has moved. Our vocabulary has shifted. Our metaphors have changed. Flip back a few pages, and we imagined the cell as a lone spaceship. Then, in the chapter "The Dividing Cell," the cell was no longer single but became the progenitors of two cells, and then four. It was a founder, the originator of tissues, organs, bodies — fulfilling the dream of one cell becoming two and four. And then it transformed into a colony: the developing embryo, with cells settling and positioning themselves within the landscape of an organism.

And blood? It is a conglomerate of organs, a system of systems. It has built training camps for its armies (lymph nodes), highways and alleys to move its cells (blood vessels). It has citadels and walls that are constantly being surveyed and repaired by its residents (neutrophils and platelets). It has invented a system of identification cards to recognize its citizens and eject intruders (T cells) and an army to guard itself from invaders (B cells). It has evolved language, organization, memory, architecture, subcultures, and self-recognition. A new metaphor comes to mind. Perhaps we might think of it as a cellular civilization.

1

Emily Whitehead, the first child treated at the Children's Hospital of Philadelphia for relapsed, refractory acute lymphoblastic leukemia (ALL). In the absence of experimental therapy or a bone marrow transplant, this form of the disease is lethal. Whitehead's T cells were extracted, genetically modified to "weaponize" them against her cancer, and reinfused into her body. These modified cells are called chimeric antigen receptor T cells, or CAR-T cells. Originally treated in April 2012, when she was seven years old, Emily remains in good health today.

2

Rudolf Virchow, standing in his pathology laboratory. As a young pathologist working in Würzburg and Berlin in the 1840s and '50s, Virchow would revolutionize the idea of medicine and physiology. Virchow argued that cells were the basic units of all organisms and the key to understanding human illness was to understand the dysfunctions of cells. His book Cellular Pathology *would transform our understanding of human disease.*

ANTONI VAN LEEUWENHOEK.
LID VAN DE KONINGHLYKE SOCIETEIT IN LONDON.

A portrait of Antonius (or Antonie) van Leeuwenhoek. A secretive, temperamental draper in Delft in the Netherlands, Van Leeuwenhoek would be among the first to visualize cells under a one-lensed microscope in the 1670s. He called the cells that he saw—likely protozoa, single-celled fungi, and human sperm—"animalcules." Van Leeuwenhoek made more than five hundred such microscopes—each a marvel of delicate engineering and tinkering. The English polymath Robert Hooke had seen cells in a section of a plant wall about a decade earlier, but no definitive portrait of Hooke survives.

4

George Palade (right) and Philip Siekevitz next to an electron microscope at the Rockefeller Institute in the 1960s. Palade's team of cell biologists and biochemists, collaborating with Keith Porter and Albert Claude, would be among the first to define the interior anatomy and function of cellular compartments—or "organelles."

5

Vierville-sur-Mer, June 6, 1944. On the Normandy coast, on "Omaha Beach," Army medics give a blood transfusion to an injured soldier. Blood transfusion—cellular therapy—would save the lives of thousands of men and women during the war.

6

A portrait of Paul Ehrlich and his collaborator Sahachiro Hata, c. 1913. Ehrlich and Hata, biochemists, would devise new drugs to treat infectious diseases such as syphilis and trypanosomiasis. Ehrlich's theory of how B cells generate antibodies would lead to a vigorous debate in the 1930s. Ehrlich would eventually be proved wrong, but his idea of an "antibody" created specifically to bind to and attack an invader would form the basis of our understanding of adaptive immunity.

7

*Santiago Ramón y Cajal in 1876. Cajal's drawings of the
nervous system, aided by a stain devised originally by Camillo
Golgi, would revolutionize our ideas about how the brain and
the nervous system work. Cajal's drawings are considered
among the most beautiful and revelatory drawings in science.*

8

Frederick Banting and Charles Best with a dog on the roof of the Medical Building, University of Toronto, in August 1921. Banting and Best devised ingenious experiments to identify and purify the hormone insulin, the central coordinator of glucose levels in the body.

PART FOUR
KNOWLEDGE

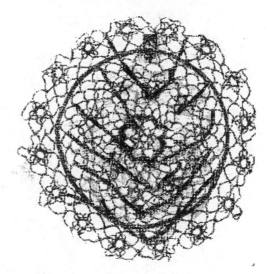

THE PANDEMIC

Into the notable city of Florence, fair over every other of Italy, there came the death-dealing pestilence, which . . . had some years before appeared in the parts of the East and . . . had now unhappily spread toward the West . . .To the cure of these maladies nor counsel of physician nor virtue of any medicine appeared to avail or profit aught . . . The mere touching of the clothes or of whatsoever other thing had been touched or used of the sick appeared of itself to communicate the malady to the toucher . . . Each thought to secure immunity for himself. Some . . . lived removed from every other and shut themselves up in those houses where none had been sick and where living was best.
— Giovanni Boccaccio, *The Decameron*

In the early winter of 2020, before our self-assuredness was upended, it seemed as if the immune system, of all the complex cellular systems in the body, was the one that we understood the best. In 2018, when Allison and Honjo were awarded the Nobel Prize for their discovery of how tumors evade T cell immunity, the Prize seemed to mark a crest in our understanding of immunity — and perhaps of cell biology in general. Powerful medicines were being created that could uncloak immunologically cloaked tumors. Fundamental mysteries remained, of course. How this system achieved the acrobatic balancing act between generating a forceful immune response against pathogens, while ensuring that same response did not turn on our own bodies — how the *Kampf* against microbial invaders did not degenerate into the civil war of horror autotoxicus — was still somewhat of a profound riddle (in Sam P.'s case, we could never control the autoimmune hepatitis induced by the cancer-rejecting immunotherapy). But the central pieces of the puzzle seemed to have fallen into place. Several years ago, I spoke to a postdoctoral researcher who was transitioning from his university position to a job at a biotech company that planned to make novel immunological therapies for cancer. He told me that researchers were increasingly imagining the immune system as a knowable machine, with movable — manipulable, decipherable, changeable — cogs, gears, and parts. I did not sense hubris in his optimism. In 2020, eight of the approximately fifty drugs approved by the

US Food and Drug Administration involved the immune response; in 2018, that number was twelve of fifty-nine; almost a fifth of *all* the human drugs being discovered had something to do with the immune system. We were stepping, it seemed, quite confidently from basic immunology to applied immunology.

And then, biblically, we tumbled.

On January 19, 2020, a thirtysomething man, just off a flight from Wuhan, China, walked into a clinic in Snohomish County, Washington, with a cough. To read that first case report, published in the *New England Journal of Medicine* in March that year, is to experience an ascending chill:

"On checking into the clinic, the patient put on a mask in the waiting room."

Who stood next to him in that room? How many people had he infected in the past days? Who was sitting across the aisle on the flight from Wuhan to Seattle?

"After waiting approximately twenty minutes, he was taken into an examination room and evaluated."

Was the doctor who examined him masked? The nurse who checked his temperature? Where are they now?

"He disclosed that he had returned to Washington State on January 15 after traveling to visit family in Wuhan, China."

On January 20, a nasal swab and an oral swab (and subsequently, stool samples) were sent to the Centers for Disease Control and Prevention. Both tested positive for a novel coronavirus: SARS-COV2.

On the ninth day of his illness — and the fifth day of hospitalization — his condition deteriorated. His oxygen levels dropped to 90 percent — decidedly abnormal for a young man with no prior lung condition. A chest X-ray revealed hazy, opaque streaks across his lung, signaling a deepening pneumonia. His liver function blood tests revealed abnormalities; high fevers came and went. He teetered on the brink of dying, but ultimately recovered.

More than two years have passed since the man with a cough walked into the Seattle clinic. As I write this, in March 2022, the world has recorded nearly 450 million infections, nearly 6 million deaths (both numbers are likely vast underestimates, given the lack of reliable reporting of testing and deaths from the virus). The contagion has leapt across the globe, leaving virtually no corner untouched. Waves of viral strains bearing new mutations have appeared, some more lethal than others — Alpha, Delta, and now, Omicron. More than sixty vaccines against the virus are undergoing clinical tests. Three have been approved in the United States, nine approved by the World Health Organization (WHO), and several others are still in development.

Wealthy nations, with mature health care and

delivery systems, have been brought to their knees. The United Kingdom has had more than 160,000 deaths. In the United States, the official number of deaths is 965,000. And the counts of the dead, the sick, the ravaged, the displaced, the bankrupted, the bereaved goes on. And on.

I cannot shake the images, or the sounds, of the pandemic. Who can? The orange body bags, piled in bunks, in makeshift morgues. The mass graves in Chile. The constant wailing of ambulance sirens outside my hospital, fusing together until all I could hear was a wall of shrieks; the Emergency Room, in the spring of 2021, packed to its brim, with stretchers spilling into the corridors; the patients gasping as they drowned in their own fluids; the ICU scrambling for more beds every day. The exhausted doctors and nurses every evening as they zombied their way across the crosswalk outside my office, with that hollow-eyed, post-shift look, and the characteristic marks on their faces, where the edges of the N95 masks had cut into their cheeks. The withdrawn, empty cities, with brown paper bags blowing on the streets. The looks of suspicion, or frank terror, when a man on the subway coughed or sneezed.

The photograph of a friend's cousin — a healthy, vigorous, fortysomething Brazilian man — on a beach in Rio de Janeiro two summers ago, his arms lifted joyfully above the water. In late July 2021, he fell sick with the virus. His pneumonia worsened. His breathing rate climbed to the thirties. I can only imagine a second image: the same man in an ICU bed straining to breathe so hard that the strap muscles on his neck are taut and visible, his lips blue. His arms are lifted again, but flailing — lifted not to express joy but to signal a desire for survival. I exchanged urgent texts, night after night, with my friend, and, for a while, felt reassured. The cousin was on a ventilator, but improving, albeit slowly. And then, the final text, received sometime late at night on April 9: "I'm sorry, but he passed."

A second wave that moved through India in April 2021 was vastly more lethal than the first. The virus had mutated into a strain now called the Delta strain — much more infectious, and possibly more lethal, than the original strain that had come from Wuhan. Delta rampaged through India, decimating its already shattered public health system and revealing the shocking absence of an organized, coordinated response. Delhi closed down, leaving millions of migrant workers stranded. My mother was a lone prisoner in her flat in the city. Over the weeks and weeks of confinement, her daily texts to me shortened into a Morse code of reassurance: *Today: OK.*

I cannot shake from my mind the picture of a migrant laborer from New Delhi on his knees outside the hospital, begging for a cylinder of oxygen for his family. A sixty-five-year-old journalist from Lucknow tweeted that he was infected, febrile, short of breath, but his phone calls to hospitals and doctors went unheeded. The tweets, ricocheting through cyberspace, spelled out an escalating catalogue of desperation. As the world watched in abject horror, he sent out pictures of his oxygen levels falling — 52 percent, 31 percent — levels incompatible with living. In the last tweet, there is an image of him, holding a pulse oximeter with blue fingers. Oxygen level: 30 percent. And then, there were no more messages.

There were days that I dared not open the newspaper. It was as if we had reinvented the phases of grief: anger turning into blame, and then helplessness. India burned so briskly that every system, every grid, collapsed, corroded, and melted.

I think, sometimes, of a legend. Bali, a demon king, has conquered three worlds — the earth, the underworld, and the heavens. A tiny man with smoky eyes and an umbrella, Vamana — Vishnu's avatar — appears before him and asks him to grant him a single wish. His arrogance inflated into munificence, Bali, the demon king, agrees. Vamana asks for something ludicrously small: a square plot of land whose edges are defined by the distance that he might cover in three strides. The man is — what — two arm lengths tall? He wants a few square feet of a kingdom that stretches to infinity? Bali laughs it off; yes, of course, the little man can have his little plot of land.

And then, as Bali watches in terror, Vamana expands. His body arches exponentially through the skies. His first step crosses the entire Earth. His second stretches across the heavens; the third arcs over the netherworld. There is no more kingdom to grant. He plants his foot on Bali's head and pushes him into the lower reaches of hell.

The analogy fails quite obviously in some parts — Vamana was a divine being, and the virus was anything but a divine intervention. Our failings, unfortunately, were all too human: a fraying, sclerotic global public health system, the absence of preparedness, misinformation that spread, virus-like, through nations, supply chain issues that made it impossible to procure protective masks and disposable medical suits, strong-men leaders of nations who turned out to be flaccid in their responses to the viral contagion.

But the foot planted on our head was real. Just when we felt that we *knew* the cell biology of the immune system, when our confidence had crested, scientist's heads were pushed into the lower reaches of hell.

When a diminutive microbe began to cross worlds, leaping from continent to continent, very little made sense. As Akiko Iwasaki, a virologist at Yale told me, coronaviruses similar to SARS-COV2 had circulated through human populations for millennia, but none had caused this kind of devastation. Some cousin viruses, like SARS and MERS, were deadlier than COV2, but they had been rapidly contained. What feature of the interaction between SARS-COV2 and human cells enabled *this* virus to precipitate a global pandemic?

Two clues came from a medical report from a German clinic that, at first glance, hardly read ominously. In January 2020 (in retrospect, how innocent, how self-assured, we seem to have been during that short-lived calm; *How much kingdom could a three-foot man claim?*), a thirty-three-year-old man from Munich had a business meeting with a woman from Shanghai. A few days later, he fell ill, with fever, headache, and flu-like symptoms. He recuperated at home, and went back to work, holding meetings with several other colleagues. A fever, a headache — and a quick recovery. An everyday infection. A common case of a common cold.

A few days later, the hospital in Munich called the man in: the Shanghai woman had fallen ill on her flight back to China. She had tested positive for SARS-COV2. But here was the puzzle: she had had no symptoms when she had met him; she had seemed perfectly well. She had only fallen ill two days later. In short, she had transmitted the virus to the man *while being presymptomatic.* No one could have told her, or the exposed man, that she had been a carrier of the virus. No isolation or quarantine based on symptoms could have stopped the virus.

The mystery deepened when the man was tested. His symptoms had defervesced by then; he had returned to work, feeling perfectly well. But when the virus was measured in his sputum, he turned out to be a roiling cauldron of contagion: there were *one hundred million* infectious viral particles per milliliter of spit; just a few coughs might fill a room with a dense, invisible, deeply contagious fog. He, too, was transmitting virus while experiencing no discernible symptoms.

As the contact tracing continued, the second ominous feature of the virus emerged: the man had infected three other people. The "contagiousness" of the virus — a key factor in determining the course of the growth of an infection — was at least three. If one person can infect three, the growth of infections is necessarily exponential. Three, nine, twenty-seven, eighty-one. In twenty cycles, the number reaches 3,486,784,401 — about half the population of the globe.

Asymptomatic/presymptomatic transmission. Exponential growth. The two critical ingredients in the recipe for a pandemic had already been established by that innocuous-seeming report. The third was soon to become apparent: unpredictable, mysterious deadliness. By then, who had not seen the grainy photograph, captured on a cell-phone camera, of the ophthalmologist from Wuhan who had dared to report the first cases? He is drenched in perspiration and gasping to breathe as he battles the pneumonia that will eventually kill him. As the infection spread, the world woke up to its macabre lethality: in Seattle, New York, Rome, London, and Madrid, the ICUs filled up and the body counts kept rising. And the questions arose just as furiously: What was the basis of presymptomatic infections? And how could a virus that caused relatively mild infections in some turn so lethal in others?

Why, you might ask, do the medical mysteries of the Covid-19 pandemic sit at the center of a book on cell biology? Because cell biology sits at the center of the medical mysteries. Everything we understood about cells and their interactions with each other — how the innate immune system responds to a pathogen; how immune cells communicate with each other; how a virus, growing so tenaciously inside a lung cell, can cause a presymptomatic infection without alerting other cells around it; how the cells of the gastrointestinal system may act as first responders to a pathogen

— has to be rethought and dissected. The pandemic demands autopsies of many kinds, but an autopsy of our knowledge about cell biology is also necessary. I could not write this book without writing about Covid.

In 2020, a group of Dutch researchers, looking for genes that might increase susceptibility to severe Covid, found the glimpse of an answer. The group identified two sibling pairs from different families, four young men, who had suffered unusually aggressive forms of the disease. Genetic sequencing revealed that one pair had inherited an inactivating mutation in a gene, TLR7 (fraternal siblings, on average, will have half their genes in common). Astonishingly, the second pair of brothers had also inherited a mutation in that very gene that appeared to decrease its activity (the exact mutation was different — but it was in the same gene).

What about the TLR7 gene might explain its involvement with such severe consequences of SARS-COV2 infection? Recall the innate immune system, which responds to patterns or signals of danger sent out by cells during the very first phases of an infection. Well, before the innate system can be activated, the cell first has to *detect* the invasion. TLR7 — or Toll Like Receptor 7 — it turns out, is one of the key detectors of viral invasion. It's a molecular sensor, built into cells, that is turned

470

"on" when the cell is infected by a virus. The activation of TLR7, in turn, galvanizes a cell's danger signals — among them, a molecule called type 1 interferon, to alert other cells to amplify their antiviral defenses, and to jump-start the immune response.

The theory is that mutations in TLR7 in both sibling pairs had somehow inactivated the protein, or decreased its function. As a consequence, the secretion of type 1 interferon — the danger signal — was blunted. The invasion wasn't detected, the alarm bell never went off, and the innate immune system never reacted adequately. Something about the functional impairment of the early, innate cellular response to viral infection had made the two pairs of Dutch brothers susceptible to the severest form of disease caused by SARS-COV2.

As scientists flocked to study SARS-COV2 and its interaction with immunity, more suggestive clues emerged. At Ben tenOever's lab in New York, researchers discovered that soon after infection, the virus "reprograms" the infected cell. In January 2020, I spoke to tenOever, a forty-year-old immunologist who works at Mount Sinai Hospital. "It's almost as if the virus hijacked the cell," he told me.

The cellular "hijack" involves an uncannily devious trick: even as it converts the cell into a factory to produce millions of virions, SARS-COV2 stops the infected cell from secreting type 1 interferon. At Rockefeller University in New York, Jean-Laurent Casanova converged on the same conclusion: the most severe cases of SARS-COV2 infection, he found, occurred in patients — typically men — who lacked the ability to elicit a functional type 1 interferon signal after infection. At times, cell biology produces the most peculiar and unexpected of results. These men with severe Covid had *pre*existing auto-antibodies against type 1 interferon — i.e., their bodies had attacked and rendered the protein nonfunctional even *before* they had been infected. These patients were *already* deficient in the type I interferon response — but were unaware of their deficiency until the virus struck. For them, Covid infection had unveiled a long-standing, but previously invisible, *autoimmune* disease — a dormant, unknowable horror autotoxicus (against type 1 interferon, the viral alarm bell) that was only revealed by SARS-COV2 infection.

The studies began to fit together, like pieces of a jigsaw puzzle: the virus was most deadly when it infected a host whose early antiviral response had been functionally paralyzed — like "a raider that had come into an unlocked house," as one writer described it. The pathogenicity of SARS-COV2, in short, perhaps lay precisely in its ability to dupe cells into believing that it is not pathogenic.

More data poured in. The infected host cell,

with its impaired ability to send out an initial danger signal, wasn't simply an "unlocked house." Rather, it was an unlocked house with not one but two dysfunctional alarm systems. It couldn't summon an early alert — type 1 interferon, among the signals — but as the house burned, the cell yanked the trigger on a powerful, second alarm, sending out a separate series of danger signals — cytokines — to summon immune cells. An uncoordinated army of cells — confused, bamboozled soldiers — surged into the sites of infection and started a carpet-bombing program. It was too much, too late. The trigger-happy immune cells poured out a fog of toxins to contain the virus. The war against the virus — as much as the virus itself — became an escalating crisis.

The lungs were bogged with fluid; debris from dead cells clogged the air sacs. "There appears to be a fork in the road to immunity to Covid-19 that determines disease outcome," Iwasaki told me. "If you mount a robust innate immune response during the early phase of infection [presumably via an intact type 1 interferon response], you control the virus and have a mild disease. If you don't, you have uncontrolled virus replication in the lung that [. . .] fuels the fire of inflammation leading to severe disease." Iwasaki used a particularly vivid phrase to describe this kind of hyperactive, dysfunctional inflammation: she called it "immunological misfiring."

■■■■

Why, or how, does the virus cause "immunological misfiring"? We don't know. How does it hijack the cell's interferon response? We have a few hints, but no definitive answer. Is it the *timing* of the response — the impairment of early phase, followed by hyperactivity of the late phase, the main problem? We don't know. What about the role of T cells that detect bits of viral protein in infected cells? Could they provide some protection against the severity of viral infection? Some evidence suggests that T cell immunity can dampen the severity of infection, but other studies don't support the degree of protection. We don't know. Why does the virus cause more severe disease in men than in women? Again, there are hypothetical answers, but we lack definitive ones. Why do some people generate potent neutralizing antibodies after infection, while others don't? Why do some suffer long-term consequences of the infection, including chronic fatigue, dizziness, "brain fog," hair loss, and breathlessness, among a host of other symptoms? We don't know.

The monotony of answers is humbling, maddening. We don't know. We don't know. We don't know.

Pandemics teach us about epidemiology. But they also teach us about epistemology: how

474

we know what we know. SARS-COV2 has forced us to point our most powerful scientific flashlights on the immune system, resulting in arguably the most intense scrutiny that this community of cells, and the signals that move between them, has ever been subject to. But perhaps what we think we understand about SARS-COV2 is limited to what we *already* know about the immune system — i.e., the known knowns. We cannot know the unknown unknowns.

And perhaps the pandemic has pointed to another gap in our understanding: perhaps other viruses, like SARS-COV2, have unexpected ways of contorting the cells of the immune system that results in their pathogenicity, and we've just ignored these deeper explanations (indeed, we know of such mechanisms in viruses such as cytomegalovirus, or Epstein-Barr virus). The story we've told ourselves about why SARS-COV2 is so devious at hijacking our immune systems is, perhaps, a totally incomplete story. Our understanding of the true complexities of the immune system has been partially shoved back into its black box.

Science hunts for truths. There's a haunting image in one of Zadie Smith's essays that involves a cartoon of Charles Dickens surrounded by all the characters that he's invented: tubby Mr. Pickwick in an ill-fitting waistcoat, adventurous David Copperfield in a top hat, bedraggled and innocent Little Nell.

Smith is writing about authors — in particular, about the out-of-body, in-another-mind experience that a fiction writer feels when she fully inhabits the mind, body, and world of a character that she's created. That familiarity, or intimacy, feels like a "truth." "Dickens didn't look worried or ashamed," Smith writes of the cartoon. "Didn't appear to suspect he might be schizophrenic or in some other way pathological. He had a name for his condition: novelist."

Imagine another character now, except surrounded by half-ghosts. Some of these "characters" — like type I interferon, the toll-like receptor, or the neutrophil — are mostly visible, except they exist in the half-light of visibility. We think we know and understand them, but we don't, really. Some only cast shadows. Some are completely invisible. Some mislead us about their identities. And there are others around us whose presence we cannot even sense. We haven't even met them, or named them, yet.

I, too, have a name for this condition: scientist. We look, we create, we imagine — but find only incomplete explanations for phenomena, even phenomena we may have (partially) discovered through our own work. We cannot inhabit their minds.

Covid exposed the humility that is required to cohabitate with these characters surrounding us. We are like Dickens, except encircled by shadows, ghosts, and liars. As one physician told me: "We don't even know what we don't know."

There's an alternative story — a triumphalist narrative — that can also be told about the pandemic. It goes this way: immunologists and virologists, building on decades of investigation into the fundamentals of cell biology and immunity, developed vaccines against SARS-COV2 in record time — some less than a year after the man from Wuhan had entered the Seattle clinic. Many of these vaccines functioned with entirely new methods of eliciting immunity — an altered chemical form of mRNA, for instance — again, using decades of knowledge of how immune cells detect foreign proteins, and how they might stave off infections.

But triumphalism fails in the face of more than six million deaths. The pandemic energized immunology, but it also exposed gaping fissures in our understanding. It provided a necessary dose of humility. I cannot think of a scientific moment that has revealed such deep and fundamental shortcomings in our knowledge of the biology of a system that we had thought we knew. We have learned so much. We have so much left to learn.

PART FIVE
ORGANS

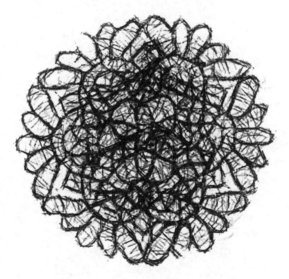

We've spoken a lot about organs, but we haven't really met one yet. Blood, which we encountered as a model of cellular cooperation and communication, is not a simple "organ." It is, rather, a system of organs: one to deliver oxygen (red cells), and another to respond to injury (platelets), and yet another to respond to infections and inflammation. Some of its systems have yet other systems within them — there's innate immunity (neutrophils and macrophages that have an inbuilt capacity to detect and kill pathogens) that cooperates with learned immunity (B and T cells that adapt and learn to make a specific immune response to the pathogen).

In biology, an "organ" is defined as a structural or anatomic unit, in which cells come together to serve a common purpose. In smaller animals, even a small collection of cells will serve the purpose. The nematode worm, C. elegans, which many biologists study, has a nervous system consisting of 302 neurons. That's about three hundred million–fold fewer than neurons found in the human brain.

As organisms grew bigger and more complex, organs, too, grew necessarily bigger and more complex. But the fundamental defining feature of organs — a common purpose, the "citizenship" of cells that Virchow had imagined — remained, and still remains, constant. In animals, organs are anatomically defined so that the cells that reside in the organ can act in concert — as citizen cells — to enable physiology.

The cells in organs, as we shall see, still utilize the basic principles of cell biology — protein synthesis, metabolism, waste disposal, autonomy. But each cell in each organ is also a specialist: it acquires a unique function that serves the organ as a whole, and ultimately coordinates some aspect of human physiology. Human organs, and their cells, thus had to acquire more and more specialized functions. A roundworm can breathe through its skin, but humans need a lung. And in mega-cellular organisms, such as humans, there are oceanic distances to cover: the pancreas sends insulin to cells in the toes with every heartbeat — a longer distance than most nematodes travel in their lifetimes.

Cellular specialization and citizenship — the hallmarks of the cell biology of an organ — result in the profound "emergent" properties of human physiology — i.e., properties that can only emerge when multiple cells coordinate their functions and work together. A heartbeat. A thought. And *the restoration of constancy — the orchestration of homeostasis.*

To understand the biology of humans, then, we need to understand organs. And to understand organs, their dysfunctions in disease, and the possibility of rebuilding them, we must understand the biology of the cells that make them work.

THE CITIZEN CELL
THE BENEFITS OF BELONGING

The crowd, suddenly there where there was nothing before, is a mysterious and universal phenomenon. A few people may have been standing together — five, ten or twelve, not more; nothing has been announced, nothing is expected. Suddenly everywhere is black with people, and more come streaming from all sides as though the streets had only one direction.

— Elias Canetti, *Crowds and Power*

For the concept of a circuit of the blood does not destroy, but rather advances traditional medicine.

— William Harvey, 1649

For months, during the early, deadening days of the pandemic in New York, I found myself unable to write. As a doctor, I was considered an "essential worker" — and so "essential work" carried on. Between February and August 2020, as the contagion swirled like a

485

virulent tornado around the city, I went to my office at Columbia, donned my requisite N95, and tended to my patients who needed care (the Cancer Center was still running, albeit with minimal staff. Somehow, we managed to keep essential chemotherapy, transfusions, and procedures on schedule). Some of my patients contracted the virus — a sixtysomething woman with pre-leukemia, another with myeloma whose stem cell transplant had to be pushed back — but thankfully, there were only two ICU admissions and no deaths. The rest recovered.

But my motions were robotic, and my mind was blank: I stared at the screen, often until one or two at night, produced a paragraph or two, and then junked them in the trash bin every morning. What I felt wasn't writer's block, but writer's languish: I wrote, yes, but everything that I put on the page seemed to lack life and energy. What preoccupied me was the collapse of infrastructure, and of homeostasis, that we had witnessed during the worst of the crisis in the United States, and then, around the world.

When my frustration reached its crest, I near-regurgitated an essay, later published in the *New Yorker*. It was part cri de coeur, part plea for change, part autopsy of what I had witnessed in the middle of the pandemic. Medicine, I wrote, isn't a doctor with a black bag. It's a complex web of systems and processes. And systems that we had thought were self-regulating and self-correcting, like a human body in good health, turned out to be exquisitely sensitive to turbulence, like the body during critical illness.

I had spent nearly a year thinking about bodies succumbing to illness, about a cellular system poised for battle against invaders. But as the spring of 2021 approached, the constant metaphors of the battle had become denervating. I wanted to think about normalcy and about restoration, about cellular systems that form the infrastructure of human physiology (and conversely, of the future repair and restoration of the human systems that had failed). I wanted to write about homeostasis and self-correction. I was exhausted by my mulling about how the body recognizes things — viruses — that do not belong. I wanted to turn to citizenship, to belonging.

The heart, of all organs in the body, epitomizes belonging. We use the word "belong" to signify attachment, or love — and the heart has been the central signifier of that feeling for millennia (although, of course, we now know that much of emotional life sits in the brain). When you say "my heart belongs to you," you signify the link between that organ and attachment.

As a child, my heart belonged to my mother. My father was a distant presence — reliable and gentle, but reserved, somehow unreachable. His mother — my grandmother — lived with us. Scarred by her move during Partition, she kept to a room by herself, cooked her own food and washed her own clothes, almost as if a house was a temporary shelter that might be snatched from her hands at any time. Her belongings, largely untouched and still packed in sheets of newspaper, sat in the steel trunk that she had ferried across the border from East Pakistan into India. Aside from a bed and a threadbare mattress, her room had nothing else; she had separated from the possibility of separation. I have no memory of her touching me. Her heart had been broken.

My transition into manhood marked a different relationship with my father. As a student at Stanford in a world before cell phones and e-mail, I began to write to him. At first, our letters were brief and stilted, but over time, they grew longer and warmer. I began to understand him in a new light. His loss of home seemed familiar: in 1946, he had been uprooted from his village when he was barely a teenager, stuffed into an overnight ferry bound for Kolkata, a city on the edge of a nervous breakdown. In the late fifties, he had moved again, as a young executive to Delhi, a city as culturally and socially alien to a young man from East Bengal as the Frisbee-slinging, Froyo-guzzling, beer-ponging life of my California dorm seemed to me. In 1989, five weeks into my first term, San Francisco was hit by the Loma Prieta earthquake — a shuddering of such magnitude that, as I stood under the lintel of my dorm room door, I saw the corridor buckle and a sine wave move through the cement, as if I were standing on the back of a serpent that had suddenly been awoken. My father heard the news and immediately wrote to me. In 1960, when he was building his first house in Delhi, an earthquake had decimated the one-story structure into which he had poured all his savings. He told me — he had told no one — that he had spent the night sitting on the foundation, surrounded by the crumbled rafters, weeping.

I longed to be back home — even for a short while. One afternoon, I picked up my mail and found a heavy packet: he had surprised me with tickets to return to Delhi in my first winter (I was supposed to stay in California until the next summer). It was a sixteen-hour flight, and I slept through it, until the fog-choked lights of the city came into view, and the plane made that elephant-screech sound of the wheel hatch opening just before landing. Since that trip, I must have flown to India four dozen times — but it is a sound that makes my heart bound with a strange joy.

The man at Customs asked me for a small bribe, and I felt like hugging him; I was home. I can still feel the boom of my heartbeat as I exited the airport. I could tell you about the neural cascade I experienced — the memories flooding back, the release of epinephrine in my blood — but while the stimulus was triggered in the brain, the experience was felt in my heart. My father was there, as he would be year after year when I returned, with a white shawl draped around himself, and an extra one to wrap around me. Coming back. Belonging.

Beyond metaphors, the heart is actually an organ where the belonging and citizenship among cells is of crucial importance. What makes heart cells special? What allows them to perform the precisely coordinated activity — second upon second, day after day — that we recognize as the heartbeat? Contemplate the heartbeat: this phenomenon that many of us might consider the epitome of the everyday — the heart will beat more than two billion times over an average person's life — is, in fact, a miraculously complex feat of cell biology. The heart is a model of cellular cooperation, citizenship, and belonging.

Aristotle, for one, thought of the heart as the first among equals — the most important citizen of all organs, the center of vitality in the body. The other organs that clustered around the heart, he proposed, were there just as heating and cooling chambers. The lungs were like bellows, expanding and contracting to keep the engine cool. The liver was a glorified heat sink, diverting the excess heat produced by the most vital of organs, lest it overheat. Galen of Pergamon propelled the idea forward: "the heart is, as it were, the hearthstone and source of the innate heat by which the animal is governed."

But the heart's centrality to human life — so much so that all the other organs were mere heating and cooling pipes for its engine — begged the question: What does this organ *do*? The medieval physiologist Ibn Sina (or Avicenna) who lived around AD 1000 tried to tackle the question in a majestic treatise that he named *al-Qanun fi'at-Tibb* — *The Canon of Medicine* (the word *Qanun* can also be translated as "Law"; Ibn Sina was looking for universal laws that governed physiology). Ibn Sina concentrated on the pulse, noting its wave-like quality, and its correlation with the heart's pulsation. When the pulse was irregular, so, too, was the heartbeat — and palpitations caused symptoms, such as fainting or lethargy. When the heartbeat became thready, so did the pulse, and the symptoms portended death. Anxiety increased the pulse concomitantly with the heartbeat. And, Ibn Sina noted, so did "love sickness" — longing, or belonging. A friend told me the story of visiting a Tibetan doctor who specialized in pulses. The doctor asked him a few perfunctory questions and then checked his pulse.

"You've gone through a terrible breakup," the doctor said. "Your life isn't going to be the same again." The Tibetan doctor was right: something about the pulse — its rapidity, or its dullness — had provided a clue about the longing and belonging. My friend's breakup, and life, had forever been uprooted.

Ibn Sina's description of the heart as a source of pulsation — in essence a pump — was one of the earliest attempts to describe its function. But it was English physiologist William Harvey, working in the 1600s, who fully described the conjoined circuitry of the heart-as-pump in the human body. Harvey trained in medicine in Padua, and then returned to Cambridge to continue his medical studies. In 1609, he was appointed a physician at St. Bartholomew's Hospital with an annual salary of £33. Short and round-faced — "eyes small, round, very black and full of spirit; his hair as black as a raven and curling" — he was a man of simple tastes. He lived in a small house in rundown Ludgate, even though his position as hospital physician gave him access to two much larger houses near the hospital. It's tempting to connect his material austerity to the austerity of his experimental methods. Using nothing more than bands and torniquets, and an occasional pinch of an artery or vein, Harvey set out to solve a problem that had confounded physiologists for centuries.

We've already encountered Harvey's

orthodoxy-dismantling, inquisitive mind in embryology and physiology: he was among the strongest critics of the idea that the embryo came "preformed" in the womb, or that blood was the warming oil of the body. But it was his seminal work on the heart and circulation that was Harvey's most significant scientific contribution. Harvey didn't have the use of powerful microscopes, so he used the simplest physiological experiments to understand the workings of the heart. He punctured the arteries of animals and found that when the blood had drained from them, the veins were also eventually emptied of blood: hence, he concluded, arteries and veins must be connected in a circuit. When he pinched the aorta, the heart swelled up with blood. When he pinched the main veins, the heart was drained of blood: hence the aorta must lead blood out of the heart, and the veins must lead blood into the heart — a conclusion so manifestly central to the understanding of circulation that it is unfathomable that it had eluded generations of physiologists.

Most importantly, when he examined the septum — the wall — between the left and right sides of the heart, he found it too thick, and without any pores in it: hence the blood from the left side must travel to the lungs first before it reentered the right side (a direct assault on the beliefs of Galen and early anatomists that blood moved through the lungs

493

before reaching the heart). When Harvey watched the heart beating, he saw it contracting and relaxing: hence the heart must be the pump that sends the blood in a circuit around the body, from arteries to veins and back.

In 1628, Harvey published his conclusions in a seven-volume series now typically referred to as *De Motu Cordis (An Anatomical Account of the Motion of the Heart and Blood)* that would upset the very foundations of anatomy and physiology of the heart. The heart, Harvey argued, was a pump that moved blood circuitously through the body — from arteries to veins and back again. These views, he wrote, "pleased some more, others less: some [. . .] calumniated me and laid it to me a crime that I had dared to depart from the precepts and opinions of all anatomists: others desired further explanation of the novelties which they said were both worthy of consideration, and might perchance be found of signal use."

We now know, in part, from Harvey's work on the anatomy of the heart, that the heart is, in fact, *two* pumps — one left and one right, laid side by side, like twins in a womb.

It's all a circle, so let's begin with the right side. The right-sided pump collects blood from the veins of the body. Exhausted and depleted, having delivered oxygen and nutrients to the organs, "venous" blood (often darker red than bright crimson) pours into the upper right chamber called the right atrium. It then

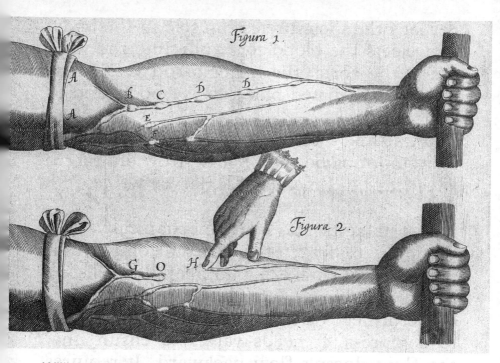

William Harvey's drawing (from De Motu Cordis*) of simple exercises, such as pinching veins and arteries to illustrate how blood flows from the veins toward the heart, and from arteries away from the heart.*

passes through a valve and is moved into the pumping chamber, the right ventricle. A powerful heave from the right ventricle pumps the blood to the lungs. This is the right-sided circuit — veins to heart to lung.

The lungs, having received blood from the right side of the heart, oxygenate the blood and clear the carbon dioxide. Replete with oxygen and cleansed, blood, now a vivid crimson, moves to the left side. It collects in the left atrium of the heart. It is then pushed into the left ventricle. It is this left ventricle, perhaps the most tireless muscle in the body, that ejects the blood forcefully into the wide arc of the aorta, the major blood vessel that carries oxygenated blood to the body, and to the brain.

Round and round in circles. "The concept of a circuit of blood does not destroy, but rather advances [. . .] medicine," Harvey wrote.

But to imagine the heart mechanically, as a pump, is to forget the central conundrum: How do you make a pump out of cells? A pump is, after all, a highly coordinated machine. It needs a signal to dilate and a signal to compress. It needs valves to ensure that the fluid doesn't flow backward. It requires a mechanism to ensure that the contracting bladder doesn't wobble without purpose or direction. An uncoordinated pump is no better than a wobbling balloon.

On January 17, 1912, Alexis Carrel, a French scientist at the Rockefeller Institute in New York, cut a small fragment of the heart of an eighteen-day old chick fetus and grew it in liquid culture. "[T]he fragment pulsated regularly for a few days and grew extensively," he noted. "After the first washing [. . .] the culture grew again very extensively." When he removed and recultivated a piece of it, he found it still capable of pulsating: in March, nearly three months after having removed it from the chick's heart, "it was [still] pulsating at a rate that varied between 60 and 84 per minute [. . .]." Ultimately, "on March 12 the pulsations were irregular, and the fragment beat for a series of 3 to 4 pulsations, and then stopped for about 20 seconds." Over the course of about three months, the piece of chick heart in a petri dish had generated about nine million heartbeats.

Carrel's experiment was widely heralded as proof that organs could live, and function, outside the body — but it also signaled an equally important idea: the cells of the heart, cultivated outside the body, had the *autonomous* capacity to pulsate rhythmically Something intrinsic about the cells made them capable of "pump-like" action — coordinated pulsation. That same year, W. T. Porter, a physiologist at Harvard, had shown that severing the nerves from a dog's heart still enabled the ventricles to pulsate autonomously — a "live" demonstration of what Carrel had shown in a petri dish.

The coordinated pulsation of the heart's cells fascinated physiologists. In the 1880s, German biologist Friedrich Bidder had noted that the cells of the heart "branch and intercommunicate, forming a continuum." They form a consortium of sorts — a citizenry of cells. The source of their contractile power seemed to lie in their togetherness, their belonging.

But how was that contractile power

generated? In the 1940s, a Hungarian-born physiologist, Albert Szent-Györgyi, began to investigate how a cell might acquire the ability to contract and relax. By then, he had already established himself as one of the most eminent physiologists of his generation: he had won the Nobel Prize for discovering vitamin C and studied how the cell generates energy. Much of what we know about the mitochondrial reactions that make energy molecules emerged from this work. He was a man of powerful convictions and itinerant curiosity. Conscripted to the medical corps during World War I, he became so disgusted with the carnage and disillusioned by the war that he shot himself in the arm, claimed to have been injured by enemy gunfire, and was therefore able to resume his scientific and medical studies. He moved from university to university, lab to lab, and city to city — Prague; Berlin; Cambridge, England; and Woods Hole in Massachusetts — studying the biochemistry of cellular respiration, the physiology of acids and bases in the body, and vitamins and biochemical reactions essential for life.

By the 1940s, his infinitely curious mind had turned to the study of cardiac muscle. The question that preoccupied him was essential to understanding the function of the heart: How was its pumping force generated? Szent-Györgyi began with Virchow's idea: if an organ is capable of contracting and dilating, then *its cells* must be capable of contraction and dilation. Sitting within each muscle cell, Szent-Györgyi mused, must be some specialized molecule, or set of molecules, that was capable of generating a directional force, thereby shortening the cell — contracting it. "In order to make a system which can shorten," he wrote, "Nature has to use thin and long protein particles." By then, one of the "thin and long proteins" had already been identified. He wrote: "The threadlike, very thin and long protein particles out of which Nature has built the contractile matter is 'myosin.'"

But a thin long protein is merely a rope. Tether the rope to the ends of a cell and you begin to have the basic elements of a contractile apparatus. But how is this system of ropes pulled tight and loosened? Szent-Györgyi and his colleagues found that fibers of myosin are intimately connected to another dense, organized network of long and thin fibers — composed mainly of a protein called actin. In short, there were *two* systems of interconnected fibers inside a muscle cell: actin and myosin.

The trick to a muscle cell's contraction is that these two fibers — actin and myosin — slide against each other, like two networks of ropes. When a cell is stimulated to contract, a part of the myosin fiber binds to a site on the actin fiber, like a hand from one rope gasping the other. It then unclutches it and reaches forward to bind to the next site — a man suspended on one rope, grasping and pulling on the other, one fist upon the next. *Clutch. Pull. Release. Clutch. Pull. Release.*

Each muscle cell has thousands of such lined ropes — bands of actin in parallel with bands of myosin.* As the ropes, lined side by side, slide against each other — *clutch, pull, release* — the edges of the cell are also yanked, and the cell is dragged into a contraction. The process requires energy, of course, and every heart cell and muscle cell is chock-full of mitochondria to supply the energy required for the two fibers to slide (a quick aside here: one peculiarity of the system is that it is the *release* of actin from myosin — not the binding of the fibers — that requires energy. When an organism dies, and the source of energy is lost, the muscle fibers, unable to unclasp their fists, are caught in a permanent grip — bound. The cellular ropes in every muscle

---

*There are three fundamental kinds of muscle cells in the human body: cardiac muscle, which constitutes the main subject of this chapter; skeletal muscle (the kind that moves your arms on command); and smooth muscle (the kind that moves involuntarily, but consistently, allowing, say, liquid in the intestines to keep moving). All three muscles use variants of the actin/myosin system, along with a smattering of other proteins, for contractility.

tighten. The body hardens and contracts into the permanent clasp of death — the phenomenon that we call *rigor mortis*).

But this describes the contractile journey of one cell. For the heart to function as an organ, all its cells must contract in the coordinated order. And this is where Friedrich Bidder's observation — that the cells of the heart muscle seem to form a "continuum" — becomes critical. In the 1950s, microscopists would discover that heart cells are connected to each other through minuscule molecular channels, called gap junctions. In other words, every cell is inherently designed to communicate with the next. Although many, they *behave* as one. When a stimulus to contract is generated in one cell, it automatically travels to the next cell, resulting in its stimulation, and ultimately resulting in contraction in unison.

What is that "stimulus"? It's the movement of ions — principally calcium — in and out of cells through specialized channels on the membranes of heart cells. In its resting state, the heart cell has low levels of calcium. When it is stimulated to contract, calcium floods into a heart cell, and it instigates contraction. And calcium's entry is a self-feeding loop: the entry of calcium releases more calcium from the heart cell, resulting in a sharp, steep spike in calcium levels. The interconnections between the cells — those "junctions" that were identified in the 1950s — carry the ionic message from cell to cell. The one becomes the many. Crowds generate power. The organ — a continuum of cells — thus behaves as a whole.

There are two final cellular elements of the heart that are essential to its function. First, there are valves between the chambers to ensure that blood does not flow backward. The cells of the atria — the collecting chambers — contract first sending blood into the ventricles. The valves between the atria and ventricles close, making a flapping noise: *Lub,* the first heart sound. And after that, the cells of the ventricle contract in a similarly coordinated manner. The outlet valves of the ventricles close: *Dub,* the second heart sound. *Lub-Dub, Lub-Dub.* The sound of a citizenry in lockstep, working together.

The final element of a pump is a rhythm generator, or a metronome. Physiologists found that specialized nerve-like cells, resident in the heart, generate paced, rhythmic electrical impulses that stimulate the contraction. Yet other nerves — fast-conducting electrical wires — carry these impulses throughout the heart, first to the atria, and then to the ventricles. Once the impulse reaches one cell, the junctions between the cells ensure that all cells contract together.

The result is a miraculous coordination.

Atrial contraction. Ventricular contraction. The heart's cells form an orchestrated citizenship. Each cell of the heart muscle maintains its identity. But each cell is so intimately connected to the next that when the impulse to contract arrives, the contraction is purposeful and coordinated. The heart doesn't wobble; its ventricles contract in one potent push. We might imagine the organ as if it behaves, almost, as a single-minded cell.

THE CONTEMPLATING CELL
THE MANY-MINDED NEURON

The Brain — is wider than the Sky —
For — put them side by side —
The one the other will contain
With ease — and you — beside —

The Brain is deeper than the sea —
For — hold them — Blue to Blue —
The one the other will absorb —
As sponges — Buckets — do —
— Emily Dickinson, c. 1862

If the heart is single-minded, then the brain is many-minded. Let's acknowledge one challenge upfront: it is impossible to tackle the function of an organ of such immense complexity in a whole book, let alone a single chapter.

But, leave function aside for a moment; let's begin with structure. During my anatomy lab at medical school, the students were divided into groups. My group, four students in total, was handed a squishy human brain that had been pickled in formaldehyde — a gift to medical science, left by a fortysomething man who died in a car accident and had chosen to donate his organs. It was an infinitely strange feeling to hold an organ, about the size and shape of a large boxing glove, and imagine it as the repository of memory, of consciousness, of speech, temperament, sensation, and feeling. Love. Envy. Hatred. Compassion. All of these had reposed in some tangle of neurons. I was holding *him*, I thought, this man whose name, or identity, I would never know. Somewhere within that organ lived the neurons that had once remembered his mother's face. Somewhere was the memory of his last moment before the car had hurtled off the road, somewhere the melody of his favorite song.

From the outside, the most extraordinary of all organs looked extraordinarily dull — a bulk of tissue sheathed in curvilinear bulges of gray matter. A cerebellum, each of its two lobes the size of a child's fist, hung below. There were protrusions on the two sides of the brain — the "thumbs" of the boxing glove, seen from the side. A severed, stem-like nub of tissue was the site where it had once connected to the spinal cord.

But as I sliced into the tissue from the side, it was as if I was opening a box of wonders. There was a seeming infinitude of structures — beltways of nerves, ventricles filled with fluid, sacs, glands, and dense clusters of nerve cells, called nuclei. The pituitary gland — one of the few unpaired glands in the body, dangled, like a tiny berry, from the middle. The pineal gland, which Descartes thought was the seat of the soul, was also nestled in the center. Each of these glands and nuclei contained a unique set of cells dedicated to some particular and often distinctive function. How this endless array of structures — and an equally endless array of cells (neurons, hormone-making cells, and glial cells, the non-neuronal cells that support the function of the nerves) — ultimately enable the brain's profound functions cannot be captured in a book on cell biology. But the function of the neuron — the most essential unit of the brain's whole — is where we might begin to understand the brain.

For several decades in the late nineteenth century, the most versatile and mystifying of cells in the body was not even considered a cell. In fact, it was not visible to most microscopists: the structure of a neuron was largely hidden. In 1873, Camillo Golgi, the Italian biologist working in Pavia, found that if he added a solution of silver nitrate to a slice of translucent neuronal tissue, a chemical reaction occurred, resulting in a black stain that accumulated within some of the neurons. Under the microscope, Golgi saw a lacy network. He thought that the network represented a continuous connection — a "reticulation," as he called it. Cell theory was itself in its infancy — Schwann and Schleiden had proposed that all organisms were collections of cells in 1838 and 1839 — and so, Golgi wondered whether the entire nervous system was a spiderweb of "cellular appendages," an "inextricable tangle" of interconnected, contiguous cellular extensions, as one writer called it. It was a mumbo-jumbo theory: as Golgi imagined it, the whole nervous system was like a fishnet, composed of wiry extensions sent out by the brain.

A rebellious young pathologist from Spain challenged Golgi's theory. A gymnast, an athlete, and an avid draftsman — "shy, unsociable, secretive, brusque," as one biographer described him — Santiago Ramón y Cajal was the son of an anatomy teacher who, in the tradition of Vesalius, took his young boy to the graveyards in his town to dissect specimens. As a child, Cajal was known for his elaborate pranks. His first "book" was on the construction of slingshots — a fusion, as it were, of his love for accuracy and his disdain for authority. He also drew compulsively — bird's eggs, nests, leaves, bones, biological specimens, anatomical structures: all forms of natural objects fascinated him, and he sketched them in his notebook. He would later call this drawing habit his "irresistible mania." Cajal attended medical school in Zaragosa and eventually moved to Valencia, where he was appointed as a teacher of anatomy and pathology. In Madrid, he chanced upon a friend who had just returned from Paris where he had learned Golgi's staining method.

Many scientists had attempted to reproduce Golgi's stain, but it was a capricious, temperamental reaction, often yielding just a blob of black-stained tissue. When it worked, it typically lit up — or rather silhouetted — the dense reticular network that had led Golgi to imagine the nervous system as an intricate connection of continuous wires. But Cajal's genius involved tinkering relentlessly with the method — again mixing accuracy with his disdain for prior authority. He titrated the nitrate to an exact dilution, cut the tissue into precise, razor-thin sections, and used the finest of microscopes to visualize the neurons stained by the "black reaction." And unlike Golgi, what Cajal saw was a radically different organization of cells. There was no tangled "reticulation" in the nervous system, no hodgepodge of wiry extensions. Rather, there were *individual* neuronal cells, with intricate, delicate anatomy, that reached out to connect with *individual* neuronal cells.

He sketched them by hand, in black ink, producing among the most beautiful drawings in the history of science. Some neurons were like thousand-branched trees, with dense arbors of extensions above, a pyramidal cell body in the middle, and a stem-like extension below. Some were like starbursts, some like many-headed hydra. Some had infinitesimally slender, many-fingered extensions. Some were compact; some stretched from the surface of the brain to the deeper layers below.

Yet, despite their unfathomable diversity, Cajal discovered, neurons often shared common features. They possessed a cell body — the soma — from which often sprouted dozens, hundreds, or even thousands of branch-like projections called dendrites. And they possessed an outflow tract — an "axon" — that extended toward the next cell. Notably, the axon of one neuron, its outflow point, was separated from the second neuron by an intervening space — eventually called the "synapse." The nervous system was wired, yes, but the "wires" consisted of cells connected to cells connected to cells, with intervening spaces between them.

Cajal used these drawings, as tenderly beautiful as they were forensically accurate, to propose a theory of the structure of the nervous system. He argued that information traveled unidirectionally in a nerve. The dendrites — those extensions that he had visualized budding out of the neuron's cell body — "received" the impulse. The impulse then moved through the cell body. And it moved out through the axon, through the synapse, to the next nerve cell. The process was then re-peated in the next cell: *its* dendrites picked up the impulse, transmitted it to the cell body, and then the impulse flowed out through the axon to the next cell. And so forth, ad infinitum.

The process of nervous conduction, then, was the movement of the impulse from cell to cell. There wasn't a single, reticular spider-web of "cellular appendages," as Golgi had proposed, or a syncytium of citizen cells, as in the heart. Rather, nerve cells "chattered" with each other — collecting inputs (via dendrites) and generating outputs (via the axon). And it was this cellular chatter — or rather, *inter*cellular chatter — that gave rise to the profound properties of the nervous system: sentience, sensation, consciousness, memory, thinking, and feeling.

In 1906, Cajal and Golgi were jointly awarded the Nobel Prize for their elucidation of the structure of the nervous system. It might have been the oddest prize in its history because it was an armistice more than an award: Cajal's and Golgi's ideas about the structure of the nervous system were precisely opposed to each other. In time, with the invention of more powerful microscopes, Cajal's theory — of discrete neurons communicating with each other, and the impulse moving from cell to cell in a directional course — would be proved right. The nervous system *was* made of wires and circuits, but the "wires" were not a contiguous reticulum, but rather individual cells that had the capacity to collect information and transmit it to another series of neurons.

It is one of Cajal's legacies that he never performed a single experiment in cell biology — or at least an experiment in the traditional sense. To see his drawings of neurons is to realize how much can be learned by just *seeing*. It is to return to characters such as Da Vinci or Vesalius who imagined drawing as thinking: an astute observer and draftsman could generate a scientific theory as much as an experimental interventionist. Cajal sketched what he saw, and his understanding of how the nervous system "worked" emanated entirely from drawing cells and drawing conclusions. Even that phrase — "drawing a conclusion" — illuminates the connection between thinking and drawing: to "draw" is not merely to illustrate, but to extract a substance, to pull out a truth. It was Cajal's "irresistible mania" — drawing truths, drawing out truths — that laid the foundations for neuroscience.

Return, for a moment, to Cajal's idea of a neuron: that it is a discrete cell, capable of transmitting an impulse — a message — to another cell. What was the message, and who was the messenger?

For centuries, scientists had believed that

nerves were hollow conduits, like pipes, and some fluid, or air — pneuma — flowing through them carried a wave of information from one nerve to the next, and from the nerve to a muscle, ultimately causing that muscle to contract. According to the "balloonist" theory, as it was called, the muscle was a balloon, and when it was filled with pneuma, it swelled like an air-filled bladder.

In 1791, an Italian biophysicist, Luigi Galvani, deflated "balloonism" with an experiment that changed the course of neurological science. The story, likely apocryphal, runs that his assistant was dissecting a dead frog with a scalpel when he accidentally touched a nerve. A nearby electrical spark reached the scalpel, and the dead animal's muscle twitched, as if it had come to life.

Astonished, Galvani repeated the experiment with several variations. He connected the frog's leg to its spinal cord using a makeshift wire, one length made of iron, and the other of bronze. When he put the two wires in contact, a current sparked through the electrodes, and, again, the frog's leg twitched (Galvani assumed that the electricity that was moving from the cord to the muscle was intrinsic to the animal — a phenomenon he called animal electricity. His colleague Alessandro Volta, fascinated by Galvani's experiment, found that the real source of electricity was not the animal, but the contact between the two metals submerged partially in the dead frog's fluids. In time, Volta would use this idea to devise the first primitive battery).

Galvani spent much of his life exploring "animal electricity" — a unique form of biological energy that he considered his most exciting discovery. But the center of Galvani's finding would turn out to be rather peripheral. Most animals — electric eels and mantas aside — don't discharge bioelectricity. It was Galvani's lesser discovery that would prove to be revolutionary. It was the idea that the signal that moved from nerve to nerve, and nerve to muscle was not air but *electricity* — the influx and efflux of charged ions.

In 1939, Alan Hodgkin, an undergraduate fresh out of Cambridge, England, was invited to work on nerve conduction with Andrew Huxley, a physiologist, at the Marine Biological Association in Plymouth. The lab was a large brick building on Citadel Hill, its corridors coursed by a bracing ocean breeze. The location was critical. From the sea-view windows that overlooked Plymouth Bay, the researchers could look at the catch coming in on fishing boats. And of all things trawled up from the ocean, there was one most precious to them: squid, which happens to possess one of the largest neurons in the animal kingdom, nearly a hundredfold larger than some of the slender, minute neurons that Cajal had drawn in his notebook.

Hodgkin had learned to dissect that neuron out of the squid at the Marine Biological Labs in Woods Hole. And the duo punctured the cell with a minuscule silver electrode, sharpened beyond a pinpoint. They learned to send impulses in and record the output, eavesdropping on an individual neuron's "chatter."

In September 1939, just as Hodgkin and Huxley were recording the impulse from an axon, the Nazis invaded Poland, plunging the Continent into war. The two scientists had finished their first recordings of electrical conduction, and they rushed their paper off to *Nature*. It was an astonishing work, with just two figures, one showing the experimental setup, with the squid axon and a sliver of a silver wire inserted into it.

It was the second figure, though, that was breathtaking. They saw the arrival of a small electrical impulse — a mini wave — followed by a large wave of charged ions moving into the neuron. The large wave subsided, and dipped, and then the system was reset to normal. Again and again, when they stimulated the axon, they saw the same rising spike of charge, and its restoration to normal. They had observed the dynamics of a nerve conducting its signal to another nerve.

The war interrupted Hodgkin and Huxley's collaboration for nearly seven years. Hodgkin,

514

the engineer-tinker, was sent off to make oxygen masks and radars for pilots; Huxley, the mathematician, was deployed to use equations to make machine guns more accurate. In 1945, shortly after the war ended, they resumed their work in Plymouth, scouring the catch for squid and delving deeper and deeper into the nervous system, finding more accurate ways to measure the flow of charge into the neuron, culminating in a mathematical model to describe the movement of ions into the neuronal cell.

Nearly seven decades later, neuroscientists are still using the Hodgkin and Huxley equations, and their experimental methods, to understand the nervous system. The broad outlines of how a neuron "chatters" are now understood. Perhaps we can turn to one of Cajal's drawings as a template to elucidate the movement of a signal through a nerve. Imagine the nerve, first, in its "resting" state. At rest, the internal milieu of the neuron contains a high concentration of potassium ions and a minimal concentration of sodium ions. This exclusion of sodium from the neuron's interior is critical; we might imagine these sodium ions as a throng outside the citadel, locked out of the castle's walls and banging at the gates to get inside. Natural chemical equilibrium would drive the influx of sodium into the neuron. In its resting state, the cell actively excludes sodium from entry, using energy to drive the ions out. The net result is that the resting neuron has a negative charge, just as Hodgkin and Huxley had discovered in their original experiment in 1939.

Turn, now, to the dendrites, the many-branched structures that Cajal drew. The dendrites are the site within the neuron where the "input" of the signal originates. When a stimulus — typically a chemical called a "neurotransmitter" — arrives at one of the dendrites, it binds to a cognate receptor on the membrane. And it is at this point that the cascade of nerve conduction begins.

The binding of the chemical to the receptor causes channels in the membrane to open. The citadel's gates are thrown ajar, and sodium floods into the cell. As more ions swarm in, the neuron's net charge changes: every influx of ions generates a small positive pulse. And as more and more transmitters bind, and more such channels open, the pulse increases in amplitude. A cumulative charge courses through the cell body.

Imagine, now, that the army of invading ions, a charge (indeed, literally so) marches past the dendrites toward the neuron's cell body — the soma — and reaches a pivotal point in the neuron, called the "axon hillock." It is here that the critical biological cycle that enables nerve conduction is set into motion. If the pulse that reaches the axon hillock is greater than a set threshold, the ions begin a self-fulfilling loop. *The ions stimulate the opening of more channels in the axon.* In biology, when a chemical stimulates the release of the very same chemical, it sparks a positive feedback loop — more becomes more. The ion-sensitive ion channels are the linchpins in axonal conduction: they are self-propagating, as if the throng is now a perpetuating crowd — bludgeoning open more gates to the citadel, letting more of themselves in. More sodium rushes in through the channels, while potassium, another ion, rushes out.

The process amplifies: the invading crowd of ions cracks open more gates, and even more sodium ions flow in. As more and more channels are unshackled open, a tidal wave of sodium ions enters and potassium ions exit, causing the large positive spike that Hodgkin and Huxley first saw in 1939. The net charge of the axon turns from negative to strongly positive. The cascade of conduction, once sparked, is now unstoppable: it moves farther and farther along the axon.* The process is self-propagating. One set of channels opens and closes, generating an electrical spike. That first spike opens another set of channels

*This mechanism of conduction within a neuron — sodium channels opening and flooding sodium in — does not apply to *all* neurons. Some neurons use other ions, such as calcium, as their mechanism to conduct their signals.

a few micrometers down the neuron, thereby producing a second spike just a short distance farther. Then a third spike another few micrometers down — and so forth, until the impulse has marched to the end of the axon.*

But once the spikes have sparked through the neuron, equilibrium must be restored. As the cell completes its spike in charge, the channels begin to close. The neuron begins to reset, pumping the sodium out and the potassium in, restoring equilibrium, and returning, ultimately, to its resting negatively charged state.

If you peer carefully into the squiggly depths of Cajal's drawings, you might discover yet another unusual feature in them. In the thinnest of slices that he cut and drew, and in the most delicate of his sketches, where the neurons do not overlap with each other, there is a minute gap between the end of one neuron, where its impulse terminates (i.e., at the end of its axon), and the beginning of the next

*Most neurons are coated by a sheath similar to the plastic insulation around a wire. The insulating sheath is interrupted every few micrometers along the length of the axon. These "unsheathed" parts of the neuron's membrane are where the ion channels are located. The electrical spikes are generated at these locations. The spike then moves a few micrometers down the length of the neuron to the next "unsheathed" location, where it generates the next spike.

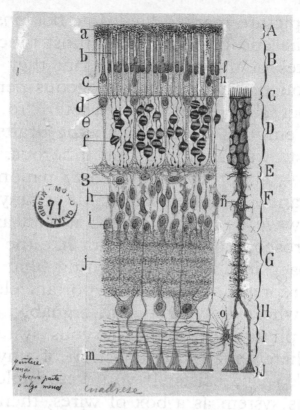

A drawing by Santiago Ramón y Cajal showing a slice of the brain with various layers of neuronal cells. Note how some neurons end in a bouton (for instance, in the layer marked [f]), which represents part of the synapse. Note also that the end of the axon often doesn't physically touch the dendrites (delicate processes of the second neuron). This empty space represents the synapse, later found to carry chemical signals (neurotransmitters) that would activate or inhibit the second neuron. These empty spaces, and their proximity to the dendritic branches of the second neuron, are especially apparent in the neurons marked in (f).

neuron, where the impulse presumably sets off a second one (i.e., at the beginning of one of its tree-like dendrites).

Look again, for instance, at the detail of part

of the picture marked "g." The *boutons* that mark the end of one nerve almost touch the dendrites of the next nerve. But they don't *quite* touch. "It takes a courageous person," the poet Kay Ryan once wrote, "to leave spaces empty" — and Cajal, the draftsman-scientist, was anything but timorous. That space — about twenty to forty nanometers in distance — is left blank. It is tiny; you could wave it away. Perhaps it's an artifact of microscopy or staining. But like the negative space in a Chinese painting, that space might represent the most important element of the whole drawing — and arguably, of the entire physiology of the nervous system. It immediately raises the question of why such a blank space exists: If you were building a nervous system as a box of wires, then what foolish electrician would leave gaps between wires? But Cajal drew exactly what he saw — the horse of observation leading the cart of theory. And yet again, as with so much in this history, it was seeing that led to unbelieving.

How does a nerve impulse, having traversed the nerve as Hodgkin and Huxley described it, move to the next nerve? In the 1940s and 1950s, the eminent neurophysiologist John Eccles, who dominated the field of neuro-transmission, argued vehemently that the only means for the signal was electrical. Neurons were electrical conductors — "wires" — Eccles posited, so why would wires ever use anything but electrical impulses to move signals from one to the other? Who ever heard of a device where the wiring changed *modes* of transmission between wire to wire? In a textbook published in 1949, Eccles's colleague John Fulton, another physiologist, wrote: "The idea of a chemical mediator released at the nerve ending and acting on a second [neuron] or muscle thus appeared to be unsatisfactory in many respects."

It might be useful to distinguish two broad kinds of problems in science. The first kind — call it the "eye in the sandstorm" problem — arises when there's such immense confusion in a field that no pattern or road map is visible. There's sand in the air everywhere you look, and a completely new pathway of thinking is needed. Quantum theory serves as a good example. In the early 1900s, as the atomic and subatomic worlds were discovered, the heuristic principles of Newtonian physics just would not suffice, and a shifted paradigm about this atomic/subatomic world was required to get out of the sandstorm.

The second is its converse: call it the "sand in the eye" problem. Everything makes perfect sense, except one ugly fact that just won't fit the beautiful theory. It irritates the scientist like a sand grain lodged in the eye — *why, why,* she asks herself, won't this one irritating contradictory fact go away?

In the 1920s and 1930s, for the English neurophysiologist Henry Dale, and his lifelong colleague, Otto Loewi, the gap between neurons had become a sand-in-the-eye problem. Yes, they agreed that transmission between neurons was electrical; there was no arguing with the signal that Hodgkin and Huxley had witnessed by eavesdropping into a neuron's impulse. But if everything was a box of wires, then what to make of the spatial interruption between nerves?

Unusual for his times, Dale, having trained in Cambridge, followed by a brief stint in Ehrlich's lab in Berlin, initially left academic posts — which he deemed too risky — to work as a pharmacologist at the Wellcome Laboratories in England. There, building on the work of John Langley and Walter Dixon, he began to isolate chemicals that had profound effects on the nervous system. Some, like the chemical acetylcholine, infused into a cat, would slow its heart rate down. Other chemicals could accelerate heartbeats. Yet others could act as stimulants to the activity of nerve cells on muscles. In 1914, Dale became the director of the National Institute of Medical Research at Mill Hill outside London. Dale speculated cautiously that these chemicals were "transmitters" of information between neurons, or between neurons and muscle cells that they enervated. Their infusion in the bodies of cats had just stimulated the nerves that enervated the heart, resulting in the heart-slowing and heart-speeding activities. And these *chemicals* restarted the next electrical impulse. Dale kept circling back to the thought. Chemicals — not just electricity — could transmit impulses from nerve to muscle and even, perhaps, from nerve to nerve.

In Graz, Austria, another neurophysiologist, Otto Loewi, also converged on the idea of chemical neurotransmitters. The night before Easter Sunday, 1920 — in the brief lull of peace between the wars — he dreamed of an experiment. He remembered very little of the dream, but perhaps it involved a muscle and a nerve in a frog. *"I awoke"* he wrote, *"turned on the light and jotted down a few notes on a tiny slip of thin paper. Then I fell asleep again. It occurred to me at 6.00 o'clock in the morning that during the night I had written down something important, but I was unable to decipher the scrawl. The next night, at 3.00 o'clock, the idea returned. It was the design of an experiment to determine whether or not the hypothesis of chemical transmission that I had uttered seventeen years ago was correct. I got up immediately, went to the laboratory, and performed a simple experiment on a frog heart according to the nocturnal design."*

Easter Sunday, a little after three in the morning, Loewi ran to his lab. First, he severed the vagus nerve of one frog, thereby isolating it from one of the principal drivers of the heartbeat. The vagus sends an impulse to slow the heartbeat — and so, as expected, the vagus-lacking frog's heart sped up. Then he stimulated the intact vagus nerve of a second frog, causing the heart to beat more slowly. This, too, was expected: stimulate the inhibitory nerve, and the heart should slow down.

But what factor in the stimulated, intact vagus nerve had stimulated the slower heartbeat? If it was an electrical impulse — as Eccles so adamantly insisted — it could never be transferred from one to another (electrical ions would diffuse and be diluted during the transfer). The trick in the experiment was in the transference: when Loewi collected the chemical substances ("the perfusate") that had emerged out of the stimulated vagus nerve, and transferred them to the *first* frog's heart — the one that had sped up — it slowed down as well. Since he'd severed the nerve, it could not have come from the frog's own vagus. It could only have come from the perfusate.

In short, some *chemical substance* — not an electrical impulse — coming out of one vagus nerve could be transferred from one animal to another to control the heart's beating rate. That chemical — a neurotransmitter — would later be identified as none other than the one that Henry Dale had identified: acetylcholine.

By the late 1940s, as an increasing body of evidence supported Dale's and Loewi's hypothesis, even Eccles was convinced. Dale and Loewi, who received the Nobel Prize in 1936, wrote that Eccles's conversion was like "the conversion of Saul on the road to Damascus when 'the sudden light shone and the scales fell from his eyes.'"

We now know that the released chemicals — transmitters — are stored in vesicles (membrane-bound sacs) at the end of the axon. Once the electrical impulse reaches the axon's end, these vesicles react to its arrival by discharging their cargo. These chemicals cross the space between one cell and the next — the synapse — and start the process of stimulation all over again. They bind to their receptors in the dendrites of the next neuron, open ion channels, and reinitiate the impulse in the second (recipient) neuron.* The signal moves

---

*A small number of neurons in animals do transmit their impulses between each other through electrical stimuli. Rather than release neurotransmitters, these neurons are directly connected electrically to each other through specialized pores called gap junctions — similar to the connecting pores found in heart cells. The proximity between the neurons is thus even greater — tenfold less than a chemical synapse. Although present, these "electrical synapses" are rare. Their principal advantage is speed — electricity moves rapidly from one cell to the next — and thus they are often found in cellular circuits where speed is essential. The sea slug (or more technically, sea

to the third cell. One chattering, thoughtful neuron has "spoken" to the next. The neuron's two countermelodies are woven together in tandem, like a child's chant: electrical, chemical, electrical, chemical, electrical.

One critical feature of this form of communication is that the synapse has the capacity not just to excite the neuron to fire — as in the example above — but it can also be an *inhibitory* synapse, making the next neuron *less* prone to excitation. A single neuron can thus have positive inputs and negative inputs from other neurons. Its job is to "integrate" these inputs. It is the integrated total of these excitatory and inhibitory inputs that determines whether a neuron will fire or not.

I have sketched the skeletal outline of how a neuron functions, and how that function relates to the building of the brain. But this is the barest of sketches. Of all cells in the body, the neuron is, perhaps, the most subtle and the most magnificent. The pared-down principle is this: we should imagine the neuron not just as a passive "wire" but as an active integrator.* And once you think of each neu-

---

hare) *Aplysia,* uses an electrical circuit to squirt ink to obscure itself from predators during its flight response.

*One philosophical and biological question arises: Why is the neuronal circuit not entirely electrical? Why *not* Eccles's idea of building a wiring system

ron as an active integrator, you can imagine
building extraordinarily complex circuits out

simply to conduct electricity, as opposed to build one
that constantly moves from electricity to chemical
signals to electricity and back again, in endless cycles?
The answer, perhaps, lies (as always) in evolution and
the development of the neural circuit. A neuronal
circuit is not just a wire transmitting signals from
the brain to the rest of the body. It is, as I wrote
above, an "integrator" of physiology. There might
be times when the heart needs to be speeded up or
slowed down. Or in a more complex realm: mood, or
motivation, might have to be regulated up or down.
If neuronal circuits were sealed in a "closed box" of
an electrical wiring system, then integrating them
with the physiology of the rest of the body would be
onerous, and potentially impossible. Furthermore,
beyond integration, chemical synapses have the
capacity to "gain" or amplify a signal or dampen it
— phenomena that make them more amenable to
the building of circuits required for the complexities
of the nervous system. Imagine your laptop: a closed
box, internally wired system. The laptop cannot
"know" when you are frustrated, or irritable, or need
to work faster, or when you need to slow down; it is a
box of electrical wiring and circuitry with no synapse
with your emotional or mental state. Organs cannot
be sealed boxes. A signal carried between neurons,
hormones, and transmitters carried by blood, or
by other neurons, must be capable of intersecting
with the other signals to modify and modulate their

of these active wires. Those complex circuits, you might surmise, could be the basis for building even more complex computational modules — those that can support memory, sentience, feeling, thought, and sensation. A collection of such computational modules could coalesce to form the most complex of machines in the human body. That machine is the human brain.

"If a subject [. . .] has a glamorous aura, if its practitioners are prizewinners who receive large grants," the biologist E. O. Wilson once advised, "stay away from that subject." For cell biologists exploring the brain, the neuron was so brazenly glamorous — so mysterious, so unfathomably complex, so functionally diverse and exorbitantly glorious in its form — that it overshadowed a companion cell that consistently lurks around it. The glial cell, or glia, was like a film-star's assistant stuck perpetually in the shadows of celebrity. Even its name, drawn from the Greek word for "glue,"

---

function, thereby integrating neuronal physiology with the rest of the body's physiology. And a soluble chemical mediator is an ideal solution. It can speed up or slow down the activity of a circuit. This is an "intelligent" laptop that is responsive and complex: tell it that you are in a bad mood, and it can give you feedback to stop sending out angry e-mails that you'll later regret. Give it a deadline, and it speeds up.

signaled a century of neglect: glial cells were considered nothing more than the glue that stuck neurons together. A small group of die-hard neuroscientists had studied these cells since the early 1900s, when Cajal had described them in slices of the brain. The rest considered them irrelevant — not the stuff, but the stuffing of the brain.

Glial cells are present all over the nervous system — in about the same number as neurons. At one point, they were thought to be tenfold more common, thereby fueling the "stuffing-of-the-brain" hypothesis. Unlike neurons, they don't generate electrical impulses, but like neurons, they are extraordinarily diverse in structure and function. Some possess fat-rich, branching extensions that wrap themselves around neurons, forming sheaths. These wrappings, called myelin sheaths, act like electrical insulation for the neurons, akin to the plastic wrapped around wires. Some are wanderers and scavengers, dedicated to clearing debris and dead cells from the brain. Yet others supply nutrients to the brain, or mop up transmitters from the neuronal synapses to reset neuronal signals.

The emergence of glial cells from the shadows of neuroscience to the center stage of inquiry represents a fascinating shift in the cell biology of the nervous system. A few years ago, I went to Harvard University to visit the lab of Beth Stevens, who has been studying glia for more than a decade. Like so many neurobiologists over the course of history, Stevens found her way to glia through neurons. In 2004, Stevens began work as a postdoctoral fellow at Stanford University to study the formation of neural circuits in the eye.

Neural connections between the eyes and the brain are formed long before birth, establishing the wiring and the circuitry that allow a child to begin visualizing the world the minute she emerges from the womb. Long before the eyelids open, during the early development of the visual system, waves of spontaneous activity ripple from the retina to the brain, like dancers practicing their moves before a performance. These waves configure the wiring of the brain — rehearsing its future circuits, strengthening and loosening the connections between neurons. (The neurobiologist Carla Shatz, who discovered these waves of spontaneous activity, wrote, "Cells that fire together, wire together.") This fetal warm-up act — the soldering of neural connections before the eyes actually function — is crucial to the performance of the visual system. The world has to be dreamed before it is seen.

During this rehearsal period, synapses — points of chemical connection — between nerve cells are generated in great excess, only to be pruned back during later development. To create a synapse, the neuron has specialized structures, often seen as tiny swellings, at its terminal end of the axon where it stores the chemicals that are emitted to transmit a signal to the next neuron. Synaptic "pruning" is thought to involve the paring back of these special structures, thereby eliminating the synaptic connection at that site — akin to removing, or cutting, the soldering joint between two wires. It's an odd phenomenon — our brains make connections in vast excess, and then we pare back the excess.

The reasons for this paring back of synapses is a mystery, but synaptic pruning is thought to sharpen and reinforce the "correct" synapses, while removing the weak and unnecessary ones. "It reinforces an old intuition," a psychiatrist in Boston told me. "The secret of learning is the systematic elimination of excess. We grow, mostly, by dying." We are hardwired not to be hardwired, and this anatomical plasticity may be the key to the plasticity of our minds.

But who does the pruning of synapses? In the winter of 2004, Beth Stevens joined the laboratory of Ben Barres, a neuroscientist at Stanford. "When I began my work in Ben's lab, little was known about how specific synapses are eliminated," she told me. Stevens and Barres focused their attention on visual neurons: the eye would be the eye to the brain.

In 2007, they announced a startling discovery. Stevens and Barres found that glial cells were responsible for this pruning of synaptic connections in the visual system. The work, published in *Cell,* garnered enormous attention, but also threw open a host of new questions. Which specific glial cell was responsible for the pruning? And what was the mechanism of pruning? The following year, Stevens moved to Boston Children's Hospital, to set up her own lab. When I visited her on an icy March morning in 2015, the lab was thrumming. Graduate students were folded over microscopes. One woman sat on her bench determinedly mashing a freshly biopsied fragment of a human brain into individual cells so that she could grow them in a tissue-culture flask.

There is something effortlessly kinetic about Stevens: as she speaks, her hands and fingers trace the arcs of ideas, forming and unforming synapses in the air. "The questions we took on in the new lab were direct continuations of the questions that I had at Stanford," she said.

By 2012, Stevens and her students had created experimental models to study synaptic pruning, and identified the cells responsible for the phenomenon. Specialized cells known as microglia — spidery and many-fingered — had been seen crawling around the brain, scrounging for debris, and their role in eliminating pathogens and cellular waste had been known for decades. But Stevens also found them coiled around synapses that had been marked for elimination. Microglia nibble at the synaptic connections between neurons and pare them away. They are the brain's "constant gardeners," as one report put it.

Perhaps the singularly striking feature of synaptic pruning is that it uses an immunological mechanism to eliminate connections between neurons. Macrophages in the immune system phagocytose — eat — pathogens and cellular debris. Microglia in the brain use some similar proteins and processes to mark synapses that are to be nibbled — except, rather than ingesting pathogens, they ingest the bits of a neuron involved in the synaptic connections. It's yet another captivating instance of repurposing: the very proteins and pathways that are used to clear pathogens in the body have been rejiggered to fine-tune connections between neurons. Microglia have evolved to "eat" pieces of our own brain.

"Once we knew about the involvement of the microglia, all sorts of questions popped up," Stevens said. "How does a microglial cell know which synapses to eliminate? [. . .] We know that synapses compete against one another, and the strongest synapse wins. But how does the weakest synapse get tagged for pruning? The lab is now working on all these questions."

The pruning of neural connections by glial cells has become the focus of intense study — and not just in Stevens's lab. Recent experiments suggest that dysfunctions in glial pruning may be related to schizophrenia — a disease where the pruning doesn't occur appropriately. Other functions of different glial cells have been linked to Alzheimer's disease, to multiple sclerosis, and to autism. "The deeper we look, the more we find," Stevens told me. It's hard to locate an aspect of neurobiology that *doesn't* involve the glial cell.

I walked away from Stevens's lab onto ice-slicked Boston streets, mentally reciting the lines of Kenneth Koch's poem "One Train Might Hide Another":

In a family one sister may conceal another,
So, when you are courting, it's best to have
    them all in view [. . .]
And in the laboratory
One invention may hide another invention,
One evening may hide another, one shadow,
    a nest of shadows.

For decades, the neuron sashayed down cell biology's runway so glamorously that it hid the glial cell. But when you are courting scientific insights, or making inventions, it's best to have all cells in view, not just the sashaying ones. The glial cell has emerged out of its "nest of shadows." Like one of its own subtypes, it has wrapped itself, sheath-like, around the entire field of neurobiology. Far from the celebrity's assistant, it is the discipline's new star.

In the spring of 2017, I was overwhelmed by the most profound wave of depression that I have ever experienced. I use the word *wave* deliberately: when it finally burst on me, having crept up slowly for months, I felt as if I were drowning in a tide of sadness that I could not swim past or through. Superficially, my life seemed perfectly in control — but inside, I felt drenched in grief. There were days when getting out of bed, or even retrieving the newspaper outside the door, seemed unfathomably difficult. Simple moments of pleasure — my child's funny drawing of a shark, or a perfect mushroom soup — seemed locked away in boxes, with all their keys thrown into the depths of the ocean.

Why? I could not tell. Part of it, perhaps, was coming to terms with my father's death a year before. In the wake of his passing, I had thrown myself manically back to work, neglecting to give myself time and space to grieve. Some of it was confronting the inevitability of aging. I was at the edge of the last years of my forties, staring into what seemed like an abyss. My knees hurt and creaked when I ran. An abdominal hernia appeared out of nowhere. The poems that I could recite from memory? I would now have to search my brain for words that had gone missing ("I heard a Fly buzz — when I died — / The

Stillness in my Room / Was like" . . . um . . . like . . . like what?). I was becoming fragmented. I was officially entering middle age. It wasn't my skin that began to sag, but my brain. I heard a fly buzz.

Things got worse. I dealt with it by ignoring it, until it had crested fully. I was like the proverbial frog in the pot that doesn't sense the incremental rise in temperature until the water starts boiling. I started antidepressants (which helped, but only moderately) and began to see a psychiatrist (which helped much more). But the sudden wave of the disorder, and its recalcitrance, mystified me. All I could feel was the "dank joylessness" that the writer William Styron describes in *Darkness Visible*.

I called Paul Greengard, a professor at Rockefeller University. I had met Greengard at a retreat in Maine several years earlier — we had recognized each other as fellow scientists, and walked for a mile on a white-pebbled, wind-swept beach, talking about cells and biochemistry — and had become close friends. He was substantially older than me — eighty-nine when we met — but his mind seemed perpetually young. We met for lunch in New York often, or took long, slow walks on York Avenue, or on the campus of the University. Our conversations ranged widely. Neuroscience, cell biology, university gossip, politics, friendships, the latest exhibition at the Museum of Modern Art, the newest findings in cancer research; Paul was interested in everything.

In the 1960s and 1970s, Greengard's experiments had led him to a novel way of thinking about neuronal communication. Neurobiologists studying the synapse had largely described the communication between neurons to be a rapid process. An electrical impulse arrives at the end of the neuron — i.e., at the axon terminal. It causes the release of chemical neurotransmitters into a specialized space — the synapse. The chemicals, in turn, open channels in the next neuron, and ions surge in, reinitiating the impulse. This is the "electrical" brain — a box of wires and circuits (with a chemical signal — a neurotransmitter — thrown in between the two wires).

But Greengard argued that there was a different kind of neurotransmission. The chemical signals sent out by one neuron also create a cascade of "slow" signals in the neuron. Neuronal signaling from one cell to the next instigates profound *biochemical* and *metabolic* changes in the recipient cell. An elaborate cascade of chemical changes is sparked off in the recipient neuron: alterations in metabolism, in gene expression, and in the nature and concentration of chemical transmitters that are secreted into the synapse. And these "slow" changes, in turn, alter the electrical conduction of an impulse from nerve to nerve. For decades, this slow cascade was considered peripheral ("Oh, he'll eventually come around," another researcher said of Greengard's work). But the biochemical alterations produced in neuronal cells — the "Greengard cascade" — is now known to permeate the brain, change the function of neurons, and dictate many of its subsequent properties.

We might, then, divide the pathologies of the brain into those that affect the "fast" signals (the rapid electrical conduction of neuronal cells), those that affect "slow" signals (the biochemical cascades that are altered in nerve cells), and those that fall somewhere in between.

Depression? When I told Greengard of the soupy fog of grief that I was experiencing, he invited me to meet him for lunch. It was the late fall of 2017. We ate at the university cafeteria — he was a slow, fastidious eater, examining each morsel on his fork as if it were a biological specimen before putting it in his mouth — and then took a walk around the campus of Rockefeller University. His Bernese mountain dog, Alpha, lumbered, drooling, by our side.

"Depression is a slow brain problem," he said.

I was reminded of the Carl Sandburg poem: "The fog comes / on little cat feet. / It sits looking / over harbor and city / on silent haunches

/ and then moves on." My brain felt perpetually fogged, as if some creature had descended on slow, silent haunches, but would not move on.

Andrew Solomon, the writer, once described depression as a "flaw in love." But in medical terms, it was a problem with the regulation of neurotransmitters and their signals. A flaw in chemicals.

"Which chemicals? What signals?" I asked Paul.

I knew that serotonin, the neurotransmitter, had something to do with it.

Paul told me the story of the origin of the "brain chemical" theory of depression. In the autumn of 1951, doctors treating tubercular patients at Sea View Hospital on Staten Island with a new drug — iproniazid — had observed sudden transformations in their patients' moods and behaviors. The wards, typically glum and silent, with moribund, lethargic patients, were "bright last week with the happy faces of men and women," a journalist wrote. Energy flooded back and appetites returned. Many patients, ill and catatonic for months, demanded five eggs for breakfast. When *Life* magazine sent a photographer to the hospital to investigate, the patients were no longer lying numbly in their beds. They were playing cards, or walking briskly in the corridors.

Researchers later discovered that iproniazid

had, as a side effect, increased serotonin levels in the brain. And the idea that depression was caused by the paucity of the neurotransmitter serotonin, in the neural synapse, gripped psychiatry. There isn't enough serotonin in the synapse, and so the electrical circuits that respond to the chemical don't get enough stimulation. The inadequate stimulation of mood-regulating neurons results in depression.

If that was all there was to depression, then increasing serotonin in the brain should solve the crisis. In the 1970s, Arvid Carlsson, a biochemist at Göteborg University in Sweden, collaborated with the Swedish pharmaceutical company Astra AB to develop a drug, zimelidine, that increased the levels of the neurotransmitter in the brain. These early drugs led to more selective chemicals that increased serotonin levels in the brain — the SSRIs, such as Prozac and Paxil.* And, indeed, some depressed patients, treated with these SSRIs, experienced profound remissions in their disease. In her bestselling 1994 memoir *Prozac Nation,* author Elizabeth Wurtzel wrote of a transformational experience. Before she

---

*Carlsson, a neurophysiologist, was already well known for his earlier work on the neurotransmitter dopamine and its effects on Parkinson's disease. His work on the chemical L-DOPA, a precursor of dopamine, led to the development of this drug to treat the movement disorder of Parkinson's disease.

began treatment with antidepressants, she floated from one "suicidal reverie" to the next. Yet, just a few weeks after starting Prozac, her life was changed. "One morning I woke up and really did want to live. . . . It was as if the miasma of depression had lifted off me, in the same way that the fog in San Francisco rises as the day wears on. Was it the Prozac? No doubt."

But the response to SSRIs was far from universally positive. And experimental and clinical results with SSRIs revealed contradictory data: in some trials, with the most severely depressed patients, there was a measurable improvement in symptoms for patients who received the drug versus those that had a placebo, while in other studies, the effect was marginal, often vanishingly so. And the time to obtain the effect — often weeks, or months — did not suggest that simply raising the level of serotonin could reset the level of some electrical circuit, and thereby cure depression. When I tried Paxil, and then Prozac, the fog in my brain did not lift. One thing was obvious: merely adjusting the level of serotonin in the synapses of the mood-regulating neurons couldn't be the simple answer.

Paul nodded in agreement. Greengard's lab at Rockefeller University had just discovered a "slow" pathway, instigated by serotonin, that might be responsible for depression. Serotonin, Greengard and other researchers had found, doesn't only act as a "fast" neurotransmitter, and depression isn't just a malfunctioning neuronal circuit that can be reset by increasing serotonin in the synapse. Rather, serotonin sets off a "slow" signal in neurons — biochemical signals that come on cat's feet — including altering the activity and function of several intracellular proteins that Greengard's lab had identified.

Paul believes that these proteins, which modify neuronal activity, are crucial to the slow signaling in neurons that regulate mood and emotional homeostasis. In his earlier work, he had shown that one such factor, called DARPP-32, is critically responsible to the way a neuron responds to another transmitter, called dopamine, which is involved in many other neurological functions, including our mind's response to reward and addiction.

"It's not just the *level* of serotonin," Paul said emphatically, jabbing his fingers in the air. The New York air was clear and bitingly cold, and his breath left a drifting trail of mist behind him. "That's way too simple. It's what serotonin *does* to the neuron. The way it changes the neuron's chemistry, and its metabolism," he said. "And that might vary from one individual to the next." He turned to face me. "In your case, there might be inputs, or genetic reasons, that makes the response more difficult to sustain or restore."

"We are looking for new drugs that will

542

affect this slow pathway," Greengard continued. He was searching for an entirely new paradigm for depression, and thereby a new way of treating this disorder.

Our walk had come to an end. He had not touched me, but I felt as if he had healed some implacable wound inside me. I waved good-bye, and watched him return to his lab. Alpha was exhausted, but Paul was energized.

Depression is a flaw in love. But more fundamentally, perhaps, it is also a flaw in how neurons respond — slowly — to neurotransmitters. It is not just a wiring problem, Greengard believes, but rather a cellular disorder — of a signal, instigated by neurotransmitters, that somehow malfunctions and creates a dysfunctional state in a neuron. It is a flaw in our cells that becomes a flaw in love.

Paul Greengard died of a heart attack in April 2019 at ninety-three. I miss him dearly.

I met Helen Mayberg at Mount Sinai Hospital in New York on a November afternoon in 2021. The wind stung my face as I walked to her office. Autumnal leaves were falling around me like snowflakes, presaging the winter. Mayberg is a neurologist who specializes in neuropsychiatric diseases and runs a center called Advanced Circuit Therapeutics. She is one of the pioneers of a technique called deep brain stimulation (DBS), in which minuscule electrodes are surgically inserted deep into very specific parts of the brain. Tiny bolts of electricity are sent through those electrodes to cells of the brain whose malfunction may be responsible for neuropsychiatric diseases. By modulating these areas of the brain with electrical stimulation, Mayberg hopes to treat the most recalcitrant forms of depression that are resistant to normal therapies. It is cellular therapy of sorts — or rather, a therapy aimed at cell circuits.

In the early 2000s, taking a radical turn from the then-current prominence of using drugs, such as Prozac and Paxil, Mayberg began to use a variety of techniques to map cellular circuits in the brain that may be responsible for depression. Deep brain stimulation had already been used to treat Parkinson's disease, and researchers had noted that it could improve the coordination of movement in affected patients. But DBS was yet to be tried in recalcitrant depression. Using powerful imaging techniques, the circuit mapping of neuronal cells, and neuropsychiatric tests, Mayberg found one area of the brain, called Brodmann area 25 (BA25), the presumptive residence of cells that seem to regulate emotional tone, anxiety, motivation, drive, self-reflection, and even sleep — the symptoms that are markedly dysregulated in depression. BA25 was hyperactive in patients with recalcitrant depression, Mayberg found. Chronic electrical stimulation, she knew, can diminish the activity of a brain area. This may sound like a contradiction but it isn't; chronic electrical stimulation of a neuronal circuit at high frequencies can depress its activity. Mayberg reasoned that electrical stimulation delivered to cells in BA25 might relieve the symptoms of chronic, severe depression.

Brodmann area 25 is not an easy place to access. If you imagine the human brain as a folded boxing glove in its punching position, BA25 sits in the deep center of the clutch, just where the middle finger might be (there's one area on either side of the brain). As one journalist described it: "In a pair of pale-pink curves of neural flesh called the subcallosal cingulate, each about the size and shape of a newborn's crooked finger, [Brodmann] area 25 occupies the fingertips." In 2003, collaborating with neurosurgeons in Toronto, Mayberg launched a trial to insert electrodes on both sides of the brain and stimulate BA25 in patients who were suffering from treatment-resistant depression. It seemed like an impossibly delicate task: tickling a newborn's fingertips to make her laugh.

There were six patients in the study: three men and three women, ranging between thirty-seven and forty-eight years old. "I remember each one of those patients," Mayberg told me. "The first was a nurse with a physical disability. She described herself as totally numb," as if permanently anesthetized. "Like many patients that I've seen before and after her, her metaphors for her illness were vertical. She was trapped inside a hole, a void. She had fallen into it. Others would talk about caves; about force fields that pushed them down into something. I hadn't realized it then, but listening to the metaphors was absolutely vital. It was the metaphors that allowed me to track whether a patient was responding or not."

To position the electrode accurately into BA25, the neurosurgeon collaborating with Mayberg, Andres Lozano, had to put a frame around the patient's head (the frame acts like a three-dimensional GPS system to track the electrode's position as the surgeon introduces it into the brain). As Mayberg tightened the clasps of the stereotactic frame, the patient looked at her blankly, registering neither fear nor apprehension. "Here she was, a woman about to have holes bored into her head and a totally untested procedure performed in her brain, and all she could register was numbness. Nothing. That's when I knew how bad it was for her."

Mayberg brought her to the operating room. "Gosh, we were so apprehensive. We had no idea *what* the stimulation might do." Would it make the blood pressure drop? Turn on a cellular circuit that neuroscientists knew nothing about? Unleash some unexpected psychosis? The surgeon bored through the patient's skull and inserted the electrodes. The position seemed right, and Mayberg turned the current on, slowly increasing the frequency.

"And then it happened," Mayberg said. "As we hit the right spot, she [the patient] suddenly said: "What did you do?"

"What do you mean?" Mayberg asked.

"I mean you did something, and the void lifted."

The void lifted. Mayberg turned the stimulator off.

"Oh, maybe I just felt something weird. Never mind."

Mayberg turned it on again.

The void lifted again. "Describe it," Mayberg urged her.

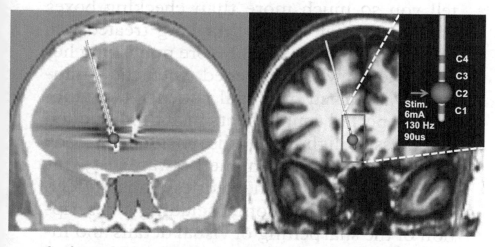

An image from Mayberg's paper showing the insertion of an electrode through the skull into Brodmann area 25 deep inside the brain. Chronic electrical stimulation of neuronal cells in this area was used to treat recalcitrant depression.

"I'm not sure I can. It's like the difference between a smile and laughter."

"That's why you have to listen to the metaphors," Mayberg told me. The difference between a smile and laughter. There was a picture in her office of a stream with a deep sinkhole in the middle, where water gushed in from all sides. "A patient sent me that picture to describe her depression." Another void, a hole. Vertical, inescapable traps. When Mayberg turned the stimulator on, the woman said she saw herself lifted out of the sinkhole and sitting on a rock above the water. She could see her former self in the hole — but she was on a rock, sitting above the hole. "These pictures, these descriptions tell you *so* much more than checking boxes on a depression scale." Mayberg treated five more patients with DBS before publishing her data. Here's what happened when the stimulator was turned on: "[a]ll patients spontaneously reported acute effects including 'sudden calmness or lightness,' 'disappearance of the void,' sense of heightened awareness, increased interest, 'connectedness,' and sudden brightening of the room, including a description of the sharpening of visual details and intensification of colors in response to electrical stimulation."

The patients were sent home, with their electrodes and batteries. At six months, four of the six continued to respond, with significant and objective measures of improvement in their mood. "The whole syndrome recovers," Mayberg later told an interviewer. "It can be extremely dramatic in some patients, while in others it takes time to become evident — as much as a year or two. Other patients appear not to be helped by [deep brain stimulation], for reasons that are still unclear."

Mayberg has treated nearly a hundred patients. "Everyone doesn't respond, and we don't know why," she told me. But in some patients, the effect is almost immediate. One woman, also a nurse, described her illness as a complete incapacity to feel emotional, or even sensory, connections. "She told me that when she held her own children she felt nothing. No sensation, no comfort, no pleasure." When Mayberg turned the DBS on, the patient turned to her and said: "You know what's strange? I feel connected to you." Another patient remembered the precise moment of the onset of her disease. "She was walking her dog along a lake, and felt that all the colors had gone. They had become black and white. Or just gray." When Mayberg turned the DBS on, the patient looked startled. "The colors just popped." Yet another woman described her response as if a season was about to turn. It wasn't spring yet, but she felt the *portent* of spring. "The crocuses. They just came out."

"There are still all sorts of mysteries that I don't understand," Mayberg continued.

"You know that depression has a psychomotor component to it — patients often cannot move. They lie in bed, they become catatonic. When we turn the DBS on, patients want to move again, but the activities that they want to do involve cleaning out rooms. Taking the trash out of the kitchen. Washing dishes. One patient, before he fell into his depression, used to be a thrill seeker. He would jump out of planes. When we turned the DBS on, he said he wanted to move again."

"What do you want to do?" Mayberg asked him.

"I want to clean my garage."

More stringent studies — randomized, controlled, multi-institutional trials — focusing on DBS for the treatment of treatment-resistant depression are ongoing. Significantly, a pivotal study (called BROADEN — for Brodmann area 25 Deep Brain Neuromodulation), initiated in 2008, was halted because early data did not show anywhere near the kind of efficacy that Mayberg had seen in her first studies. In 2013, when data were available on about ninety patients who had had DBS for at least six months, their depression scores were no better than the control group — patients who had undergone the surgery, but without the stimulator turned "on" (and worse: some patients with the implant suffered multiple complications of the surgery.

550

Some had infections, some had intolerable headaches, some reported *increases* in depression and anxiety). The trial's sponsor, a company called St. Jude's (since then, acquired by Abbott) halted it. As one journalist wrote: "The experience, which was searing, has sent [Mayberg] back to her first research principles: scrutinizing the criteria by which potential candidates for [deep brain stimulation] are selected; determining ways to improve implantation procedures to accommodate surgical teams less experienced with the procedure; improving methods of tweaking the device once implanted in a patient; and, most importantly, performing research to determine why DBS might not work in certain patients, and how to identify them before committing to surgery. The converse is also being studied: figuring out who is likely to be helped, and helped most rapidly, before the surgical procedure is performed."

Mayberg thinks there are a variety of reasons as to why the BROADEN study went wrong. "We have to find the right patient, the right area, and the right way to monitor the response. There's a lot here that we still have to learn." Some of her harshest critics remain unconvinced. ("Electroceuticals are in, and pharmaceuticals are out," a blogger wrote, with biting sarcasm that was not lost on his readers.)

But intriguingly, over many months, the

patients in the halted study who chose to keep their DBS devices "on," began to experience potent and objective responses. In a paper published in *Lancet Psychiatry* in 2017, when patients were tracked for two years rather than the six months used in the initial analysis, 31 percent had experienced a remission — nearing the remission rates that Mayberg had documented in her initial studies. And so there's renewed enthusiasm for DBS for the treatment of chronic, severe depression. "We just have to do the study the right way," Mayberg said. The field has gone through its own cyclical mood disorder: hopelessness, followed by ecstatic (and perhaps premature) optimism, then a relapse into despair. Finally, there is renewed but cautious hope again. Mayberg, it seemed to me that November afternoon, had begun to feel the portent of a season turning. There were no crocuses in the gardens outside Mount Sinai West — this was November, after all — but I knew that they would bloom in February.

Meanwhile, deep brain stimulation — "cell circuit" therapy, as I like to think about it — is being attempted for a variety of neuropsychiatric and neurological disorders, including obsessive compulsive disorder (OCD) and addiction, among others. The point is this: the electrical stimulation of cellular circuits is trying to become a new kind of medicine. Some of these attempts might succeed; some might fail. But if these attempts garner even a measure of success, they will generate a new kind of person (and personhood) — humans implanted with "brain pacemakers" to modulate their cellular circuitry. They will presumably wander about the world with rechargeable batteries in fanny packs and move through airport security saying: "I have a battery in my body with an electrode through my skull that sends impulses into the cells of my brain to regulate my mood." Maybe I will be among them.

THE ORCHESTRATING CELL
HOMEOSTASIS, FIXITY, AND BALANCE

Every cell has its own special action,
even though it derive[s] its
stimulus from other parts.
— Rudolf Virchow, 1858

Now we will count to twelve
and we will all keep still.
This one time on the face of the Earth,
let's not speak any language,
let's stop for one second,
and not move our arms so much.
— Pablo Neruda, "Keeping Still"

Most of the cells that we've encountered thus far talk to each other locally. Aside from the cells of the immune system, where a signal from one cell can summon distant cells to the site of infection or inflammation, we haven't heard much about cellular chatter that can reach across the vast expanses of an organism's body. A nerve cell whispers through the synapse to the next nerve cell. Heart cells are

554

so physically conjoined that an electrical impulse within one spreads to the other through junctions between the cells. There are lots of murmurs, but very few shouts.

But an organism cannot depend on local communication alone. Imagine an event that affects not just one organ system but the whole body. Starvation. Chronic illness. Sleep. Stress. Each individual organ might mount a particular response to this event. But — to return to Virchow's idea of the body as a cellular citizenship — the messages between organs must be orchestrated. Some signal, or impulse, must move between cells, informing them of the global "state" that the body is inhabiting. The signals move from one organ to the next, carried by blood. There must be a means for one part of the body to "meet" a distant part of a body. We call these signals "hormones," from the Greek *hormon* — to impel, or to set some action into motion. In a sense, they impel the body to act as a whole.

Tucked away in a bend in the abdomen, elbowed between the stomach and the swirls of intestines, is a leaf-shaped organ — "mysterious, hidden," as one pathologist described it. It has two lobes — called its "head" and its "tail" — connected by a body. The Alexandrian anatomist Herophilus, who lived around 300 BC, was likely among the first to identify it as a distinct organ — but he didn't name it.

(Admittedly, it's hard to attribute a discovery without a name.) The name *pancreas* appears in the medical literature in Aristotle's writing — "a so-called pancreas," he wrote, somewhat dismissively — but the word still carried no hint of its function. It was simply labeled "pan" (all) and "kreas"(flesh) — an organ of all flesh. At some point in his anatomical dissections, Galen — four hundred years after Herophilus — noted that the pancreas was filled with secretions. But he, too, was uncertain about its function — although that rarely stopped Galen from hazarding a guess. "As the vein, the artery and the nerve join behind the stomach, all these vessels are easily vulnerable at the site of their division. . . . Nature therefore has wisely created a glandular body, called pancreas, and placed beneath and around it all organs, filling the empty spaces so that none of them may tear without a support."

Centuries later, Vesalius drew among the most detailed diagrams of the organ, placing it in relation with the stomach and the liver. He noted that it looked like a "big glandular body" — and therefore must be designed to secrete *something,* as glands always do — but then, like Galen, Vesalius returned to the idea that it existed mainly as a supportive structure to keep the stomach from crushing blood vessels against the spine. In short: a cushion filled with some fluid. A glorified pillow.

There was, it seemed, only one dissenter to the pancreas-as-cushion story, and his logic was based on simple anatomical reasoning. Gabriel Fallopius, a biologist in sixteenth-century Padua, couldn't make sense of it all: in animals that walked on four legs, he argued, how could a cushion that sat behind the stomach be of much value. "[I]t would be totally useless to animals that walked in prone position," Fallopius wrote. But his perceptive line of reasoning, like the organ that he was thinking about, was soon forgotten.

The discovery of the function of pancreatic cells began, inauspiciously, with a quarrel between two anatomists that ended in a murder. The senior of the two, Johann Wirsung, was a highly respected German professor of anatomy in Padua. On March 2, 1642, at a hospital attached to the San Franciscan church, Wirsung dissected the abdomen of a hanged criminal to remove the pancreas. Several assistants helped with the autopsy, including his student, Moritz Hoffman. As Wirsung extracted the pancreas and probed the organ further, he found a previously unnoticed feature: it had a duct that ran through it — later called the main pancreatic duct — heading out toward the intestines. Wirsung published a series of medical drawings describing his discovery, and sent the plates to the leading anatomists of his time, with little comment about the function of such a duct (although one might have asked: What possible reason might an anatomical cushion have a *duct* running through it — unless the channel was carrying something?).

Wirsung's claim to his anatomical discovery may have stoked an old rivalry. On the evening of August 16, 1643, just a little more than a year after Wirsung had sent off news of the breakthrough identification of the pancreatic duct, he was walking in an alley outside his house in Padua when he was accosted and shot to death by a Belgian assassin. The reasons for this strange and brutal end to his life remain a matter of speculation — but at least one possible motive stands out. Moritz Hoffman, Wirsung's star pupil, was embroiled in a bitter dispute with his mentor. Hoffman claimed that he had shown Wirsung the existence of the pancreatic duct in a bird and Wirsung had used Hoffman's findings to identify the same duct in humans, giving no credit to his student. The master anatomist, Hoffman alleged, was actually a masterful plagiarist.

One might have thought that Wirsung's assassination would send a chill through the field of pancreatic anatomy — I cannot think of another murder incited by a duct — but interest in the function of the pancreas was ignited. If the pancreas wasn't the stomach's cushion, then what was it doing? What was

558

that duct buried inside it carrying? On March 25, 1848, a Saturday morning, Claude Bernard — the physiologist in Paris who coined the concept of "homeostasis" — performed a crucial experiment. It wasn't an easy time to concentrate on science. Revolutions were swirling around Europe. The French king had just abdicated. Armies were out on the streets, but Bernard was sequestered in his lab. He was more concerned with the restoration of equilibria in the body, with how cells could maintain a steady state (unlike Virchow, he wasn't particularly interested in keeping the state steady).

He extracted pancreatic "juice" out of a dog and added a lump of candle fat to it. In about eight hours, he found, the juice had emulsified the fat — broken it down into small particles — such that a layer of milky droplets floated above it. Building on prior work done by other physiologists, Bernard found that pancreatic juices, secreted by pancreatic cells, would also break down starches, and proteins — in essence, converting complex food molecules into simpler, digestible units. In 1856 Bernard published *Mémoire sur le Pancréas,* in which he detailed his idea that the pancreas released these juices to enable digestion. The duct that Wirsung had discovered, then, was the central channel for these juices: it relayed them into the digestive system, where they broke down complex food molecules into simple ones. He had, at last, found the function of the gland.

But the world must also be measured by eye. By the time Bernard was done with his physiological studies on the pancreas, cell theory was in its full glory, and microscopists had already trained their lenses on the micro-anatomy of the pancreatic gland. And in the winter of 1869, when the physiologist Paul Langerhans looked at thinly cut slices of pancreatic tissue through a microscope, he found that the organ hid yet another surprise. As expected, he found the ducts that Wirsung had described, surrounded by large, distended, berry-shaped cells that would later be identified as the ones producing the digestive juices — "acinar" cells, as they would eventually be called (*acinus* is Latin for "berry"). But as Langerhans's eyes turned his lens to look beyond the acinar cells, he found a second cellular structure. Nestled within the pancreas, and distinct from the acinar cells, he discovered little islands of cells that stained bright blue with a cellular dye. These cells looked entirely different from those that produced the digestive juices.* They were often widely separated from each other, seeming to float like archipelagoes of islands in the sea of the

*Islet cells produce a range of other hormones, including glucagon, somatostatin, and ghrelin.

560

pancreatic tissue. In time, the clusters would be called islets of Langerhans.

The field was again rife with questions and speculations about the function of these islands of cells. The pancreas, it seemed, was a gland that kept on giving.

In July 1920, Frederick Banting was a surgeon working in a suburb of Toronto. His practice was small and floundering, and he often sat alone in his clinic without any cases. That July, he earned $4; in September, $48 — barely enough for him to afford basic comforts, let alone continue his clinic. He drove a beat-up, fifth-hand car which ran for about two hundred fifty miles, and then fell into pieces. As debt and doubt mounted that fall, Banting took a job as a medical demonstrator — an assistant to the main lecturer — at the University of Toronto.

Late one evening in October 1920, he read an article in the journal *Surgery, Gynecology and Obstetrics* describing the development of diabetes in patients who had developed various diseases of the pancreas, including stones that had clogged up the ducts carrying digestive juices. The author noted how some of these diseases, especially the ones that caused plugged ducts, led to the degeneration of the acinar cells — the ones that produced the digestive enzymes. But the odd thing was that while the *acinar* cells typically shriveled and degenerated early when the ducts got clogged,

561

the *islet* cells managed to survive much longer. Diabetes, the author noted almost parenthetically, did not typically develop until the islet cells of Langerhans had finally degenerated.

Banting was intrigued. The function of the islands of cells was unknown; perhaps they had some relationship with diabetes. A disease of sugar metabolism — when the body cannot sense or adequately signal the presence of sugar, causing sugar to build up in blood and spill out in urine — diabetes was a mystifying illness. Banting tossed and turned through a sleepless night pondering the idea. Perhaps the pancreas — two-lobed — was, in fact, two-minded. Generations of physiologists, Bernard most prominently among them, had concentrated exclusively on its external function — the secretion of digestive juices. But what if the islet cells secreted a second chemical — an *internal* substance — that sensed and regulated glucose? The dysfunction of these cells would make the body unable to sense glucose and send the levels of the sugar spiking sky-high in the blood — the fundamental hallmark of diabetes. "I thought about the lecture and about the article and I thought about my miseries and how I would like to get out of debt and away from my worry," Banting wrote. He jotted down the vague outlines of an experiment.

If he could separate the "external" and "internal" functions — the secretions of the acinar cells from the secretion of the islets — he might find the substance responsible for sugar control, the key to understanding diabetes.

"Diabetus," he wrote that evening.

"Ligate the pancreatic ducts of dogs. Keep dogs alive till acini degenerate leaving Islets.

"Try to isolate the internal secretion of these to relieve glycosuria [sugar in the urine, a sign of diabetes]."

Karl Popper, the eminent historian of science, once recounted the story of a man in the Stone Age asked to imagine the invention of the wheel in some distant future. "Describe what this invention will look like," his friend asks. The man struggles to find words. "It'll be round and solid, like a disk," he says. "It will have spokes and a hub. Oh, and an axle to connect it to the other wheel, also a disk." And then the man pauses to reconsider what he's done. In anticipating the invention of the wheel, he has *already* invented it.

In later years, Banting would describe his notes from that October evening much like the invention of the wheel. As far as he was concerned, he had already discovered the hormone that controls sugar, later to be named insulin.

But where might he perform the experiment to prove it? Goaded into confidence by some combination of anxiety and intrigue, he soon plucked up the courage to approach one of the most senior professors in Toronto, a serious, scholarly Scotsman named John Macleod, to try an experiment on dogs.

The initial meeting, on November 8, 1920, was a disaster. They met in Macleod's office, where a desk was covered in sheaves of paper, and the professor distractedly flipped through some of them as they talked. Macleod had worked on sugar metabolism for decades and was a towering figure in the field, compassionate but exacting. He wasn't impressed. Perhaps he had expected Banting to have in-depth knowledge about diabetes and the metabolic response to sugar; instead, he found an insecure young surgeon with little research experience, talking about an organ he seemingly knew little about, and a stumbling, incoherent plan to explore it. Nonetheless, Macleod agreed to let Banting try his experiment with some dogs in his lab. Banting pestered him relentlessly; the experiment *had* to work. In the end, Macleod assigned one of two students to help Banting. The students tossed a coin to see who would work with Banting first. A gifted young researcher, Charles Best, won the toss.

Banting and Best began their main experiments in the broiling heat of the summer of 1921, operating on dogs in a dusty, unused, tar-roofed lab on the top floor of the Medical

Building. On May 17, Macleod showed Best and Banting how to perform a pancreatectomy on dogs — a two-step process that was far trickier than had been described in journal articles. The lab was barely equipped, and the heat was oppressive. Banting, dripping with perspiration, cut off the sleeves of his lab coat. "We found it next to impossible to keep the wound clean in very hot weather," he complained.

The experiment that Banting had devised was simple in principle, but devilishly complex in practice. In some dogs, they would surgically stitch the ducts of the pancreas closed until the acinar cells had atrophied and died but the islet cells remained, following the protocol of the surgical paper that Banting had read. In a second set of dogs, they would remove the whole pancreas — no acini, no islets — and so no islet "substance." By transferring the secretions between the two groups — one with islets and the ones without — they would identify the function of the islet cells, and the material substance secreted by them.

The first attempts were failures. Best killed the first dog with an overdose of anesthetic. The second died from bleeding. Then a third from an infection. It took several attempts before Banting and Best could get a dog to survive long enough to perform the very first stage of the experiment.

Late that summer, with the temperature

still rising, dog 410 — a white terrier — had its whole pancreas removed. As expected, it began to experience mild diabetes, with blood sugar levels about twice the normal level. It was far from the most extreme case, but Banting and Best decided it was good enough. The next step was crucial: they ground up the pancreas from a dog that had intact islet cells left and injected the extracted juice into the terrier. If the "islet substance" existed, it should reverse the diabetes. An hour later, the terrier's sugar normalized. They injected a second dose — and again, the sugar returned to normal.

Banting and Best repeated the experiment over and over. Remove the pancreatic extract from a dog, leaving the islets intact. Inject the extract into a diabetic dog and measure the blood sugar level. After multiple attempts, they began to gain confidence that something secreted by the islet cells was causing the blood sugar to go down. They concocted a name for a substance they had only seen in abstract. Isletin, they called it.

Isletin was a difficult substance to work with — temperamental, unstable, unpredictable. As its name suggested: insular. But Macleod began to believe that Banting and Best had actually found something important — even if the signal was weak. He soon assigned another scientist to the project — James Collip, a young Canadian biochemist who had already proved himself to be a savant of biochemical extractions. Collip's job was to purify the elusive substance, isletin, out of pancreatic extracts that Banting and Best had made.

The first attempts were crude — and their effects disappointing. Collip worked with liter upon liter of soupy, ground-up pancreatic sludge, trying to follow the hint of the sugar-reducing activity that Banting and Best had seen in dogs. Eventually he had a first sample — dilute and impure, but a preparation extracted from the pancreas nonetheless.

The crucial test of the extract was to determine if it might reverse diabetes in a human patient. It was a tense clinical experiment. The subject was Leonard Thomson, a fourteen-year-old boy in the throes of a severe diabetes crisis. Sugar poured out of his urine. His body, cachectic and starving, was a heap of skin and bones. He drifted in and out of a coma. In January 1922, Banting injected him with the crude extract, but the result was disappointing. The boy had a mild, nearly undetectable response that soon flickered away.

Banting and Best felt defeated: their first human experiment had ended in failure. But Collip forged ahead with purer and purer extractions. If the "substance" existed anywhere in the pancreas, he would find a way — some method — to purify it. He sourced new solvents, found novel methods of distillation, varied temperatures, and changed concentrations of alcohol to solubilize the material, until he'd found a highly purified extract.

On January 23, 1922, the team returned to Thompson. The boy, still desperately ill, was injected again with Collip's highly purified extract. The effect was immediate. The sugar in the blood dropped dramatically. It disappeared from the urine. The sweet, fruity smell of ketones in his breath, an ominous warning of a body in severe metabolic crisis, dissipated. The boy, semicomatose, woke up.

Banting now wanted more of the extract to treat more patients. But Collip, a latecomer, refused to give the team the protocol for purification; after all, hadn't *he* solved the puzzle? Psychologically and physically strained to the point of breaking, Banting, the man who had hunted this substance, Ahab-like, for four years walked into Collip's lab and grabbed his overcoat. He threw Collip onto a chair, and clasped his hands around Collip's neck, threatening to throttle him. Had Best not intervened at the right moment and pulled the two men apart, the pancreas would have been responsible for not one, but two murders.

In the end, a fragile truce was reached between Collip, Best, Macleod, and Banting. They licensed the purified substance to the university and set up a lab to produce more of it to treat patients. Its name was changed from isletin to insulin. A larger clinical trial proved an equal and dramatic success: the sugar levels in patients injected with insulin dropped precipitously. Semicomatose children, with ketoacidosis, woke up. Cachectic, wasted bodies gained weight. Soon, it was evident that insulin was the master regulator of sugar metabolism — the hormone responsible for sensing sugar and sending a signal to the cells throughout the body.

In 1923, just two years after Banting and Best had performed their first experiment, Banting and Macleod were awarded the Nobel Prize for the discovery of insulin. Banting was so perturbed at the choice of Macleod and the exclusion of Best that he declared that he would split his prize privately with Best. Macleod countered that he would split his half with Collip. History has, perhaps appropriately, pushed Macleod, who had alternated between skeptical and supportive through the whole project, to the background. The discovery of insulin is now widely attributed to Banting and Best.

Insulin, we now know, is synthesized by a particular subset of islet cells in the pancreas — beta cells — and its secretion is stimulated by the presence of glucose in the blood. It then travels all over the body. Virtually every tissue responds to insulin: the presence of sugar means that the extraction of energy, and everything that flows from energy — the synthesis of proteins and fats, the storage of chemicals for future use, the firing of neurons, the growth of cells — can proceed. It is, perhaps, among the most important of the "long range" messages that acts as a central coordinator and orchestrates metabolism all through the body.

Type 1 diabetes, which affects several million patients around the globe, is a disease in which immune cells attack the beta-islet cells of the pancreas. Without insulin, the body cannot sense the presence of sugar — even if there is enough of the chemical in the blood. The cells in the body, imagining that the body has no sugar, begin to scramble around for other forms of fuel. The sugar, meanwhile, all readied up but with nowhere to go, spikes threateningly in the blood, and spills into the urine. Sugar, sugar everywhere — but not a molecule in cells to satiate them. It is one of the defining metabolic crises of the human body — cellular starvation in the presence of plenty.

For decades since the discovery of insulin, the lives of millions of type 1 diabetics have been transformed. When I trained in medicine in the 1990s, patients used to check their blood sugar levels using drops of blood tested on monitors, and inject themselves with the right dose of the drug based on a chart. Now, there are implantable monitors that check your blood glucose continuously — continuous glucose monitors, or CGMs — and pumping machines that automatically dispense the correct dose of insulin. It is a closed loop system.

But the dream of diabetic researchers is to make humans with bioartificial pancreases. If beta cells could somehow be cultured inside an implantable sac, and inserted into a human being, then the cells might function autonomously: sensing glucose, secreting insulin, perhaps even dividing to form more beta cells. Such a device would require a supply of blood to bring nutrients and oxygen in, and an exit to send the insulin out. And most importantly, it would have to be protected from the immune attack — i.e., the autoimmune killing of the islet cells by the person's immune system — that had started the diabetes in the first place.

In 2014, a team led by Doug Melton at Harvard published a method to take human stem-like cells and coax them, step by step, to form insulin-producing beta cells. Melton began his academic career as a developmental and stem cell biologist, studying the signals that an embryo uses to make organs and how stem cells respond to these signals.

Then both of Melton's children developed type 1 diabetes. When Melton's son, Sam, was six months old, he began shaking and vomiting — ultimately becoming so sick that he was rushed to the hospital. His urine was dense with sugar. Melton's daughter, Emma, born a few years earlier, also eventually developed the illness. For a while, as Melton told a journalist, Melton's wife *was* their children's pancreas — pricking their fingers four times a day, checking glucose levels, and injecting them with the right dose of insulin. But over the years, that personal saga has turned Melton into a diabetes researcher on a compulsive quest to make human beta cells and implant them into the body — a bioartificial pancreas.

Melton's strategy was to recapitulate human development. Every human begins his or her life as a single pluripotent cell (i.e., a cell capable of giving rise to all tissues in the body) and eventually buds out a pancreas fully capable of sensing sugar and developing insulin-producing islet cells. If it could be done in utero, Melton believed it could be done in a petri dish with the right factors and steps. Over the next two decades, many scientists in Melton's lab worked to coax human pluripotent stem cells to form islet cells. But they would inevitably get stuck in the penultimate stage before becoming mature.

One evening in 2014, a postdoctoral researcher named Felicia Pagliuca stayed up late in Melton's lab, performing experiments. Her husband had already called her, asking her to come home for dinner, but Pagliuca had one last experiment to finish. She added a dye to the stem cells that she had coaxed along the islet cell pathway, hoping it would turn blue — a sign that they were making insulin. At first, she saw a faint blue tinge — but then it grew darker and darker. She looked again, and then again to confirm that her eyes weren't misleading her. The cells had made insulin.

Melton, Pagliuca, and their team reported their success that year. The cells that they had generated, Melton's team wrote, "express markers found in mature beta cells, flux calcium in response to glucose [a sign that they've detected the sugar], package insulin into secretory granules, and secrete quantities of insulin comparable to adult beta cells in response to multiple sequential glucose challenges in vitro." They are as close as researchers have gotten to making human beta cells that survive and function and can be grown into millions of cells.

The insulin-secreting cells made out of stem cells have now found their way to clinical trials. One strategy is to take these millions of islet cells and infuse them directly into a patient's body while giving the patient immunosuppressants to prevent the rejection of the cells. One of the first patients to receive the infusion, a fifty-seven-year-old type 1 diabetic from Ohio named Brian Shelton, seems to have achieved sugar control — a crucial first step in measuring the effectiveness of the whole strategy. More patients are being rapidly enrolled in the trial.

A potential next step is to encapsulate these cells into a device that is immune-protected, stable in the body, and that can act as entry and exit slots for nutrients. A team involving Jeff Karp, also at Harvard, is devising tiny, implantable machines that might achieve these goals.

Sometime in the future we might encounter a new kind of diabetic patient with no injections, batteries, or beeping monitors (instead, the batteries and monitors will be worn, like those getting deep brain stimulation for Parkinson's disease or depression). After so many winding errors and misconceptions, one murder, one throttling, one Nobel Prize split into four — and that unforgettable moment of a blue stain spreading on a cluster of cells — we may have solved the conundrum of the two-minded organ and made it into a bioartificial self. Once that neo-organ integrates into our bodies, the pancreas — the central coordinator of metabolism, the maker of the hormone to which all tissues respond — would live up to its Greek name. It would become part of us, a new form of "all flesh."

You go out for dinner one evening. Perhaps in Venice, Italy — a resplendent restaurant near the Giardini, the public gardens of the city near the banks of the Bacino di San

Marco. You begin with baccala manticato, the whipped salt-cod concoction that the Venetians stole from the Portuguese and appropriated into a national culinary monument. There's a mound of toasted bread and a gigantic bowl of rigatoni to follow, and enough Chablis to fill a small canal.

Perhaps you don't realize as you walk back that a cellular cascade has been activated. Leave aside digestion for a moment. It is the *metabolic* cascade — and the restoration of chemical balance — that is the small miracle of cell biology that unfolds in your body as you walk back to your hotel.

The carbohydrates from the bread and rigatoni are digested into sugars — ultimately, into glucose. The glucose is picked up from the intestines, absorbed into blood and moved into circulation. When the blood reaches the pancreas, it senses the spike in glucose, and sends out insulin. The insulin, in turn, moves the sugar from the blood into all the cells of your body, where it can be stored, if needed, or used for energy, as needed. The brain is the ultimate recipient of these signals: if the sugar drops too low, it reacts by sending out converse signals. Yet other hormones, secreted by different cells, send signals to release stored sugars into the blood. The stores come from liver cells, which respond, at least transiently, by releasing their stockpiles of stored glucose to restore equilibrium.

But what about all the salt? Your body was just assaulted by sodium chloride. If there was no restoration, day upon day, your blood would slowly become ocean water, with the salinity of the canal that you just sat next to. And so, perhaps unaware of it, you feel a pang of thirst. You drink one, two, perhaps three glasses of water. And now a second metabolic sensor kicks in. To understand how the salt gets dispensed, we need to understand the cell biology of another orchestrating organ — the kidney.

Deep within the kidney lies a multicellular anatomical structure called the nephron. Each nephron — first identified by cellular anatomists in the late 1600s — can be imagined as a mini-kidney. The nephron is the site where the blood and kidney cells meet, and the first drops of urine are generated. The circulation of blood carries the excess salt, dissolved in plasma, to the kidneys. The blood vessels split and split further to form finer and finer-walled arteries. Finally, the thinnest arteries whirl around themselves to form a thin-walled nest of cells — so delicate and porous that the liquid, noncellular part of blood — plasma — can leak out of the vessels into the nephron — that is, into the mini-kidney.

The liquid then moves through a membrane that surrounds the vessels and, finally, through a wall of specialized kidney cells that form a fenestrated barrier. Each one of these transitions — out of the blood vessel, in through a membrane, and through the walls of the kidney cell — serves as a filter. Large proteins and cells are selectively left behind, only allowing small molecules, such as salts, sugar, and metabolic waste products to pass through. The liquid — urine — then moves into a collecting bowl, and then into a system of cell-lined tubes called the renal tubules. The tubules connect to pipes that drain into larger collecting tubes, like tributaries joining to form a river, until they converge on the large duct — the ureter — that brings urine into the bladder.

Back, then, to the sodium that you consumed. The excess sodium causes a hormonal system, regulated by the kidney and the adrenal gland, which sits just above the kidney, to decrease its signal. The cells in the tubule respond to these changes by excreting the excess sodium into urine, thereby discarding the salt and returning the sodium level to normal. The salt is also detected by specialized cells in the brain that monitor the overall concentration of salts in the blood, a property called osmolality. These cells, sensing a high osmolality, send out another hormone to cause the cells in the kidney to retain more water. As more water is absorbed into the body, the sodium level in the blood is diluted, and the concentration is reset — albeit at the cost of retaining more water overall. You might find your feet swollen the next morning — but you might well argue that the baccala was worth the shoes that don't fit.

What, then, of the *non*-waste products? Why don't we lose essential nutrient molecules or sugars every time we make urine? The sugar and other essential products are *reabsorbed* into the body by the cells in the collecting duct through special channels. The answer returns us to one of the strange strategies often used by cells: we generate excess, and then pare it back to restore normalcy.

And the alcohol? The last cell type in this trio of orchestrating cells (or quarto, if you count the brain) are the cells of the liver — hepatocytes. The liver cell is functionally specialized for both storage and waste disposal, secretion, protein synthesis, among dozens of many other functions. But waste disposal is so essential for the body — and the liver so deeply specialized for it — that it's worth its own focus.

We think of metabolism as a mechanism to generate energy. But flip it around, and it's also a mechanism to generate waste. The kidney dispenses some of this, as above, through urine. But the kidney is not a detoxifying plant: its master plan for waste is to merely wash it away down a sewer.

Liver cells, in contrast, have evolved dozens of mechanisms to detoxify and dispense waste. In one system, it generates a sacrificial molecule that attaches itself to a potentially toxic one and renders it inactive; both the sacrificial molecule and the toxin are then broken down further until the poison is detoxified. For other waste products, it destroys the chemical using specialized reactions. Alcohol, for instance, is detoxified in a series of reactions, until it is broken down into a harmless chemical. There are even specialized cells within the liver that eat dead or dying cells — red cells, for instance. Reusable products from the dead cells are recycled. Others are dispensed into the intestines or excreted by the kidney. Liver cells, in short, are also part of the "orchestra" of regulation and constancy — except, unlike pancreatic islet cells, they perform their regulation locally. The pancreatic cell maintains metabolic constancy, the kidneys salt constancy. The liver maintains chemical constancy.

In the early spring of 2020, the labs were closed because of Covid metastasizing through New York and the world. I was seeing limited numbers of patients in the hospital — partly because, yet unvaccinated (the vaccines were yet to be approved), I feared transmitting an infection to my chemotherapy-receiving patients whose immune systems could not battle a lethal virus. I still tended to the sickest, the most vulnerable. The oncology wing of the hospital went on heroically, kept alive by nurses.

When not in the hospital or the lab, I spent my weekends in a house on a bluff that overlooked Long Island Sound. By the early light of morning, with the geometric crosshatches of sunbeams streaming across the lawn like rays of light coming out a prism, I would watch two ospreys that had come to nest there. They would fly up above the ocean, and then, miraculously, seem to sit still in midair — even while capricious gusts of wind might arrive from any direction. The writer Carl Zimmer described the same phenomenon in bats. Their miraculous fixity in midair, he wrote, was another form of homeostasis at work.

The liver, pancreas, brain, and kidney are four of the principal organs of homeostasis.* The pancreatic beta cells control metabolic homeostasis through the hormone insulin. The kidneys' nephrons control salt and water, maintaining a constant level of salinity in the blood. The liver, among many of its functions, prevents us from being soused in toxic products, including ethanol. The brain

*Note that I wrote "principal." Every cell in every organ in the body has some form of homeostasis. Some are unique, while some are common to all cells, as we discussed in part one.

coordinates this activity by sensing levels, sending out hormones, and acting as a master orchestrator of balance-restoration.

Stillness. Counting to twelve. *"Now we will count to twelve and we will all keep still."* Perhaps the most underrated of our qualities.

In the end, of course, we will all be blown out of position by some fierce gust of pathological aberration in one of these systems of cells. But the four guardians of homeostasis, working together, like systems of wing and tail feathers, adjusting ever so slightly as the winds change direction, keep the organism in position. When those systems succeed, there is fixity. There is life. When they fail to operate, the delicate balance is thrown off. The osprey can no longer stay still.

PART SIX
REBIRTH

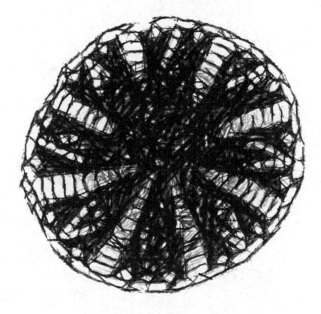

Old age is a massacre," Philip Roth wrote. But in truth it is a maceration — the steady grind of injury upon injury, the unstoppable decline of function into dysfunction, and the inexorable loss of resilience.

Humans counter this decline by two overlapping processes — repair and rejuvenation. By "repair," I am referring to the cellular cascade that begins upon injury. It is typically marked by inflammation, followed by the growth of cells to seal the damage. "Rejuvenation," on the other hand, refers to the constant replenishment of cells, typically from a reservoir of stem or progenitor cells, in response to the natural death and decay of cells. Both diminish — either in the number of stem cells or in their function — drastically with age. The rate of repair decelerates. The reservoir of rejuvenation decays.

One of the remaining mysteries of cell biology is why, in adulthood, some organs can repair, and some rejuvenate, while others lose both capacities. Blood-forming stem cells can

585

completely regenerate a blood system. But the death of a neuron almost never regenerates a neuron in its place. Other organs mix and match the two processes. Bone, perhaps, is among the most complex — it uses repair and rejuvenation to combat decay. Cells that can repair bone remain throughout adulthood — albeit much diminished in function with age. But the cells that form cartilage in joints decay dramatically with age. My mother broke her ankle and the crack healed, if slowly. But the joints in her knees are irreversibly swollen, never to return to the suppleness of her childhood, when she clambered so easily up guava trees.

Finally, we turn to a cell type that defies decay — the cancer cell, or rather, various cancer cells. Is that because some cancers behave like organs that possess reservoirs of rejuvenation — cancer stem cells? Or is it just that the cells keep giving rise to cells — as they do when an organ repairs itself after injury. Is cancer a disease of repair or rejuvenation, or both?

Yet another persistent mystery of cancer is why some malignant cells grow in certain organs, while refusing to grow in others. Is there something about the surrounding milieu of cells that support or reject cancers? The nutrients they supply?

Clearly, there is an understanding of the cellular ecology of cancer that we are missing. And so, we end our story of cells by borrowing concepts from ecology. We've learned about

586

cells, systems of cells, organs, and tissues. But there is yet another layer of organization left to learn: the ecosystems of cells. It is the music that drives the complexity of cellular physiology — and, conversely, the playlist of malignant pathology — that remains one of the unsolved puzzles of cell biology.

THE RENEWING CELL
STEM CELLS AND THE BIRTH
OF TRANSPLANTATION

"He not busy being born is busy dying"
[. . .] You are busy being born the whole
first long ascent of life, and then,
after some apex, you are busy
dying: that's the logic of the line.
— Rachel Kushner, *The Hard Crowd*

Stem cells don't simply transform
themselves into other cells
(a process called differentiation) to build
what the body needs and then,
their work done, quietly disappear. They are
more than the progenitors
of other cells. They also replicate
themselves—in an unrefined,
undifferentiated state—so that they can
stick around to answer
the call later when the blood
system needs rebuilding.
— Joe Sornberger, *Dreams*
and Due Diligence

On August 6, 1945, at about eight fifteen in the morning, thirty-one-thousand feet above the Japanese city of Hiroshima, an atomic bomb nicknamed Little Boy was dropped from an American military aircraft, a B-29 bomber nicknamed the *Enola Gay*. The bomb took about forty-five seconds to descend, and then detonated in midair, nineteen hundred feet above the Shima Surgical Hospital, where nurses and doctors were at work, and patients still in their beds. It released about the energetic equivalent of fifteen kilotons of TNT — about thirty-five thousand car bombs going off at once. A circle of fire, more than four miles in radius, spread out from the epicenter, destroying everything in its wake. The tar on the streets boiled. Glass flowed like liquid. Houses were flicked into oblivion, as if by a giant, incinerating hand. Outside the stone steps of Sumitomo bank, a man or woman who was vaporized instantly left a shadow of herself on the stone that had been blistered white by the conflagration.

The waves of death that followed had three crests. About seventy thousand to eighty thousand people — nearly 30 percent of the city's population — were broiled to death almost instantly. "I was trying to describe the mushroom [cloud], this turbulent mass," one of the tail gunners of the aircraft wrote: "I saw fires springing up in different places, like flames shooting up on a bed of coals [. . .] it looked like lava or molasses covering the whole city, and it seemed to flow outward into the foothills where the little valleys would come into the plain, with fires starting up all over."

Then came a second wave — from radiation sickness (or "atomic bomb sickness" as it was initially called). As the psychiatrist Robert Jay Lifton noted, "Survivors began to notice in themselves a strange form of illness. It consisted of nausea, vomiting, and loss of appetite; diarrhea with amounts of blood in the stools; fever and weakness; purple spots on various parts of the body from bleeding into the skin . . . inflammation and ulceration of the mouth, throat, and gums."

But there was yet a third wave of devastation to come. Survivors who received the lowest doses of radiation began to develop bone marrow failure, resulting in chronic anemias. Their white cell counts sputtered, then declined and collapsed over a few months. As the scientists Irving Weissman and Judith Shizuru put it, "those who died from the lowest lethal dose of irradiation almost certainly died of hematopoietic [blood production] failure." It wasn't the acute death of blood cells that killed these survivors. It was the inability to sustain the constant *replenishment* of blood; a collapse of homeostasis of blood. The balance between regeneration and death had been tipped. To paraphrase Bob Dylan: the cells not busy being born were busy dying.

Macabre as it was, the bombing of Hiroshima provided proof that the human body possesses cells that continuously generate blood, not just in the moment, but for prolonged periods of time, through adulthood. If these cells are killed — as they were in Hiroshima — the entire blood system would eventually falter, unable to balance the rate of natural decay with the rate of rejuvenation. In time, these cells, capable of rejuvenating blood, would be termed "blood-forming" — or "hematopoietic" — "stem and progenitor cells."

Our understanding of stem cells was born of a paradox: an unfathomably violent attack in an attempt to restore peace at the end of an unfathomably violent war. But stem cells are themselves a biological paradox. Their two principal functions seem, at face value, to be precisely opposed to each other. On one hand, a stem cell must generate functional "differentiated" cells; a blood stem cell, for instance, must divide to give rise to the cells that form the mature elements of blood — white cells, red cells, platelets. But on the other hand, it must also divide to replenish itself — i.e., a stem cell. If a stem cell achieved only the former function — differentiation into mature, functional cells — the reservoir of replenishment would eventually be exhausted. Over the course of adulthood, our blood counts would keep falling year after year, until there were none left. In contrast, if it only achieved its own replenishment — a phenomenon termed "self-renewal" — there would be no production of blood.

It is the acrobatic balance between self-preservation and selflessness — self-renewal and differentiation — that makes the stem cell indispensable for an organism, and thereby enables the homeostasis of tissues such as blood. Cynthia Ozick, the essayist, once wrote that the ancients believed that the moist track of slime left behind by a snail in its trail was part of the snail's self. Bit by bit, as the slime rubs off, the snail is depleted, until the organism disappears altogether. A stem cell (or in the snail's case, a slime-producing cell) is a mechanism to ensure that the moist track of slime — i.e., new cells — are generated constantly and that the snail does not rub itself into oblivion.

Allow a peculiar analogy. It's tempting to think of the stem cell as an ancestral great-great-great-grandfather or -grandmother. Its progeny give rise to more progeny, resulting in a vast lineage that arises from a single great-great-great-grandfather cell.

But to be a true *stem* cell, this must be the oddest of great-great-great-grandfathers. It must also give birth to a copy of itself that can maintain the replenishment of the lineage.

This great-great-great-grandfather, besides birthing a child (that will go on to establish an enormous lineage), must also birth a copy of itself — an eternally alive twin. And once this self-renewing great-great-great-grandfather is born, the process of regeneration can become limitless. There's a mythic quality to such a setup — and indeed, in myths, one often finds attempts by powerful kings or Gods to make backup twins (dolls, voodoo objects, souls secretly stored in animals, twin person-hoods locked in amulets) to reboot themselves and their clan should something terrible happen. Like most real stem cells, these mythic doubles usually lie dormant — quiescent — until an injury wakes them up. And then they awake and reseed the whole clan. It results not in birth but rebirth.

Do all adult organisms have stem cells? Do such cells exist in every tissue — or in just some? In science, as in fashion, trends are intensely in vogue for a moment, and then discarded the next. In 1868, the German embryologist Ernst Haeckel had proposed that all multicellular organisms arose out of a single cell — the first cell. By logical extension, that first cell must have the properties of differentiating into every cell type — blood, muscle, guts, neurons. It was Haeckel who used the term *Stammzellen* — stem cell — to describe this first cell. But Haeckel's use of

594

the term "stem cell" was still loose: the first cell certainly generated a whole organism, but did it generate a copy of itself?

For a while in the 1890s, biologists debated whether such a totipotent cell — one capable of giving rise to every tissue in the body — might be hidden away somewhere in the adult organism (in a sense, females possess a *precursor* of such a cell — the egg. Once fertilized, it can give rise to all the tissues of a new organism, although, sadly, it doesn't regenerate the mother). In 1892, the zoologist Valentin Hacker, studying the *Cyclops,* a multicellular freshwater flea named after the Greek monster because its body resembles a single eye, found a cell within it that divided into two. One daughter cell gave rise to layers of tissue that formed part of the organism, while the other became a germ cell — a cell capable, in the future, of giving rise to all the tissues of the organism, and thus a stem cell. Hacker also termed these cells *Stammzellen,* borrowing the term from Haeckel. But unlike Haeckel, Hacker's use of the term was more precise: here was a first cell that divided to give rise to one daughter that gave rise to the body of the *Cyclops,* and another cell, Hacker proposed, that could generate a new *Cyclops* all over again.

But mammals? Of all organs and tissues found in mammalian organisms, the one place to look for such cells might be blood. Red blood cells, and some white blood cells (neutrophils, for instance) are constantly dying and being replenished; if a stem cell were to exist, where else would it be but in blood? The cytologist Artur Pappenheim, studying the bone marrow in the late 1890s, had found islands of cells where multiple cellular types of blood were being regenerated — as if a single central cell was capable of generating multiple cell types. In 1896, the biologist Edmund Wilson used the phrase "stem cell" to describe a cell capable of differentiation and self-renewal, just as Hacker had observed in *Cyclops*.

As the idea of a "stem cell" gathered popularity in biology in the early 1900s, it became more hierarchically defined. A totipotent cell could give rise to all types of cells, including every tissue in the organism (including the placenta, the umbilical cord, and the structures that nourish and protect the embryo). Lower down in the rung of renewal was a "pluripotent" cell — a cell capable of giving rise to *nearly* all cell types in the organism (i.e., all tissues in the fetus — brain, bones, guts — except the ones that form the placenta and the supportive structures that link the fetus to the mother). And even lower still in the hierarchy was a "multipotent" cell — a cell capable of giving rise to all cell types in a *particular kind* of tissue, such as blood, or bone.

Between the 1890s and the early 1950s,

some biologists argued that the various elements of blood — white cells, red cells, and platelets — all emerged out of the same "multipotent" stem cells resident in the bone marrow. Others suggested that each cell type emerged from a unique stem cell. But without formal proof in either direction, interest in this mystical blood stem cell fell out of vogue. By the 1950s, references to stem cells had largely vanished from the biological literature.

In the mid-1950s, two Canadian researchers, Ernest McCulloch and James Till, launched a collaboration to understand the physiology of blood cell regeneration after radiation exposure. An unlikely duo, Till and McCulloch came from very different backgrounds. McCulloch — stout, short, hefty — was the scion of an "Old Money Toronto" family, as one biographer described it. He had a lively, wandering intellect: "[h]e thought tangentially, often playing connect the dots." McCulloch trained in internal medicine at the Toronto General Hospital. He was recruited, briefly, as the head of hematology at the Ontario Cancer Institute in 1957 but found himself bored with the humdrum practice of medicine and soon left to become a full-time researcher.

Till, in contrast, was a tall, skinny farmhand from Saskatchewan with a PhD from Yale in biophysics. He had an arrow-like mind, was mathematically inclined, and relentless about details. To McCulloch's inventive madness, he brought method. There was complementarity, too, in their interests and expertise. Till had trained in radiation physics; he knew how to calibrate radiation and measure its effects on the body (he had been a student of the notoriously exacting Harold Johns, who had worked on the effects of cobalt radiation). McCulloch was a hematologist, with an interest in blood and its genesis.

In 1957, when they had launched their collaboration, Toronto was a sleepy, regional town. Scientific news trickled in. But in the aftermath of the bomb, there had been an international stir to study whether bodies, and organs, could be protected from the lethal effects of irradiation. Till and McCulloch were particularly interested in the effect of radiation on blood. But how might they quantify that effect? When they exposed a mouse to a large dose of radiation, they found that the genesis of blood would stop in about two and half weeks, and the mouse would die — akin to the victims of the third wave of death in Hiroshima. The only way to rescue the mouse was to transfer marrow cells from another mouse into it. By transferring cells from another mouse's bone marrow (the organ where blood is generated) Till and McCulloch could rescue the irradiated mouse, and the animal would resume its genesis of blood. It was this crude assay — the rescue of a near-dead animal — that would open a new frontier in stem cell biology.

On a wintry Sunday afternoon in December 1960, a few days before Christmas, Till left his house in Toronto to look at the results of an experiment in his lab. The experimental setup was simple: mice had been irradiated with doses high enough to kill their intrinsic genesis of blood, and transplanted with marrow from other mice. Each mouse had received a different number of marrow cells — a titrated dose — to rescue it from death.

Till sacrificed the mice and prepped them for autopsies, looking methodically through each organ. Marrow. Liver. Blood. Spleen. Superficially, there was not much to see. But when Till looked carefully at the spleen, he found tiny white bumps — colonies. Mathematically minded, he counted the total number of colonies in each mouse and plotted them on a graph. The "bumps" lined up nearly exactly with the number of transplanted marrow cells. The more cells they transplanted, the more colonies were formed. What could that mean? The simplest answer was that these colonies were not just a random count of transplanted cells that happened to have encountered the spleen, but rather a quantitative measure of a special kind of cell. This cell had to possess the intrinsic property to form colonies in the spleen — a sign of regeneration — and it had to exist in a fixed ratio in the bone marrow (hence the more cells that were transplanted, the more bump-like colonies were produced).

Till and McCulloch would soon discover that each bump — a colony — was a regenerating nodule of blood cells. But not any regenerating nodule. These colonies were making *all* the active elements of blood — red cells, white cells, and platelets. And they were extraordinarily rare: about one colony for every ten thousand cells from the marrow.

Till and McCulloch published their data in a paper with a prosaic title ("A Direct Measure of the Radiation Sensitivity of Normal Mouse Bone Marrow Cells"; note that there was not even a passing mention of "stem cells") in an academic journal of radiobiology. "You have to remember that it was a fairly small group that was interested in this kind of work at that point," Till wrote. "This was way before all the excitement about what has happened over the next decade or so." But instinctively, Till and McCulloch knew that their result had revealed a principle of enormous significance: a tiny fraction of transplanted marrow cells, like intrepid founders crossing an ocean on a makeshift boat, had migrated to the spleen and set up isolated colonies to regenerate blood — *all* the major cellular elements of blood. As Joe Sornberger, the science writer, described it, "The paper represented an entirely new way of looking at how the body makes blood, not to mention presenting a raft of potential implications for other biological rethinks — as in, if that's true for blood, then how does the body make heart muscle, or brain tissue? But it did not immediately knock the scientific world off its axis and went largely unnoticed by the larger biological community."

Working with Lew Siminovitch and Andrew Becker in the early 1960s, Till and McCulloch deepened their studies on these colony-forming blood cells. First, they established that some colonies were producing all three kinds of cells — red cells, white cells, and platelets — the definition of a "multipotent" cell. A year later, they proved that each colony had arisen from a single "founder" cell. And finally, when they isolated the colonies of cells from the spleen and transplanted them into radiated mice, they found that they could recapitulate their capacity to generate additional multipotent colonies — the hallmark of self-renewal.

They had, in effect, discovered a cell capable of giving rise to not just one but multiple lineages of blood cells — red cells, white cells, and platelets: the blood-forming, or hematopoietic stem cell. Irving Weissman, now the director of Stanford's Stem Cell program, was a student when he read Till and McCulloch's first paper on radiation sensitivity. "The real discovery," he said later, "was to turn things around from 'the bone marrow is a black box;

we don't know anything about it,' to 'the bone marrow has discrete cells that can make multiple different cell types.'"

Weissman remembers how the experiment ricocheted through the world of cell biology. Till and McCulloch had "reset how people think of blood, the vital source of life." "Until the Till and McCulloch experiments, people thought that every distinct cell type in blood came from a unique parental cell," he continued. "But Till and McCulloch proved just the opposite. The red cell 'mother' and the white cell 'mother' and platelet 'mother' cell," Weissman told me, "all arose from the same stem cell. And these stem cells kept giving rise to more and more cells — red, white, platelets — until an entirely new blood system had been created. The effect on the field of bone marrow transplantation was extraordinary. If transplanters could find this cell, they could regenerate the entire blood system." They could build a human with new blood out of that stem cell.

And so Weissman went looking for that cell. Where did these stem or progenitor cells reside? What was their behavior, their metabolism, their size, shape, color? Inspired by the Till and McCulloch experiments, Weissman began to use a technique developed at Stanford by the wife-and-husband team of Leonore and Len Herzenberg, called flow cytometry, to purify the cells. Simplified to its essence, flow cytometry is like coloring cells with crayons — each cell a different permutations of colors (one: blue and green; another: green and red) based on the permutations of proteins on their surface. The "crayons" are antibodies, carrying chemicals that fluoresce in different colors, that recognize the different proteins on the surface of a cell. A machine can be used to separate cells based on their staining by different permutations of colors.

Weissman went through dozens of permutations and finally found one combination of markers that could purify mouse blood stem cells taken from bone marrow. As Till and McCulloch had predicted, they were rare — with a frequency of fewer than one in ten thousand cells — but exquisitely potent. Eventually, as Weissman's technique was refined and more markers were added, researchers could isolate a *single* blood stem cell and regenerate the entire blood system of a mouse. And they could extract a single such cell out of *that* mouse and regenerate the blood of a second mouse. In the early 1990s, Weissman, and other researchers, used the same technique to identify human blood–forming stem cells.

Mouse and human hematopoietic stem cells are similar looking. They are small, round cells with compact nuclei. In their resting state, they largely remain dormant — i.e., they rarely divide. But put into the right milieu of chemical factors, or given the right internal signals in the bone marrow, they begin a ferocious program of cell division (in the 1960s, the Australian researcher Donald Metcalf was among the first to find these chemical "factors" that enable the growth of particular kinds of cells that emerge from the stem cell). A single stem cell can produce *billions* of mature red and white cells — and an entire organ system of an animal.

In the spring of 1960, a six-year-old girl named Nancy Lowry fell ill. She was dark-eyed and dark-haired, with eyebrow-skimming bangs. Her blood counts began to fall; her pediatricians noted that she was anemic. A biopsy of her bone marrow revealed that she had a condition called aplastic anemia, a form of bone marrow failure. Yet Barbara Lowry, Nancy's identical twin, was perfectly fine. Barbara's counts were normal, with no evidence of bone marrow failure.

The marrow produces blood cells, which need regular replenishing, and Nancy's was rapidly shutting down. The origins of this illness are often mysterious — an infection, or an immune reaction, or even a reaction to a drug — but in its typical form the spaces where young blood cells are supposed to be formed gradually fill up with globules of white fat.

The Lowrys lived in leafy, rain-slicked

Tacoma, Washington. At Seattle's University of Washington Hospital, where Nancy was being treated, the doctors had no clue what to do next. They tried transfusing her with red blood cells, but the counts would inevitably collapse again. One of them knew of a physician-scientist named E. Donnall "Don" Thomas, who had tried to transplant marrow between humans. Thomas worked in Cooperstown, New York. The Seattle doctors contacted him for help.

In the 1950s, Thomas had attempted a new kind of therapy in which he infused a leukemia patient with marrow cells from the patient's healthy identical twin. There was fleeting evidence that the blood stem cells from the donated marrow cells had "engrafted" into the patient's bones, but the patient had swiftly relapsed. Thomas had tried to refine the blood stem cell transplant protocol on dogs, with some marginal success. Now the Seattle doctors persuaded him to try again in humans. Nancy's marrow was faltering, but no malignant cells were occupying it. The Lowrys, fortuitously, were identical twins, with perfect "histocompatibility" — bone marrow could be transferred from one to the other without rejection. Would the blood stem cells from one twin's marrow "take" in the other twin?

Thomas flew to Seattle. On August 12, 1960, Barbara was sedated, and her hips and legs were punctured fifty times with a large-bore needle to extract the crimson sludge of her bone marrow. The marrow, diluted in saline, was then dripped into Nancy's bloodstream. The doctors waited. The cells homed their way into her bones and gradually started to produce normal blood. By the time she was discharged, her marrow had been almost completely reconstituted. Nancy's blood, in a sense, belonged to her twin.

Nancy Lowry lived through one of medicine's first successful bone marrow transplants. It was the quintessential instance of cellular therapy brought to life — her twin's *cells,* not a drug or a pill, had been Nancy's "medicine." In Toronto, Till and McCulloch were characterizing blood stem cells through their discoveries in mice. At Stanford, Weissman would eventually learn how to purify them from human bone marrow. In Seattle, Donnall Thomas had brought these blood stem cells into medical use. He had "enlivened" them in humans.

In 1963, Thomas moved to Seattle for good. Setting up his lab first at the Seattle Public Health Service Hospital and then, a dozen years later, at the newly established Fred Hutchinson Cancer Center — the Hutch, as doctors called it — he was determined to use marrow transplantation in the treatment of other diseases, notably leukemia. Nancy and Barbara Lowry were identical twins, and a noncancerous blood disease in one had been curable by cells from the other, a vanishingly rare occurrence. What if a disease involved malignant blood cells, as with leukemia? And what if the donor wasn't a twin? The promise of transplantation had been hindered by the fact that our immune systems are inclined to reject matter from other bodies as foreign; only identical twins, with perfectly matched tissues, can sidestep the problem.

Thomas saw a way around this. First, he would try to eradicate the malignant blood cells with doses of chemotherapy and radiation so high that the functioning marrow would be destroyed, purged of both cancerous and normal cells. That would usually be fatal, but the donor marrow stem cells from an identical twin would then replace it, generating healthy new cells.

The next problems arose from trying an "allogeneic" transplant (*allo,* from the Greek word for "other"), transplanting marrow from someone who *wasn't* an identical twin. In 1958, the French pioneer of bone marrow transplantation, Georges Mathé, had transplanted bone marrow from a series of donors to some Yugoslavian researchers who had accidentally received toxic doses of radiation and had thereby developed fulminant bone marrow failure. The donor cells had engrafted briefly — but then eventually disappeared. But soon after transplantation, Mathé had observed quite the opposite of what he had expected: an acute wasting disease appeared in the bodies of the Yugoslavian researchers.

Mathé deduced that this wasting disease was caused by immune response in the donor marrow attacking the body of the transplanted patients. *The guest was attacking the host.* This response is the consequence of an ancient system for maintaining the sovereignty of organisms (and rejecting invading cells) — except in bone marrow transplants, the direction of sovereignty is reversed. Like a mutinous crew forced onto an unfamiliar ship, the donor's immune cells recognize the body around them as foreign and attack it. The other (that is, previously the graft) becomes the self, and the self, ipso facto, becomes the other.

Other pioneers in organ transplantation had learned that these forces of rejection could be blunted if the donor and the host were reasonably well matched (recall our discussion of the discovery of histocompatibility genes — genes that govern whether a recipient will accept a graft from a host). There were now tests to help predict compatibility (or tolerance) and to improve the chances that allogeneic marrow cells would engraft. And various immune-suppressing drugs had been developed to further dampen the host's resistance, thereby allowing the allograft (i.e., from a foreign donor) to be accepted by the body, or the guest from attacking the host.

Over the next few years, Thomas assembled a group of doctors who pushed the frontiers of bone marrow transplantation. There was a tall, German-born rowing fanatic named Rainer Storb, who focused on tissue typing and transplant therapy; his wife, Beverly Torok-Storb, was an astute clinician. A diminutive, Siberian-born soccer enthusiast named Dr. Alex Fefer had shown that immune systems could turn against tumors in mice (hence the donor's immune system could kill the leukemia); and Don's wife, Dottie Thomas, who ran the day-to-day affairs of the lab and the clinic, and whom everyone called "the mother of bone marrow transplantation."

Thomas, who won a Nobel Prize for these studies, later described them as "early clinical successes." But for the nurses and the technicians in Seattle who cared for the patients — not to mention the patients themselves — the experience could be harrowing. "Of the hundred patients with leukemia who were transplanted in those early years, eighty-three died within the first several months," one of the doctors told me.

The final cataclysm, in this biblical array of plagues, happened when white blood cells produced by the donor's marrow mounted a vigorous immune response to the patient's body — the phenomenon called graft-versus-host disease that Mathé had discovered in his early transplants. It was sometimes a passing storm, and sometimes a chronic condition. In either acute or chronic form, the disease could be fatal.

But as Fred Appelbaum, part of the team of doctors who performed the first bone marrow transplants for leukemia, and other researchers found when they analyzed the data, the immune attack on the self — graft-versus-host — could also be an immune attack on the leukemia. Those that survived this cataclysm were also those that were most likely to beat the leukemia. It was the most definitive proof that a "rebooted" immune system — from a foreign donor — could engraft into a body, and then reject a cancer, resulting in the cures of deadly variants of blood cancers.

It was a stunning but sobering result: the poison was the cure. When I reminded Appelbaum of those early transplants, I caught a wistful look in his eye, as if he was remembering each of the patients. He has a gentle patrician air about him, with a humility acquired from years of failure. He thought back to the years when no one had survived — and then to the years when, one by one, the team had witnessed long term survivors of cell therapy for lethal diseases. They had succeeded — but at such a steep price.

I met the Thomases, Don and Dottie, at a conference in Chicago. They had grown frail and thin, and held on to each other like two playing cards propped against each other; remove one and the other would fall. I walked up, among a throng of admirers, to pay my regards to the twin parents of cellular therapy.

Don ambled up to the podium to speak. Once famous for his looming stature, he stooped as he spoke, pausing between sentences. The convention hall was packed to its edges — nearly five thousand hematologists had gathered to hear the talk — and the air was thick with reverence. Don reminisced about the early days of transplantation, and the heroic efforts — and the equal heroism of the first patients — that had ultimately led to the first allogeneic bone marrow transplants.

In 2019, I flew to Seattle to interview the nurses who had worked in the bone marrow transplant ward during the first years of its inception. Most had retired, but some had stayed in touch with the hospital. I sat in a conference room a few floors above the glistening new labs where patient's cells were being readied for gene therapy trials, such as the one that had resulted in Emily Whitehead's cure with CAR-T cells.

The nurses hugged and kissed each other as they came in. They remembered their nicknames, and the names of all the patients who had been treated in those early years. Some of them broke down in tears. This was an impromptu reunion.*

"Tell me about the first patients," I asked.

"The very first was a chronic leukemia patient," a nurse, A.L., told me. "His name was Bowlby [. . .] An elderly man," she said, and then corrected herself. "No, no, he was just in his fifties. He died [. . .] from an infection. Number two was a young man with leukemia, and then a little girl. Both of them died."

They remembered Don and Dottie, the Storbs, Appelbaum, and Fefer — the stalwarts and pioneers of early cell therapy. "Every morning, one of them would be at rounds, holding each patient's hand, asking how the night had been," one said.

"In 1970, we had a young boy with leukemia," another nurse said. "He was ten. He survived and went through college — about ten years — but he struggled with lung infections. And then he died."

I asked about what the hospital looked like, what it felt like.

"There were twenty beds," another nurse, J.M., said. "The nurse's station was in the ice room. I remember it was small. It was close. Everyone was pulling for everyone.

*I have deliberately concealed the names of the nurses. This is not to diminish their vast contributions to bone marrow transplantation, but to protect their identities and respect their privacy.

"There was one child who wanted to hear the same story every night. About a boy who went into a cave and killed a bear." And so, night after night, while chemo dripped through his veins, he was sent to sleep with that story.

The place where the patients were doused with radiation — to kill their blood cells and make space for the new marrow — was a cavernous cement bunker a few miles away. The dogs, used for transplant experiments, were housed next to it, so the patients, locked in the cement chamber, had to hear their incessant barking as they were being irradiated.

Initially, the entire dose of marrow-killing radiation was given in one slug.* "Halfway through the radiation, the patients got so nauseous that it was unbearable," one of the nurses said. "They vomited and vomited. We had to open the bunker doors and go in and tend to them. There weren't really any strong antinausea medicines then [. . .] so we went in with water, tubs, and mouth wipes and wet towels. And there was this seven-year-old boy . . ."

She choked up. Another woman stood up to hug her.

*Later, the dose would be fractionated over several days, greatly reducing the nausea. New antiemetics, such as Zofran and Kytril, also vastly improved the waves of nausea caused by the radiation.

"Tell them about the pilot," one of the nurses urged.

The pilot was Anatoly Grishchenko. In 1986, when the nuclear reactor blew up in Chernobyl, Grishchenko was sent on a helicopter to dump sand and concrete to entomb one of the open vents of the reactor that was spewing toxic radioactive gas — in essence, to convert the factory into a cemented sarcophagus. He was supposedly covered head to toe with lead shielding, but the radioactivity penetrated his body, all the way to his bone marrow.

In 1988, he was diagnosed with a pre-leukemic disease. In 1990, the disease bloomed into a full-blown leukemia. A woman in France was found to be a near-perfect match. A doctor from the Hutch flew to Paris to supervise the extraction of the bone marrow, and jetted the marrow overnight to Seattle, where Grishchenko underwent a bone marrow transplant.

"But he didn't make it," the nurse told me. "We watched over him for days, but in the end, the leukemia came back."

And so it went. "We had one survivor from 1970. Three from '71. And in '72, we had a few. We didn't have a lot of long-term survivors — but some of them did make it to twenty, thirty, forty years. By the mid-eighties, we began to really see long-term survivors. Tens, twenties, dozens of them, living five, or ten years after the transplant."

■ ■ ■ ■

Downstairs, in the lobby of the Hutch, there was a spiral sculpture that represented the seemingly relentless and steady progress of transplantation. I took a closer look, watching the numbers spiraling upwards year by year — five, twenty, two hundred, a thousand, up to several thousand by 2021. And the cure rates for deadly diseases had also increased: in one study, patients with acute myeloid leukemia had between a 20 percent and 50 percent chance of survival five years post transplantation.

One of the nurses had come down to look at the sculpture with me. She put her hands on my shoulders.

"It wasn't so easy back then," she said. The smooth spiral line, she knew, was, in fact, a jagged record of failures punctuated with infrequent successes. But then the successes accumulated. Thousands of bone marrow transplants are performed each year now, for dozens of diseases. Their success varies, but it is now one of the mainstays of cellular therapy. In my own clinic, I can think of throngs of patients with deadly variants of leukemia that have been cured by bone marrow transplantation.

The nurse ran her hands along the smooth line of the curve and smiled. I thought of Grishchenko in his helicopter, suspended in midair and surrounded by a fog of toxic plutonium. Of the boy who went into a cave to kill a bear. I could sense the terrifying fear of the young child in the cement chamber, bent over with nausea, while the dogs barked next door. I thought of the nurses with wet towels, and of those who stayed overnight, those who kept vigil against infections, those who held the patients' hands all day, and watched them as if they were their own children. As the nurses left the hospital, many of the doctors and the staff stood up as they passed. It was a silent recognition of their many, many contributions. I realized that I had tears in my eyes.

Cell therapy for blood diseases had a terrifying birth.

Stem cells have been found in diverse organs, and in diverse organisms. But more than any other type of stem cell, the two that remain the most fascinating, and most controversial, perhaps, are the embryonic stem cell (ES cell) and its even stranger cousin, the induced pluripotent stem cell (iPS cell).

In 1998, James Thomson, an embryologist working at the Wisconsin Regional Primate Research Center, procured fourteen human embryos that had been discarded from IVF procedures. The experiment that he was about to perform, he knew, was inherently controversial, and he had consulted two bioethicists, R. Alta Charo and Norman Fost, before even launching it. The human embryos were cultured in an incubator until they had reached the blastocyst stage, when the embryo forms a hollow ball. The blastocyst normally grows within the uterus, but it can also be cultured under special conditions in a petri dish. The ball has two distinct structures. There is a veil-like outer shell that will eventually form the placenta and the structures that attach the embryo to the mother's body. Furled inside the shell is a tiny bulge of inner cells that will form the embryo.

Thomson extracted these inner cells and cultured them on a "feeder" layer of mouse cells that would supply the human embryonic cells with nutrients and support (this is a common technique in cell culture. Some cells are so fragile, especially in the first days of their transition to cell culture, that they cannot live by themselves. They need feeder, or helper cells, to "nurse" them in these initial stages). Over the course of several days, five human cell lines grew out of the embryos — three "males" and two "females." They proliferated for months in cell culture, with no obvious genetic damage, and with no change in their growth potential.

Injected into immunocompromised mice, the cells made a gamut of layers of mature human tissue — gut, cartilage, bone, muscle, nerves, and elements of skin. The cells were clearly capable of self-renewal in a petri dish, and capable of differentiating into multiple (possibly all) kinds of human tissue.* They were termed "human embryonic stem cells," or h-ES cells. One of these cells, called H-9 — a "female," with XX chromosomes — has become the standard ES cell. It has been grown in thousands of incubators in hundreds of labs around the planet and been subject to tens of thousands of experiments.

I've grown H-9 myself and watched the cells grow continuously. I've also watched them differentiate into various mature cell types, including bone and cartilage. Even today, the existence of this cell line astonishes me: I cannot look through a microscope at a flask containing these cells without a small shiver, something akin to a nervous wistfulness

*A technical point: the ES cells derived by Thomson came from the inner cell mass (that will eventually form the embryo) and not from the outer wall of cells (that forms the placenta, the umbilical cord, and other structures that are called extra-embryonic). These ES cells are not totipotent since the placenta, to take one example, is derived from the outer wall of cells and not the inner cell mass. More recent work has shown that under some culture conditions, a fraction of ES cells can remain totipotent — in other words, capable of giving rise to extra-embryonic tissues. However, most investigators consider human ES cells pluripotent not totipotent, since they give rise to all tissues except the extra-embryonic tissues.

about the future. In principle, the existence of these ES cells prompts a strange thought experiment: What if one could turn time backward and inject them — a little curl — back into the cellular womb of the blastocyst from whence they came, and implant that ball back into a human womb? Perhaps we would need to mix them with some other cells of the inner cell mass — but would they now, returned to their origin, form a human? What would we name her, this new kind of cellular being? Helen-9? If a genetic change was introduced into H-9 in the petri dish, would that human now potentially bear that change — and pass it on to her children? And if the H-9 cells in the human produced an egg, and then an embryo, would we witness a new circle of life — from embryo to blastocyst to ES cell to human to embryo?

Thomson's paper, published in *Science* in 1998, set off an immediate firestorm. Many scientists sided with Thomson, who believed the inherent value of human ES cells: these cells would enable us not just to understand human embryology more deeply, but would also become therapeutically invaluable tools. As Thomson wrote pointedly toward the end of his seminal paper:

Human ES cells should offer insights into developmental events that cannot be studied directly in the intact human embryo but that have important consequences in clinical areas, including birth defects, infertility, and pregnancy loss. [. . .] Human ES cells will be particularly valuable for the study of the development and function of tissues that differ between mice and humans. Screens based on the in vitro differentiation of human ES cells to specific lineages could identify gene targets for new drugs, genes that could be used for tissue regeneration therapies, and teratogenic or toxic compounds.

Elucidating the mechanisms that control differentiation will facilitate the efficient, directed differentiation of ES cells to specific cell types. The standardized production of large, purified populations of [. . .] human cells such as cardiomyocytes and neurons will provide a potentially limitless source of cells for drug discovery and transplantation therapies. Many diseases, such as Parkinson's disease and juvenile-onset diabetes mellitus, result from the death or dysfunction of just one or a few cell types.

But critics, mostly from the religious right, would have none of it. They argued that human embryos had been destroyed — defiled — during the production of these cells, and that embryos constituted humans. That these IVF-produced embryos were yet to acquire sentience, had no organs, were no more than a ball of undifferentiated cells that would otherwise have been discarded anyway, hardly placated them; it was their *potential* to form future humans that made them currently human, Thomson's critics argued. In 2001 President George W. Bush, pressured by opponents of ES cell research, passed a law restricting federal funding to research involving ES cells that had already been derived (such as H-9); any attempts to make new ES cells could not be federally supported. In Germany and Italy, too, research on human ES cells was highly restricted and, in some cases, banned.

For about a decade, a select few human ES cell lines was all that researchers had to explore human embryology and tissue differentiation from embryonic stem cells. And then, in 2006 and 2007, the field underwent yet another radical twist. The question that had raged through the field in the early 2000s was this: *Was there something about stem cells that made them special?* Why could a skin cell or a B cell, say, not awake one morning and decide to become an ES cell — squirming its way upriver, turning back time to its origin?

The question sounds absurd at first glance. Until the 1990s, no embryologists that I knew had thought of embryology as a two-way street. Go forward and you get a human, with all his or her mature cells — nerves, blood, liver cells. Go backward, and you take a mature cell — nerve, blood, liver — and turn it into to an embryonic stem cell. "It seemed like pure madness," one researcher told me.

But there was one fact that kept the fantasy of the "two-way street" alive — at least for a small group of embryologists. The sequence of DNA in all the cells (i.e., the genome) is identical in nearly all our cells*; it is the *subset* of genes that is turned "on" and "off" in a heart cell, or a skin cell, that determines its identity. What if we could change that pattern — turn stem cell genes "on" and "off" in a skin cell? Would the skin cell now turn into a stem cell — capable of making not just skin, but bone, cartilage, heart, muscle, and brain cells — i.e., every cell in the body? And what was stopping a skin cell from doing just that?

In 2006, working in Kyoto, Japan, a stem cell researcher named Shinya Yamanaka took adult mouse tail-tip fibroblasts — spindle-shaped, run-of-the-mill cells that are found in various forms throughout the body — stuffing, as far as the stem cell world goes — and introduced four genes into these cells.

---

*We now know that the genomes of individual cells in the body can be slightly altered by mutations as an organism matures. Humans, in short, are chimeras of genomically nonidentical cells. The biological significance of these differences is still to be determined.

Yamanaka hadn't stumbled on these genes by accident: he had spent years studying and selecting Oct3/4, Sox2, c-Myc, and Klf4 for their unique ability to "reprogram" the properties of adult cells to resemble stem cells. In the late 1990s, he had started with *twenty-four* genes, comparing the effect of each gene, and each permutation, experiment after experiment, combining one with the next and then adding another, until he had narrowed the relevant genes down to the critical four. (Each of these genes encodes a master-regulatory protein, a molecular switch that flicks dozens of other genes on and off.) Each of them, he had established, plays a critical role in maintaining the stem cell state of human and mouse ES cells. What might happen if he took an adult *non*-stem cell — a humdrum fibroblast — and forcibly induced it to express all four of these master-regulatory genes that give stem cells their identity, in combination?

One afternoon, Kazutoshi Takahashi, a postdoc in Yamanaka's laboratory looked down the scope at fibroblasts in which he had forcibly expressed the four critical genes. "We have colonies!" the postdoc shouted. Yamanaka rushed over. Colonies indeed. The cells — normally spindle-shaped and prosaic-looking — had changed their morphology, turning into glowing ball-shaped clusters. Chemical changes had appeared in their DNA, Yamanaka would later find out;

the proteins that fold and package DNA into chromosomes changed. Even the metabolism of the cells had altered. The fibroblasts had turned into stem cells. As with ES cells, they renewed themselves in culture. And injected into immunocompromised mice, they, too, formed multiple kinds of human tissue — bone, cartilage, skin, neurons. *All of this derived from a skin fibroblast, a cell that is fully developed, and has no seeming function except to act as scaffolding to maintain the integrity of skin tissue or to repair a wound.*

The result was an utter shock to biologists — a Loma Prieta that shook the Earth plates of the stem cell world. I remember a senior chemical biologist from my department returning from a seminar in Toronto where Yamanaka had just presented his data visibly ruffled, breathless with disbelief. "I just cannot believe it," he told me after he'd returned from the talk. "But the result has been reproduced over and over again. It's *got* to be true." Yamanaka had made a stem cell out of a fibroblast — a transition thought to be impossible in biology. It was as if — *presto!* — he had turned biological time backward. He had changed a fully grown adult not just into a baby, but into an embryo.

In 2007, Yamanaka used this technique to turn *human* skin fibroblasts into ES-like cells. The next year, Thomson, of human ES fame, substituted two other genes for c-Myc and

Klf4, and again altered human fibroblasts into embryonic stem cells (the expression of c-Myc, in particular, to create ES-like cells was considered a potential challenge because it happens to be a cancer-causing gene, and biologists worried that these ES-like cells would eventually turn cancerous). The field termed these induced pluripotent stem cells, or iPS cells — *"induced"* because they had been changed, using genetic manipulations, from mature fibroblasts into induced pluripotent cells.

Since Yamanaka's discovery, for which he won the Nobel Prize in 2012, hundreds of labs have started working on iPS cells. The allure is this: you take your *own* cell — a skin fibroblast, or a cell from your blood — and you make it crawl backward in time and transform it into an iPS cell. And from that iPS cell, you can now make any cell you'd like — cartilage, neurons, T cells, pancreatic beta cells — and they'd still be your own. There would be no problem with histocompatibility. No immune suppression. No reason to worry about the guest turning immunologically against the host. And in principle, you could repeat the process infinitely — iPS into beta cell back to iPS cell into beta cell (to be fair, no one has tried this yet). The recursiveness, though, sets up yet another fantasy of a new human: one in whom every degenerated organ, or tissue, could be regenerated, and regenerated again, ad infinitum.

I sometimes think of the Greek story of the Delphic boat. The boat is built of many planks. Bit by bit, the planks decay and are replaced by new ones, until all of them are new. But has the boat changed? Is it even the same boat?

These musings seem metaphysical today. But they may soon become physical. And as we build new parts of humans out of iPS cells — and many scientists already have — and then try to make new parts out of those new parts, recursively, I also think of Ozick's snail. Saved from rubbing itself into oblivion, it leaves a trail of metaphysical questions behind as it moves into uncertain, unknown realms. By the end, all of it has been rubbed off, and replaced. Is it even the same snail?

THE REPAIRING CELL
INJURY, DECAY, AND CONSTANCY

*Tenderness and rot
share a border.
And rot is an
aggressive neighbor
whose iridescence
keeps creeping over.*
— Kay Ryan, 2007

Dan Worthley, an Australian postdoctoral researcher, came to my lab across many oceans, some physical and some metaphysical. He was a gastroenterologist by training — a subject that I knew very little about. He had come to Columbia University in New York to work with Tim Wang on colon cancer and the regeneration of colon cells (Wang, a professor at Columbia, is an old friend and collaborator), and to study colorectal cancer.

Standard techniques of modern genetic engineering in mice now allow us to take a single gene and alter it, such that the protein it encodes is labeled with a fluorescent marker. The protein now becomes a glow-in-the-dark beacon; you can use a microscope to detect wherever or whenever the protein is physically present. Imagine if you did this for the cyclin genes that control the cell cycle: you would see the cell begin to glow as one particular cyclin protein was made and then disappear once that protein was degraded. If you took the same approach for the actin — the protein that makes up the skeleton of the cell — virtually the whole animal would become a glow-in-the-dark mouse. The T cell receptor would only glow in T cells. Insulin would glow in pancreatic cells. The glowing proteins, incidentally, come from jellyfish; genetically speaking, a little bit of this mouse comes from a creature that wobbles and pulsates in the depths of an ocean.

Worthley had genetically manipulated a mouse and engineered a gene — called Gremlin-1 — using this technique. Whenever Gremlin-1 protein was made in a cell, that cell would become fluorescent, and thereby visible, by microscopy. Based on prior findings, Dan expected Gremlin-1 to come alight in cells in the colon. And true to form, he found it in a particular cell type in the colon. But some combination of innate curiosity and meticulousness had urged him to look for these Gremlin-1-labeled cells in other tissues. One place that the cells had lit up was in cells in the bone. And that's where our relationship began.

■ ■ ■ ■

If a catalogue of neglected, but vital, human organs was ever compiled — or, perhaps, some ratio of the "real-world" importance to "scientific neglect" of an organ was enumerated — bone may well rank among the first in both lists. Medieval anatomists thought of bones as glorified hangers for the skin, or scaffolding for the body's innards (although Vesalius, bucking this trend, drew elaborate figures of skeletons; several of his plates represent the detailed anatomies of various bones). At Massachusetts General Hospital, where I was a resident in the early 2000s, orthopedic residents jokingly, if wryly, called themselves boneheads. And who can forget the tragicomic soliloquy of "Bonehead Bill," from Robert Service's wartime poem — a soldier who is trained to maim and kill without thinking: "Me job's to risk me life and limb / But . . . be it wrong or right"?

Yet the skeleton turns out to represent one of the most elaborately complex of cellular systems. It grows to a point, and then knows when to stop growing. It heals itself continuously throughout adult life and repairs itself acutely after injury. It responds with sensitivity to hormones; it potentially even *synthesizes* its own hormones.* Its central cavities — the

*Research led by Gerard Karsenty and colleagues at Columbia has made the case that bone doesn't

marrow — are the white-walled nurseries for the genesis of blood. It is the locus of osteo-arthritis and osteoporosis, two of the major illnesses of aging, implicated in millions of deaths among the elderly around the world. And it is my personal nemesis: my father's fall, the fracture in his skull, and the bleeding that emanated from it, was ultimately responsible for his death.

But back to Dan and his bones. One morning in the summer of 2014, Dan took the elevator down to my lab — his own lab was three floors above mine — with a box filled with bone slices. I could lie and say that I was instantly intrigued. But I wasn't; researchers from all sorts of labs would descend on postdocs in my lab (and on me) entreating us to look at their samples, asking if there was something interesting in the bones, and it was a constant sap on their (and my) time. I politely asked Dan to come another time.

But Dan was unrelenting. Short, intense, energetic, and driven by a single purpose, he was like an Ozzie-made hand grenade.

---

simply respond to hormones but produces hormones. Early experiments show that one such protein, called osteocalcin, made by bone cells, appears to modulate sugar metabolism, brain development, and male fertility, although some of these findings still await confirmation.

He knew of my interest in bones. As an oncologist, I treat leukemia, a disease that originates in the bone marrow, where blood-forming stem cells live. And for decades, I've worked on how bone cells and blood cells interact: why, for instance, aren't there blood stem cells in the brain, or in the intestines? What is so special about bone? As a field, we've discovered some answers: cells resident in the bone marrow send particular signals to blood stem cells that maintain their function. Over these years, I've also learned to understand the anatomy and physiology of bones. There's the now-popular idea that you acquire a particular form of expertise when you've performed an activity — throwing a baseball, say — for more than ten thousand hours. In cell biology, it translates, well, into seeing: I have looked, through a microscope, at more than ten thousand specimens of bone.

Scarcely had a week passed when Dan was back again, lurking in the corridors with the same combination of obsequiousness and de-terminedness, and with his blue box of slides. He was indifferent to my indifference. I sighed and decided to take a look.

I darkened the room, and the microscope flickered on, sending a diffuse, blue-green fluorescence across the room. Dan paced in the back of the room like a caged animal, muttering about *Gremlins*. The slices were beautifully cut on a microtome, revealing the classic histology of bone.

Superficially, bone might look like a chunk of hardened calcium, but it is, in fact, made of a multiplicity of cells. The most familiar are cartilage cells — technically called chondrocytes — and there are two unfamiliar-sounding cell types. The second is the "osteoblast" — the cell that deposits calcium and other proteins to form a calcified matrix in layers, and then get trapped in its own deposit to form new bone. It is the bone-making, bone-depositing cell: typically, the osteoblasts thicken and lengthen the bone (my mnemonic for its name is the letter "b," for "*b*one making").

The third is an osteoclast: these are large cells with multiple nuclei that are bone-eaters. They chew away on the matrix, or punch holes in it, removing and remodeling bone, like constantly pruning gardeners (I remember it because it contains the letter "c," for "bone chewing"). The dynamic balance between osteoblasts and osteoclasts — bone makers and bone chewers — is one mechanism by which bone maintains homeostasis. Take away osteoblasts and new bone cannot be laid down. Make osteoclasts — the bone chewers — defective, and the bone turns thick — "stone bone," as early pathologists called it — tough seeming, but hard to repair. The cavities inside contract, crowding out space for the marrow, precipitating a disease called osteopetrosis.*

But bone doesn't just thin and thicken. It grows longer. And there's a cellular mystery in the growth of bone. We've met collections of cells that make organs grow larger. But how might collections of cells move directionally that makes an organ grow *longer*? Early anatomists, including Marie-Francois-Xavier Bichat, noted that bone begins, in early development, as a matrix of glutinous cartilage. Then it deposits calcium and hardens into the structure that we recognize as bone and begins to grow longer. But the main change in length occurs from the ends of the bones; the "middle" remains relatively constant. In the mid-1700s, the surgeon John Hunter hammered two screws into a growing, adolescent bone. Their distance, he noted, was unchanged. But had he hammered the screws to the two *ends* of bone, he would have watched bone growing longer — the screws separating from each other in time, like two ends of an

---

*I have only catalogued the cells in the bone. There is a much vaster catalogue of cells that reside in bone marrow. These include blood stem cells and blood progenitors. There are stromal cells that are thought to play a supportive role for blood stem cells. There are neurons, and fat-storing cells — adipocytes — and cells of blood vessels (endothelial) cells that bring blood into and out of the marrow.

elastic band traveling farther and farther away as the band lengthens. In short, there are cells at the ends of a bone — but not in its middle — that generate new cells that lengthen it.

There's a special place in the bone, just where its head — the fist-shaped end of long bones — meets the shaft. Somewhere at that junction, buried within the bone, there's a structure called a "growth plate." If you roll your fingers into a fist and imagine your lower arm as the shaft of a long bone and the fist as its end, then the growth plate would sit somewhere near your wrist.

The growth plate exists in children and adolescents — you can sometimes see it as a white line in an X-ray — but gradually closes in adults. You can imagine the growth plate as the kindergarten for young bone cells. It is the growth plate that gives rise to mature cartilage cells, and to osteoblasts. The young cartilage cells, and then the bone-forming osteoblasts, shoot off the growth plate, and migrate to the area adjacent to the head of the bone, depositing new matrix and calcium between the head and the shaft — which, in turn, lengthens the bone.

And that is where Dan's slides came into play. The existence of the "growth plate" had been known for decades. But how was bone's growth sustained, especially during that fierce burst of adolescence when a young man or woman might shoot up inch by inch every week? Fully mature cartilage — hypertrophic cartilage — we know, doesn't grow or divide — so what cells were birthing bone cells, week upon week? Was there a reservoir of skeletal cells that kept giving rise to the young cartilage cells and bone cells? The cells that Dan had lit up in his mouse sat precisely at the growth plate, in a neat, slightly curvilinear row, like a set of perfectly formed teeth. I looked, and looked again. I was now massively intrigued.

There's a moment in the life of a team of scientists — typically two researchers — when language disappears. Something akin to that moment passed between me and Dan. Language — or at least traditional language — vanished. We exchanged instincts. The pheromones of ideas passed between us, often wordlessly. At night, I would stay up late, pacing, thinking of the next experiment we should perform. The next morning, I would arrive at the lab to discover Dan had already done it.

The first series of experiments were simple. What *were* these cells? Where did they live? During what period of time? Dan's first experiment had lit up Gremlin-1-expressing cells in the growth plate of a young mouse. In a fetal mouse, he found them gathered brightly precisely where new bone and cartilage were being formed. Imagine a tiny foot,

635

or a minuscule finger, emerging. The cells were right there, dividing fiercely.

And then, as he followed them further, an astonishing thing happened: the cells migrated from the tips of the neonatal bone to the growth plate — the part where the shaft meets the long bone — and organized themselves into a neat layer just at the growth plate. As the mouse grew older, and the lengthening of bone was completed, the cells gradually grew fewer and fewer in number. Something about these cells, then, had to do with the formation of bone.

But what, precisely? The molecular beacon that Dan had created has another special property. You can use the beacon to follow the fate of a cell as it divides. It takes some additional engineering, but you can ensure that when one cell makes the Gremlin-1 protein (and thereby turns fluorescent), then its daughter cells will also become fluorescent, and the daughters of the daughters will also glow in the dark and so forth, ad infinitum. The technique is called lineage tracing — as if you could somehow find every family member of an enormous family, even if they had dispersed in time and space. It is a molecular way of illuminating an entire family tree.

Dan performed this experiment in a very young mouse. And when he traced Gremlin-expressing cells, he found that they gave rise to young cartilage. That intrigued me

— cartilage-forming cells had always been a bit of a mystery. But as he looked at the tissue over longer and longer periods of time, the family tree got more complex. The cells to light up next were plump, bloated cells of mature cartilage. And then osteoblasts — bone-forming cells began to light up. And finally, there was a totally unknown cell type, a cell with spindly fibers extending outward whose function we don't yet know, that we called reticular cells. Perhaps most notably, the original Gremlin-labeled cells — the ones that had come first — didn't go away, at least in young mice. In short, Dan had found *that* cell — the cell, sitting in the growth plate, that gives rise to cartilage cells which then matures into osteoblasts — the two principal components of bone. We called them osteo (bone), chondro (cartilage), and reticular cells — or "OCHRE" cells.

Dan published his paper with me and Tim Wang in *Cell* in 2015. At the very same time, Chuck Chan — a brilliant postdoctoral researcher (and now assistant professor) working with Irv Weissman at Stanford, also discovered a skeletal stem cell.

Rail thin and tall, Chan looks like a punk rocker; he comes to the lab as if he's been pulled out of an all-night rave. His experimental discipline, though, is unnerving. Chan, Weissman, and a surgeon-turned-scientist Michael Longaker had ground up bones and used Weissman's favorite technique — flow cytometry — to purify populations of skeletal cells that gave rise to cartilage and bone. Their paper came out with ours, back-to-back in the journal *Cell*. The similarities — genetic, physiological, histological — between our two cells was striking. For a while, we engaged in a friendly battle over how to name the cells. But OCHRE, which also happens to be a color that I particularly like, seems to have stuck.

Dan's and Chuck's original papers also left a trail of questions. It's still unknown whether these Gremlin-marked cells *first* give rise to young cartilage — an intermediate state — and then to osteoblasts. Or do they give rise to both at the same time? Are there intrinsic or extrinsic factors that bias that decision? How is the balance — this homeostasis — maintained? And do these cells self-renew? Early results, involving the transplantation of these cells into mouse bones, *do* suggest that they renew themselves. Gremlin-marked cells, then, would fit the bill for true skeletal stem cells — capable of differentiating into multiple cell types, and capable of self-renewal. The OCHRE cell — a putative skeletal progenitor or stem cell — might be the discovery that I, and my lab, are the proudest of. They represent a potential answer — a theory — that solves two age-old mysteries. How does bone grow during adolescence? Well, because a special population of cells, sitting at the growth plate at the two ends of bone, shoots off cartilage and osteoblasts that allows bone to lengthen. And why does it stop growing? Because this population diminishes over time, until early adulthood, when very few are left.

But wait. The plot was to take yet another twist. In Texas, Sean Morrison — a former trainee of Weissman's, and arguably the most tenacious stem cell biologist I know — found yet another cell type, resident inside the bone marrow, that could make osteoblasts and deposit bone. Unlike the Gremlin-labeled cells, Morrison's cells (call them LR cells, for one gene that they express) are born later in adulthood, and give rise, predominantly, to the bone that is deposited along the long shafts — not the growth plate, but the long tube of bone between the two plates. They don't give rise to cartilage cells or reticular cells. If you fracture the long shaft of bone, somewhere in the middle, the LR cells swing into action, generating bone-making cells that repair the injured long bone.

What a mess, you might say, but it's quite the opposite. Bone is not just an organ with a single supply of rejuvenating cells; it is a chimera of rejuvenation. It has at least *two* sources for two sites. There are growth-plate resident OCR (or OCHRE) cells, which form lengthening bone. They arise early in development and then gradually decay with age. And there are LR cells that arise later in adolescence and adulthood that participate in the maintenance of *thickness* of long bones, and bone fracture repair.

Morrison's data, then, represent a potential solution to a third mystery. Why can bone thicken in adults — and repair fractures — when the growth plate has diminished and gone? Well, possibly, because there's a reserve population of different cells — not resident in the growth plate, but inside the bone marrow — that performs this function. The firstborn cells (i.e., the ones that Dan found) build and lengthen the bone during fetal life, and then assume a more limited role of maintaining the growth plate during adult life, we think. The later-born cells (the ones that Morrison found) march in like a second army that fixes fractures and maintains the integrity of the bone. This "two-army" solution uncouples the function of bone making and bone maintenance. Why two armies? We don't know.

Dan returned to Australia in 2017, leaving me bereft — but then he lobbed another (much-appreciated) hand grenade across the ocean. Jia Ng — short, intense, with Dan's concentrated energy and single-mindedness — came to the lab in 2017 to study the Gremlin-marked cells. If Dan had asked the physiological question (how does bone and cartilage grow?), Jia was interested in its pathological obverse (how does it decay?).

Osteoarthritis is a disease of cartilage degeneration. The old dogma ran that the constant grind between bones eats away at the lubricating lining of cartilage at the head of a bone — a femur, say. Cartilage cells at the joint surface die, and then the bone below the joint begins to wear away. And so Jia began to study mice with osteoarthritis using the techniques that Dan had pioneered in the lab.

The first surprise concerned real estate: location, location, location. We had been so blinkered by the skeletal stem cells sitting at the growth plate, birthing new cartilage and bone, that we had missed a second site where they were also located. When we looked again, with new eyes, Gremlin-labeled OCHRE cells were also present in a thin, veil-like layer just above the head of the bone. They sat glimmering, enticingly, at the joint where two bones meet, exactly where osteoarthritis originates.

It would be hard to describe the exhilaration of the days that followed. I would swallow my cup of coffee in the mornings, pack my notebook, speed on the highway to the lab and rush to the microscope room where Jia would have set out the slides from the previous night's cuts (she worked late; I worked early). The scope would turn on, and I would look and count. *Seeing.*

Jia returned to Dan's experiment of lineage tracing — taking an indelible molecular tattoo and marking a cell, its daughters, its great-granddaughters, and so forth. And, as with Dan's experiments, the result was a surprise: when she first labeled the cells, they were located in a thin, veil-like layer right at the joint surface. And then, as the first weeks went by, they began to form layer upon layer of cartilage at the joint. In a month's time, we saw the appearance of bone cells below the cartilage.

But what might happen to these cells during arthritis? We wrote a grant together, proposing that the Gremlin-marked stem cells (or OCHRE cells) would act as a regenerative reservoir. When mice got arthritis, we reasoned, the OCHRE cells would try to regenerate the lost cartilage — much like stem or progenitor cells act in other tissues when the tissue is depleted or injured. Osteoarthritis was the forme fruste — the frustrated form — of a tissue trying to repair itself, but failing.

In the lore of science, much has been written about the joy of being precisely right about a hypothesis or a theory. In the early 1900s, Einstein's proposal of the constancy of the speed of light would spectacularly validate earlier experimental observations made by Albert Michelson and Edward Morley. ("If the Michelson-Morley experiment had not brought us into serious embarrassment, no one would have regarded the relativity theory as a [halfway] redemption," Einstein would write later.) But there's a second kind of joy in science: the peculiar exhilaration of being precisely wrong. It is an equal and opposite sensation of joy: when an experiment proves a hypothesis wrong, and the truth turns, as if on a pivot, to point *exactly* in the opposite direction.

Three weeks after Jia had introduced arthritis in the mice (there are many ways of achieving this, including using a mechanism to weaken one of the femoral joints. The induced injury is mild, and the mice almost always recover), we returned to the microscope to examine the bone slices. We had expected the OCHRE cells, lit up by the fluorescent protein, to be briskly proliferating, trying to buffer the injury. The same blue green light flooded the room.

We had been precisely wrong. In young mice that had no induced injury, the expected layer of Gremlin-marked OCHRE cells sat intact on the joint surface — the same glimmering line of cells. In injured mice, the cells — rather than becoming hyperactive and dividing to save the joint, as we had predicted — were dead or dying. *The injury had killed the stem cells* — to the point that they could no longer keep up the genesis of cartilage.*

*This work is still being reviewed by scientists.

I turned the microscope off, and an inner light flickered on. Osteoarthritis, perhaps, was a disease of stem cell loss. The cells that were being worn out — in its first stages — were the cartilage-making stem cells, and they could no longer keep up the genesis of cartilage. The balance between growth and degeneration had been disrupted. What the injury had thrown off was the capacity of cartilage at the joint to maintain its internal balance — between the growth of new cartilage (via the stem cells) and the decay of old cartilage (via age and injury).

There were many, many experiments that followed to nail the point. Toghrul Jafarov, a deliberative postdoc from Canada, picked up on Jia's work. Using remarkably dexterous techniques, he learned to forcibly kill the Gremlin-marked cells by injecting a chemical

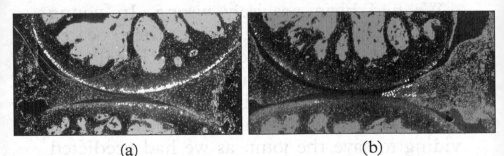

(a)   (b)

(a) A young mouse showing the cells marked with Gremlin lit up by the fluorescent protein. (b) The same joint after an arthritis-inducing injury, with the gradual death and disappearance of the Gremlin-expressing cells. The images are from Jia Ng's work.

into the knee joint — in essence, Jia's experiment in reverse (if osteoarthritis originates in the death of Gremlin-labeled cells, then does killing Gremlin-labeled cells cause osteoarthritis?). The mice, remarkably, developed osteoarthritis. Even young, healthy, mobile mice, otherwise normal, began to lose the integrity of the joint. They hobbled, until the cells restarted the growth of cartilage.

Jafarov kept pushing the direction of the experiments. He inactivated a gene that was crucial to the maintenance of Gremlin-expressing cells, and thereby killed the cells genetically. Again: the mice developed osteoarthritis, this time even more severe than any we had seen. (I gasped when I saw these bones. There were sections of the bone in which the cartilage had eroded to the point that the end of the bone looked like a mountain that had been gashed by dynamite. The raw "rocks" of the bone below were exposed — naked and crumbling.)

He purified Gremlin-positive cells from animals, grew them in tissue culture, and transplanted them into mice. They divided, grew more Gremlin-marked cells (albeit few in number) and then started making bone and cartilage all over again. He added a drug that would increase the number of Gremlin-marked cells into the joint space. The mice were protected from osteoarthritis.

Toghrul, Jia, Dan, and I sent our data for

publication in the winter of 2021. We proposed a radically new hypothesis about osteoarthritis. It isn't merely a degeneration of cartilage cells, caused by grind and tear. It is, first, an *imbalance* caused by the death of Gremlin-marked cartilage progenitor cells that cannot generate adequate bone and cartilage to keep up with the demands of the joint. And so we have a theory to answer the fourth age-old mystery: Why doesn't cartilage in joints get repaired, just as a bone fracture does, in adults? Because the repairing cells die during the injury.

Injury and repair share a border — except, as we age, injury, and the weariness of regenerative capacity, keeps infiltrating, creeping over the hedge. Osteoarthritis is a degenerative disease that arises from a regenerative disease. It is a flaw in *rejuvenative* homeostasis.

What general principle can one draw out of these experiments? One of the most unusual conundrums of cell biology is that while the early genesis of organs seems to follow a relatively ordered pattern,* the maintenance and

---

*The inner cell mass of the embryo, as I mentioned earlier, splits into three layers, followed by the formation of the notochord and the invaginations of the neural tube. The embryo is organized into various compartments, and the subsequent formation of organs along the body axis is governed by extrinsic

repair of tissues in adulthood seems idiosyncratic and peculiar to the tissue itself. If you cut the liver in half, the remnant liver cells will divide and grow the liver back to nearly its full size — even in adults. If you fracture a bone, osteoblasts will deposit new bone and repair the fracture — although the process slows down dramatically in older adults. But there are other organs where damage, once done, is permanent. Neurons in the brain and the spinal cord, once they've stopped dividing, don't divide to regenerate neurons* (they are "post-mitotic" — i.e., no longer able to divide). When certain kidney cells die, they don't come back.

Cartilage in the joints — as Dan, Jia, and Toghrul discovered — sits somewhere in between. Fully mature cartilage cells in the joint are largely post-mitotic in adult mice. But in young mice, there *is* a reservoir of cells that can generate cartilage; it diminishes radically with age and injury, until it vanishes altogether.†

---

signals that induce cells to adapt fates and intrinsic factors in cells that integrate these signals.

*There are rare instances of neuronal regeneration documented in animals and humans. The vast proportions of neurons, though, never divide or regenerate after injury.

†A recent paper by Henry Kronenberg and colleagues suggests that a fraction of mature cartilage cells can — given the right signals — "wake up" and start

It's as if every organ, and every cell system, has chosen its own kind of Band-Aid for repair and regeneration. Birds do it, bees do it — but they do it in ways and manners that are bird and bee (or liver and neuron) specific. Yes, there *are* some general principles: organs have resident "repair" cells that can sense injury and aging. But the idiosyncrasies of repair in each organ suggest that the individual cellular Band-Aids were cobbled together and remain unique to every organ. To understand injury and repair, then, we have to do it organ by organ and cell by cell. Or perhaps there's a *general* principle of repair that we are still missing — akin to the general principles of cell biology that researchers have found in other systems of cells.

In cell biological terms, then, it might be easier to imagine injury, or aging, for that matter, more abstractly, as a furious battle between a rate of decay and a rate of repair, with each rate unique for every individual cell, and individual organ. In some organs, injury overwhelms repair. In some organs, repair keeps apace with injury. In yet other organs, there's a delicate equilibrium between one rate and another. The body, in its steady state, seems to be maintained — suspended — in constancy.

---

dividing again. Whether these cells are similar to the ones that Dan, Jia, and Toghrul have found remains to be seen.

Don't just do something, stand there. But standing there, standing still, is not a statis but a frenetically active process. What appears as "stillness" — stasis — is, in fact, a dynamic war between these two competing rates. "At death you break up," Philip Larkin wrote, "the bits that were you / Start speeding away from each other for ever / With no one to see."

But death isn't a flying apart of organs. It is the withering grind of injury set against the ecstasy of healing. Tenderness, as Ryan puts it, combatting rot.

The central corporals in this pitched battle are cells — cells dying in tissues and organs, and cells regenerating tissues and organs. Return, for a moment, to the notion of homeostasis — the maintenance of a constancy in the internal milieu. We first evoked this idea to understand how the cell maintains its internal fixity. We then used it to understand how a healthy body adjusts to metabolic and environmental changes — salt load, waste disposal, sugar metabolism. We apply it, now, to the maintenance of balance between injury and repair. Death — the most absolute of absolutes — is, in fact, a relative balance between forces of decay and rejuvenation. If you tip the balance in one direction — when the rate of injury overwhelms the rate of recovery or regeneration, you fall off the edge. The osprey, buffeted by the changing winds, cannot remain suspended in midair.

THE SELFISH CELL

THE ECOLOGICAL EQUATION
AND CANCER

*Those who have not trained in chemistry or
medicine may not realize
how difficult the problem of cancer really
is. It is almost — not quite,
but almost — as hard as finding some
agent that will dissolve away
the left ear, say, and leave
the right ear unharmed.*
— William Woglom, 1947

We return in the end, full circle, to the cell
capable of infinite rebirth: the cancer cell.*
No cell's birth or rebirth has been studied as
intensely, or as passionately. And yet, despite

---

*Of course, there is no such thing as a singular "cancer
cell." Cancer is a diverse group of diseases, and even
a single cancer can have multiple types of cells. What
I have tried to do here is to distill some general
principles shared by most cancer cells. In later pages,
we will more clearly encounter how cancer cells differ
from each other, even within one patient.

650

decades of investigation, our attempts to thwart both the birth and rebirth of cancer have themselves been thwarted. Some features of the nature and the mechanisms of cancer's origin, regeneration, and spread have become clear. Much, though, still remains mystifying.

To understand the malignant division of a cancer cell, we might begin with the division of normal cells. Consider a cut in your hand. We might describe the response to the cut as a cascade of cellular events that restores the status of a tissue after injury — homeostasis at work. Blood leaks out. Platelets and clotting factors, induced by the tissue damage, gather around the wound. Neutrophils, sensing a danger signal, accumulate at the site as first responders to infection; they stand guard to ensure that pathogens don't get a chance to breach the boundaries of the self. A clot forms, and the wound is temporarily plugged.

And then healing begins. If the wound is shallow, the two ends of skin appose against themselves. If the wound is deep, fibroblasts from under the skin — the spindle-shaped cells that exist in virtually every tissue — crawl in to deposit a protein matrix underneath the wound. And then skin cells proliferate over the matrix to cover the wound, occasionally leaving a scar. Once they touch each other, the cells stop dividing. It takes a host of cells to coordinate this process. The wound has healed.

But here's the cell biological conundrum: What makes the skin cells start growing? And, more pertinently for cancer, what makes them *stop*? Each time we cut ourselves, why don't we grow a new appendage, like a tree growing a branch?

Part of the answer takes us back to the beginning of this book — to the genes that Hunt, Hartwell, and Nurse discovered that control the division of cells. When the cut occurs, signals from the wound, and from the cells that respond to it — intrinsic and extrinsic cues — activate a cascade of genes to make the repairing cells begin dividing. And when the healing is complete, and the skin cells touch each other, another set of signals inform the cells to exit the cycle. You might imagine these signals as accelerators and brakes in a car: when the road is wide open (immediately after the injury), the car accelerates, but when the traffic thickens, cell division gradually slows to a stop. This is regulated cell division, and it happens many millions of times every day in every human body. It is the basis of the development of an organism from a single cell. Why don't some embryos explode to twenty times their size? That is the basis for embryogenesis. Why don't we grow new limbs every time we get a cut? It is the basis for the continuous repair and regeneration of an organ. Why did Nancy Lowry, transplanted with her sister's cells, not burst with blood? That is the basis of our understanding of how blood stem cells make new progenitors but seemingly stop when normal blood counts have been restored.

But cancer is, in a sense, a disorder of internal homeostasis: its hallmark is that cell division is dysregulated. The genes that control these accelerators and brakes are broken — i.e., mutated — such that proteins that they encode, the regulators of cell division, no longer function in their appropriate contexts. The accelerators are permanently jammed, or the brakes fail permanently. More typically, it is a combination of both events — jammed accelerator genes and snapped brakes — that drives the dysfunctional growth of a cancer cell. The cars speed through the traffic jam, piling up on each other and causing tumors. Or they frantically move into alternate routes, causing metastasis. I don't mean to give cancer cells personalities. This is a Darwinian process, requiring natural selection: the cells that succeed are the fittest for survival. They are naturally selected to be the cells most adapted to grow and divide, in circumstances where they do not belong, and in tissues where they do not belong. Natural selection creates cells that disobey every law of belonging, except for the laws that they have created for themselves.

■■■■

As I described above, the "malfunctions" in the accelerator or brakes genes are caused by mutations — changes in DNA (and therefore changes in proteins) that dysregulate their function, so that they are typically permanently "on," or permanently "off." The jammed "accelerators" are called oncogenes; the "snapped" brakes are called tumor suppressors. Most of these cancer-causing genes are not genes that directly command the cell cycle (although some are). Rather, many of them are the commanders of the commanders: they recruit other proteins that further recruit others, until a malignant cascade of protein signals within a cell ultimately whips the cell into a kind of mitotic frenzy — to keep dividing without control. Cells pile on cells, invading tissues where they do not belong. They break the laws of cellular civility, of citizenship.

Beyond controlling cell division, many of these genes have diverse functions — activating or repressing the expression of other genes. Some of the genes co-opt the metabolism of the cell, enabling it to use nutrients to drive the malignant rebirth of the cancer cell. Some of them alter the normal inhibition that cells achieve once they contact each other; cancer cells heap on each other, even when normal cells would stop dividing.

One astonishing feature of cancer is that any individual specimen of cancer has a permutation of mutations that is unique to it. One woman's breast cancer can have mutations in, say, thirty-two genes; the second woman's breast cancer can have sixty-three, with only twelve overlapping between the two. The histological, or cellular, appearance of two "breast cancers" may look identical under the pathologist's microscope. But the two cancers may be *genetically* different — they behave differently and may require radically different therapies.

Indeed, this heterogeneity of "mutational fingerprints" — the set of mutations carried by an individual cancer cell — reaches down to the individual cell level. The woman's breast tumor which contained thirty-two mutations in its bulk? It may have an individual cancer cell that has twelve of the thirty-two mutations, and, sitting just adjacent to it, a cell with sixteen of the thirty-two — with some overlapping and some not. *Even a single breast tumor, then, is actually a collage of mutant cells — an assemblage of nonidentical diseases.*

We still don't have simple methods to understand which of these mutations is driving the pathological features of the tumor (driver mutations) and which are just encrusted on DNA as a consequence of the tumor-gathering mutations as it divides (passenger mutations). Some, such as c-Myc, are so common across multiple cancers that they are almost certain to be "drivers." Others are unique to particular forms of cancer, to leukemia, or a particular variant of lymphoma. For some mutant genes, we understand how they enable dysregulated, malignant growth. For some, we don't yet know.

When I went to meet Sam P. in the hospital, in May 2018, I was asked to wait outside. He was nauseated and excused himself to use the bathroom. He composed himself, and a nurse helped him back to bed.

It was nearing twilight, and he turned a bed lamp on. He asked the nurse if we might speak alone.

"It's over, isn't it?" he said, looking straight at my face, his brain boring a direct hole into the core of my brain. "Be honest," he said.

Was it really over? I mulled over the question. Here we had the strangest of cases — some of his tumors were responding to the immunotherapy, while others remained obdurately recalcitrant. And every time we increased the dose of the immune drugs, an autoimmune hepatitis — horror autotoxicus of the liver — would push us back. It was as if each individual, metastatic tumor had acquired its own program of rebirth and resistance, each battening down in its own niche in his body, each behaving as if it were an independent community of colonists trapped in its own island. We were fighting on multiple fronts at the same time — winning some, losing others. And every time we applied an evolutionary pressure on the cancer — an immunotherapeutic drug, say — some cell escaped the pressure and set up a new recalcitrant colony again.

I told him the truth. "I don't know," I said. "And I will not know, right till the very end." The nurse came in again to change the beeping IV line, and we changed the subject. One rule of cancer that I've learned is that it is like a fixated interrogator: it does not allow you to change subjects — even when you think you can.

Months ago, while he had been working at the paper, I had watched him and a bunch of his friends put together a playlist of music. I had borrowed the playlist for a party I was hosting. It had turned out to be my favorite set of songs.

"What are you listening to now?" I asked. For a while, the emollience of casual small talk soothed the tension, and a sense of normalcy descended on the room. Two guys talking about playlists. Rock and roll; hip-hop; rap. We spoke for another hour. And then, it was as if I had reached a place where the unavoidable questions could no longer be avoided. The fixated interrogator was back.

"Any advice, doc?" he asked. "What happens at the end?"

What happens at the end? It's a question

both ancient and unanswerable. I reached back to think about patients who had fought his kind of amorphous battle — winning, losing, winning — to think of what they had needed in their final weeks. I asked him to think of three things he could achieve. Forgive someone. Be forgiven by someone. And tell someone that he loves them.

Some truth had moved between us. It was as if he had understood why I had come to see him.

He was caught, off guard, by another wave of nausea. The nurse was called in and brought a basin. "Till next time," he said. "Next week?"

"Till next time," I said firmly.

I never saw Sam again. He died that week. I don't believe in rebirth, but some Hindus, among others, do.

The peculiar thing about the rebirth of a cancer cell is that the genetic programs that enable cancer cells to sustain malignant growth are shared, to some extent, with stem cells. If you look at genes turned "on" and "off" in leukemia stem cells, say, there is a striking overlap of that subset of genes with normal blood stem cells (making it, yet again, nearly impossible to find a drug that will kill cancer but spare stem cells). If you look at genes turned "on" and "off" in bone cancer cells, you find a similar subset of genes turned "on" and "off" in skeletal stem cells. And the overlaps continue: among the four genes that Shinya Yamanaka had turned "on" to transform normal cells into embryonic-stem-like cells (the iPS cells that won him the Nobel Prize), there is one called c-Myc — the very gene which, when dysregulated, becomes one of the main drivers of multiple forms of cancer. The relationship between cancer and stem cells, in short, is turning out to be far too uncomfortably close.

This begs two important questions. First, *do stem cells turn cancerous?* And conversely, does the population of cancer cells within the body have a *sub*population of cells within it that is responsible for cancer's continuous regeneration, just as blood and bones have reservoirs of stem cells within them? Is *that* the secret to cancer's continuous regrowth — a secret, specialized subpopulation of cells that acts as its regenerative reservoir? The first is a question of origin: *Where do cancer cells come from?* The second is the question of regeneration: *Why do malignant cells keep growing, while other cells have controlled, restricted growth?*

These questions continue to fuel fierce debates among oncologists and cancer biologists. Take the first question. Stem cells, or their immediate descendants, can certainly be made to turn cancerous in model systems. Researchers working with blood have shown that the introduction of a *single* gene into a mouse's blood stem cell descendant can create a lethal leukemia. That gene — in fact, a mutation that creates a fusion between two genes — encodes a many-fingered protein that can turn such a plethora of genes on and off, in cascade after cascade, that it can drive a stem cell toward an aggressive leukemia. Further mutations also accumulate as the cell progresses toward the leukemia.

But the converse is much harder to achieve: Can you take a fully mature, differentiated cell — a perfectly good citizen cell — and make it into a malignant actor? Yes you can, but with a lot of genetic shoving — i.e., by adding a gamut of extremely powerful cancer-goading genetic signals to the cell. Remember the glial cells that we met as accessories to the nervous system? They are fully mature; they don't grow uncontrollably. In a study performed in 2002, scientists led by Ron DePinho, then at Harvard (and now in Texas), took one such mature glial cell in a mouse, expressed powerful cancer-causing genes in it, and transformed it into a glioblastoma, a lethal brain tumor. Does the phenomena happen in real life? We don't know.

What about the second question? Do cancers have stem cells that act as a reservoir to keep them growing indefinitely? In Toronto, John Dick's group has demonstrated that a tiny fraction of the bulk of leukemia cells in the marrow are capable of regenerating the whole leukemia from scratch — just as a rare population of blood cells can repopulate blood (Dick has termed these "leukemia stem cells"). In other words, in some cancers, there is a "hierarchy" in which a small subset of cancer cells is uniquely capable of proliferating extensively and driving disease progression, while the rest of the cancer cells have little or no capacity to proliferate. These cancer stem cells are like the roots of an invasive plant. You cannot remove the plant without removing the roots — and by the same logic, you cannot kill the cancer without killing the cancer stem cells.

But the theory that *all* cancers have stem cells has its challengers. In Texas, Sean Morrison has argued that the cancer stem cell model is not relevant to some cancers, like melanoma, where most cells are capable of proliferating extensively and contributing to disease progression. The cells retain the ability to proliferate extensively, possessing stem-like properties. For these cancers, therapies must eliminate as many cancer cells as possible to have a chance of succeeding.

And there may be yet other cancers in which adherence to the cancer-stem-cell model varies from patient to patient. For example, in some breast cancers and brain tumors there might be cancer stem cells and non–stem cells, while in other breast and brain cancers there may be no such hierarchy. The normal laws of physiology — of stem cells — don't apply, because of the immense fluidity that cancer cells can accomplish by just flicking the switches of some genes.*

"Look," Morrison told me, "this is all going to become more complicated. Some cancers, including myeloid leukemias, really do follow a cancer-stem-cell model. But in some other cancers, there is no meaningful hierarchy, and it will not be possible to cure a patient by targeting a rare subpopulation of cells. The field has a lot of work to do to figure out which cancers, or even which patients, fall in each category."

This much, though, is certain: some cancer cells and stem cells "reprogram" the cell in profound ways. Genes are turned on and off inside the cells to enable their continuous rebirth. The difference is that in cancer, the program is perpetually locked — because the fixity of its mutations doesn't allow the cell to change its program of continuous division. In normal, healthy stem cells, the program is malleable, because the cell can differentiate — into osteoblasts, or cartilage cells, or red blood cells and neutrophils. Stem cells

---

*To be clear, cancer cells don't possess sentience, or a brain, that makes them flick switches on and off. It is *evolution* that selects for the cells that have turned on certain genes that enable continuous regrowth.

can change the programs of identity; as I said before, they balance selfishness (self-renewal) with self-sacrifice (differentiation). The cancer cell, in contrast, is trapped — imprisoned in a program of perpetual rebirth. It is the ultimate selfish cell.

Worse, if you apply an evolutionary pressure — a medicine to target a particular gene — there's enough heterogeneity and fluidity in cancer cells that allows them to select a different gene program to resist the medicine. A cell with a resistant mutation can grow out. A cell with a slightly altered genetic program (this is what I meant by the "fluidity" of cancer's genetic program). A cell in a different metastatic site, beyond the drug's reach, can activate a new genetic program to resist being detected and eliminated.

For the last few decades, we've tried to target specific genes, or specific mutations in cancer cells, to try to attack cancer. Some of them have been remarkable successes — Herceptin, for instance, for Her-2 positive breast cancer, or Gleevec, for a kind of leukemia called CML. But trials of other gene-targeted mutations (personalized cancer medicine) have proved more modestly successful or have failed altogether. Partly because the cells acquire resistance. Partly, because of the heterogeneity of cancer cells. Partly, because the commonalities between cancer cells and normal cells, particularly stem cells, puts a natural ceiling on the drug before the medicine becomes toxic to the body. It is cell biology's version of what Kant might have called the terrifying sublime.

As I left Sam's room in the hospital, I thought of his playlist. Imagine that all the genes in the cells — the entirety of its genome — is a fixed, preselected playlist. Stem cells can choose which songs to play, and in what order, as they change from self-renewal to differentiation. When they self-renew, they play a certain set. When they differentiate, they play a different set.

In cancer, the fixity of mutations won't allow the order of the songs to change. The accelerators are stuck in the on position and the brakes stuck in the off position. Consequently, unlike with normal stem cells, the body has little capacity to regulate their activity. The playlist is set. The same series of songs get played over and over, like a malevolent tune that one cannot get out of one's head. And when you apply a selective pressure, such as a drug or immunotherapy, it switches to a new list of genes, or even scrambles the songs — remixing a crazed jumble of hip-hop and Chopin, say — in its playlist, such that malignant cells can escape the drug. And then repeat: now the cancer cell has a *new,* fixed malignant tune that it cannot get out of its head.

When the comprehensive list of genes that

drive the growth of cancer cells were first identified in the mid-2000s, there was an exuberance that we had unlocked the key to cures for cancer.

"You have a leukemia that has mutations in Tet2, DNMT3a, and SF3b1," I would tell a bewildered patient. I would look at her triumphantly, as if I'd solved the Sunday crossword puzzle.

She would look at me as if I was from Mars.

And then she would ask the simplest question: "So, does that mean that you know the drugs that are going to cure me?"

"Yes. Soon," I would say, with exuberance. For the linear narrative ran thus: isolate the cancer cells, find the altered genes, match it with medicines that target those genes, and kill the cancer without harming the host.

And so researchers ran two kinds of trials to prove this idea right (how could it *not* be right?). The first, called "basket" trials, would put different cancers (lung, breast, melanoma) that happened to share the same mutations into the same basket and treat them with the same drug. *After all: Same mutation, same drug, same basket, same responsiveness — no?* The results were sobering. In one landmark study published in 2015, 122 patients with several different types of cancer — lung, colon, thyroid — were found to have the same mutation in common, and thus were treated with the same drug, vemurafenib. The drug worked in some cancers — there was a 42 percent response rate in lung cancer — but not at all in others: for example, colon cancer had a 0 percent response rate. And even most of the responses did not last, landing patients back at square one after a fleeting remission.

The second kind of trial was the obverse — an umbrella trial. Here, one kind of cancer, say lung cancer, was checked for different mutations, and each lung cancer with a particular mutational set was put under a different umbrella. Each individual lung cancer, under its own "umbrella," was given a different set of drugs specific for its particular combination of mutations. *After all: Different mutations, different umbrellas, different therapies, and therefore specific responses — no?* It didn't work, either. A major trial, called BATTLE-2, also generated sobering data, and most of the cancers hardly responded. "Ultimately," one reviewer commented dejectedly, "the trial failed to identify any new promising treatments."

"We biomedical scientists are addicted to data, like alcoholics are addicted to cheap booze," Michael Yaffe, a cancer biologist from MIT, wrote in the journal *Science Signaling*. "As in the old joke about the drunk looking under the lamppost for his lost wallet, biomedical scientists tend to look under the sequencing lamppost where the 'light is brightest' [because that's where it's easiest to see] — that is, where the most data can be obtained as quickly as possible. Like data junkies, we continue to look to genome sequencing when the really clinically useful information may lie someplace else."

Sequencing is seduction. It is data, not knowledge. So where is the "really clinically useful information" sitting? Somewhere, I believe, in an intersection between the mutations that the cancer cell carries and the identity of the cell itself. The *context*. The type of cell it is (lung? liver? pancreas?). The place where it lives and grows. Its embryonic origin and its developmental pathway. The particular factors that give the cell its unique identity. The nutrients that give it sustenance. The neighboring cells on which it depends.

Perhaps a new generation of cancer therapies will get us over this addiction. For decades, we have imagined cancer as the consequence of an individual malignant cell. The "cancer cell" has turned into an icon of the malignant behavior of the disease, of cellular autonomy gone rogue (there is even a scientific journal called *Cancer Cell*). The cancer cell has become the locus of our attention. Kill the cell, and we defeat cancer. "This tumor is invading the brain," one surgeon says to another in an operating room. (By contrast, whoever says that the cold catches you?) Subject, verb, object: cancer is the autonomous actor, the aggressor, the mover. The host — the patient — is the hushed audience-member, the afflicted victim, the passive onlooker. The context she provides, the particular behavior of *her* cancer cells, their location, their slippery mobility, her immune response to it; why would any of this matter?

But in Sam's case, each metastatic site of cancer behaved differently; his body was far from a passive onlooker. The behavior of the metastasis in his liver was not the behavior in the outer lobe of his ear. And some of his organs were mystifyingly spared, while others were densely colonized.

The question gets to the heart of what makes cancer metastasis survive in some places, while other sites — the kidney and spleen, notably — never seem to attract metastasis. Perhaps cancer cells, like organs and organisms, should also be imagined as a *community* — and that, too, a community that can only take up residence in a particular place and at a particular time. The metaphors of cancer are changing. Cancer as a cooperative assembly. Cancer as an ecology gone wrong. Cancer as a malevolent pact between a rogue cell and the environment which it co-opts — an armistice between the cell and the tissue in which it can flourish. "Cancer is no more a disease of cells than a traffic jam is a disease of cars," the British physician and cancer researcher D. W. Smithers wrote in the *Lancet* in 1962. "A traffic jam is due to a failure of the normal relationship between driven cars and their environment and can occur whether they themselves are running normally or not." Smithers had overstepped in his provocation. The uproar that ensued was clamorous and immediate (Bob Weinberg, one of the most influential cancer researchers, told me that it was "utter nonsense"). But Smithers — pushing provocation, to be sure — was trying to focus attention away from the cancer cell to the *behaviors* of these cells in their real environments.

And so, we are inventing new metaphors for the disease. Forget mutations. Attack metabolism. Some cancer cells, for instance, become highly dependent on ("addicted to," is the medical jargon) particular nutrients and particular metabolic pathways. In the 1920s, the German physiologist Otto Warburg discovered that many cancer cells use a fast and cheap method of glucose consumption to generate energy. Malignant cells prefer oxygen-free fermentation rather than the deep-slow burn that we had encountered in mitochondria, even when there is plentiful oxygen. Normal cells, in contrast, almost always use a combination of slow and fast burn — oxygen-dependent and -independent — mechanisms to generate energy. What if this unique metabolic quirk of malignant cells could be used to drive an inroad to killing cancer?*

*No one knows why cancer cells prefer this fast and cheap (but highly inefficient) mechanism of producing energy. After all, oxygen-dependent respiration (aerobic respiration) generates thirty-six molecules of ATP, while oxygen-independent fermentation (anaerobic respiration) generates only two such molecules — an eighteenfold difference. Why would a cancer cell use an inefficient energy-generating system when much more energy could be extracted, and resources were not limiting (a leukemia cell, for instance, is literally bathed in blood; there are enough nutrients and oxygen to use aerobic respiration)? Part of the answer may lie in the fact that the use of oxygen-dependent reactions to generate energy create toxic by-products — highly reactive chemicals that are harmful to cells that then need to be dispensed and cleansed. Toxic by-products of oxygen-dependent respiration include chemicals that induce mutations in DNA, which, in turn, activate an apparatus in cells to stop dividing (remember the G2 checkpoint, when cells check to ensure the quality of its DNA). Cancer cells may have evolved to "make the best of it" — in essence, sacrificing energy efficiency to keep away from these toxic by-products. This is one of many hypotheses; others have argued for other reasons for the preference of cancer cells for fermentation. Recent work from some researchers, such as Ralph DeBerardinis, has shown that the Warburg effect — i.e., the cancer cell's use of the non-

Another clinical trial that my group is running with a team from Cornell University and with Lew Cantley, now at Harvard, hopes to pursue the unnervingly universal way that cancers depend on sugar or protein metabolism that is different from normal cells. Working with Cantley, we've discovered that some (but not all) cancers use insulin — whose release is prompted by glucose — as a mechanism of resistance to an otherwise potent anticancer drug. In other words, cancer cells are, indeed, toxified by this drug — except, like wily criminals, they learn to use insulin to circumvent the drug's toxicity. Which raises the question of the unique dependence of cancer cells — mutations aside — on some particular nutrients. If we disabled the particular ways cancer cells use nutrients, and then unleash drugs

mitochondrial pathway to generate energy — may be exaggerated by the artificial conditions that we use to grow cancer cells in the lab compared to how they grow in the real body. When we culture cancer cells in the lab, we typically add very high levels of glucose to the culture, and this may shift metabolism toward the non-mitochondrial pathway. This said, the Warburg effect is still real: some "real" cancers growing in humans — not in the lab — use the non-mitochondrial pathway as their major mechanism to generate energy, but we may have overestimated the extent of the effect.

on the malignant cells, would they finally become "resensitized" to the drug? Or deplete the body of Proline, an amino acid that some cancers are addicted to, and thereby choke them nutritionally?

Or concentrate on the evasion of immunity. Jim Allison and Tasuku Honjo used the idea that all cancers must, at some point, find ways of resisting the immune system. Uncloak cancer's cloak, and you have a therapy that doesn't seem to depend on the immune system. Starve cancer's blood vessels, an idea championed by the researcher Judah Folkman in the 1990s. Create engineered T cells, à la Emily Whitehead, to attack her leukemia.

But first, understand the cancer cell's physiology *as a cell* in the context that it grows — the way we understand every other cell: the organ it lives in, the supportive cells it surrounds itself with, the signals it sends, the dependencies and vulnerabilities that it possesses.

There are mysteries beyond mysteries. Engineered T cells are potently active against leukemias and lymphomas, but fail against ovarian and breast cancer. Why? The kind of immunotherapy used in Sam's case eliminated the tumors in his skin, but not in his lungs. Why? As one of my own postdoctoral researchers discovered, our method of insulin depletion, via diet, decelerated endometrial and pancreatic cancers but *accelerated* the development of some leukemias in mice. Why? We don't know what we don't know.*

*Given this chapter's focus on the cancer cell, its behavior, migration, and metabolism, I have consciously chosen not to address cancer prevention and early detection. Some of these topics were covered in my earlier book *The Emperor of All Maladies: A Biography of Cancer* (2016) and more recent advances in prevention and early detection will be updated in a future edition.

THE SONGS OF THE CELL

I do not know which to prefer,
The beauty of inflections
Or the beauty of innuendoes,
The blackbird whistling
Or just after.
— Wallace Stevens, "Thirteen
Ways of Looking at a Blackbird"

In his 2021 book on ecology and climate, *The Nutmeg's Curse: Parables for a Planet in Crisis*, Amitav Ghosh recounts the story of an eminent professor of botany who accompanies a young man from a local village to guide him through a rain forest. The young man is able to identify each of the various plant species. His acumen stuns the professor, who compliments him on his knowledge. But the man is dejected. He "nods and replies with downcast eyes. 'Yes, I've learned the names of all the bushes, but I've yet to learn the songs.'"

Many readers might read the word *song* as metaphorical. But in my reading, it's far from a metaphor. What the young man laments is that he hasn't learned the *interconnectedness* of the individual inhabitants of the rain forest — their ecology, interdependence — how the forest acts and lives as a whole. A "song" can be both an internal message — a hum — and, equally, an external one: a message sent out from one being to another to signal interconnectedness and cooperativity (songs are often sung together, or to one another). We can name cells, and even systems of cells, but we are yet to learn the *song*s of cell biology.

This, then, is the challenge. We've divided the body into organs and systems — organs that perform discrete functions (kidneys, hearts, livers) and systems of cells (immune cells, neurons) that enable these functions. We've identified the signals that move between them — some at short range and some at long range. That's already a radical advance from Hooke and Leeuwenhoek, who first envisioned the body as agglomerations of unitary, independent living blocks. It brings us closer to Virchow, who imagined the body as a citizenship.

But there are still gaps in our understanding of the interconnectedness of cells. We are still living in a world where we imagine the cell, as Leeuwenhoek did, as a "living atom" — unitary, singular, and isolated, a spaceship floating in body-space. Until we leave that

675

atomistic world, we will not know, as the English surgeon Stephen Paget asked, why the liver and spleen are the same size, are anatomical neighbors, possess virtually the same flux of blood — and yet one (the liver) is among the most frequent sites of cancer metastases, while the other (the spleen) rarely has any? Or why patients with certain neurodegenerative diseases — Parkinson's among them — have a markedly lower risk of cancer. Or why, as Helen Mayberg told me, the patients who describe their depression as an "existential ennui" (her words) typically do not respond to deep brain stimulation, while those that describe themselves as "falling into vertical holes" often do. Like the dejected man in the rain forest, we have learned the names of the bushes, but not the songs that move between the trees.

Years ago, a friend told me a story that still resonates. He was walking with his grandfather, visiting from Cape Town, South Africa, when the grandfather stopped at a random apartment building in Newton, Massachusetts, where many first- and second-generation immigrant Jews had settled. My friend's great-grandfather had emigrated to South Africa from Lithuania. The grandfather walked up to the building and wanted to look at the names printed by the doorbells of the apartments. "But grandpa," my friend protested, "we don't know anyone who lives in the building." The grandfather paused and smiled. "Oh no," he said, "we know *everyone* who lives in this building."

To build new humans out of cells, we need knowledge that is not just names, but the interconnectedness between names. Not addresses, but neighborhoods; not identification cards, but personalities, stories, and the histories that accompany them.

Perhaps, as we near the end of this book, we might pause to reflect on one of the most potent philosophical legacies of twentieth-century science — and its limitations. "Atomism" argues that material, informational, and biological objects are built out of unitary substances. Atoms, bytes, genes, I had written in an earlier book. To this we might add: *cells*. We are built of unitary blocks — extraordinarily diverse in shape, size, and function, but unitary nonetheless.

Why? The answers can only be speculative. Because, in biology, it is easier to evolve complex organisms out of unitary blocks by permuting and combining them into different organ systems, enabling each to have a specialized function while retaining features that are common across all cells (metabolism, waste-disposal, protein synthesis). A heart cell, a neuron, a pancreatic cell, and a kidney cell rely on these commonalties: mitochondria to generate energy, a lipid membrane to define its boundaries, ribosomes to synthesize its proteins, the ER and Golgi to export proteins, membrane-spanning pores to let signals in and out, a nucleus to house its genome. And yet, despite the commonalities, they are functionally diverse. A heart cell uses mitochondrial energy to contract and act as a pump. A beta cell in the pancreas uses that energy to synthesize and export the hormone insulin. A kidney cell uses membrane-spanning channels to regulate salt. A neuron uses a different set of membrane channels to send signals that enable sensation, sentience, and consciousness. Think of the number of different architectures you can build with a thousand differently shaped Lego blocks.

Or perhaps we might reframe the answer in evolutionary terms. Recall that unicellular organisms evolved into multicellular organisms — not once, but many independent times. The driving forces that goaded that evolution, we think, were the capacity to escape predation, the ability to compete more effectively for scarce resources, and to conserve energy by specialization and diversification. Unitary blocks — cells — found mechanisms to achieve this specialization and diversification by combining common programs (metabolism, protein synthesis, waste disposal) with specialized programs (contractility in the case of muscle cells, or insulin-secreting capacity in pancreatic beta cells). Cells coalesced, repurposed, diversified — and conquered.

But powerful as it might be, "atomism," we've learned, is reaching its explanatory limits. We can explain much about the physical, chemical, and biological worlds through evolutionary agglomerations of atomistic units, but those explanations are straining at their leashes. Genes, by themselves, are strikingly incomplete explanations of the complexities and diversities of organisms; we need to add gene-gene interactions and gene-environment interactions to explain organismal physiology and fates. Decades ahead of her time, the geneticist Barbara McClintock called the genome a "sensitive organ of the cell." The words *organ* and *sensitive* reflected ideas totally foreign to geneticists in the fifties and sixties. Arguing against the atomistic gene-by-gene approach favored by geneticists, McClintock proposed, the genome could only be interpreted as a whole — as a "sensitive organ" — that was responsive to its environment.

By that same logic, cells, by themselves, are incomplete explanations for organismal complexities. We need to factor in cell-cell interactions, and cell-environment interactions — ushering in holism in cell biology. We possess rudimentary terms for these interactions — ecologies, sociologies, "interactomes" — but still lack models, equations, and mechanisms to understand them. I return, often, to think of disease as a violation of the social compacts between cells.

Part of the problem is that the word *holism* has become scientifically defiled. It has become synonymous with the mushing of everything we understand into a malfunctioning, soft-bladed (and soft-headed) blender. To rephrase Orwell: one equation good; four equations bad.

And then things got worse. A variant of postmodern scientific thought threw the equations, with the blackboards they were written on, into the trash; the baby went with the bathwater. But that, too, is an equal and opposite nonsense: a Newtonian ball thrown into Newtonian space *does* follow Newtonian laws. The laws that govern the ball are as real and tangible as they were during the conception of the universe. By the same logic, a cell, and a gene, are real. It's just that they aren't "real" in isolation. They are fundamentally cooperative, integrating units, and together, they build, maintain, and repair organisms. I cannot help you hold both ideas in your head at the same moment. But perhaps some experience with non-Western philosophies might help here: "cooperative" and "unitary" — selfless and self-regarding — are not mutually exclusive ideas. They exist in parallel.

Universal principles satisfy us — *one equation good* — because they satisfy our belief in an ordered universe. But why must "order" be so soldierly, so singular, so *unifest* (as opposed to manifest)? Perhaps one manifesto for the future of cell biology is to integrate "atomism" and "holism." Multicellularity evolved, again and again, because cells, while retaining their boundaries, found multiple benefits in citizenship. Perhaps we, too, should begin to move from the one to the many. That, more than any other, is the advantage of understanding cellular systems, and, beyond that, cellular ecosystems. We need to know everyone who lives in this building.

In January 1902, just as the *danse-macabre* of German sectarian division, based on the pseudoscience of racial and biological anthropology, was beginning to whir around him, Rudolf Virchow, running from appointment to appointment, missed his footing while stepping off an electric streetcar on Leipziger Street in Berlin. He fell off, injuring his thigh.

He had fractured his femoral bone. By then, he had become weak and debilitated — "a little, yellow-skinned, owl-faced, spectacled man," as an assistant wrote, "with peculiarly piercing yet slightly veiled eyes, which were conspicuously lacking in eyelashes. The eyelids were parchment-like and thin as paper [. . .] He was eating a roll and butter as we entered, and beside his plate stood a cup of café-au-lait. This was his lunch; his only refreshment between breakfast and dinner."

A cascade of cellular pathology was

unleashed. The fractured hip was likely the consequence of brittle bone, and the fragility of bone the consequence of aging bone cells unable to maintain or repair the structural integrity of the femur.

He spent the summer recovering, but then there were further setbacks: an infection from a weakened immune system (another cellular alteration) that then cascaded into heart failure (a malfunction of cardiac cells). System by system, the societies of cells that had held a man in place fell apart. He died on September 5, 1902.

Virchow continued to work on his understanding of cellular physiology and its converse, cellular pathology, right until his moment of death. The many seminal ideas sparked by his work, and their many offshoots over the ensuing decades, are his lasting legacy and the lessons of this book. His founding tenets of cell biology have expanded into at least ten that I can enumerate, but there will be more as we deepen our understanding of cells:

1. All cells come from cells.
2. The first human cell gives rise to all human tissues. Ipso facto, every cell in the human body can be produced, in principle, from an embryonic cell (or stem cell).
3. Although cells vary widely in their form and function, there are deep physiological similarities that run through them.

4. These physiological similarities can be repurposed by cells for specialized functions. An immune cell uses its molecular apparatus for ingestion to eat microbes; a glial cell uses similar pathways to prune synapses in the brain.
5. Systems of cells with specialized functions, communicating with each other through short- and long-range messages, can achieve powerful physiological functions that individual cells cannot achieve — for example, the healing of wounds, the signaling of metabolic states, sentience, cognition, homeostasis, immunity. The human body functions as a citizenship of cooperating cells. The disintegration of this citizenship tips us from wellness into disease.
6. Cellular physiology is thus the basis for human physiology, and cellular pathology is the basis for human pathology.
7. The processes of decay, repair, and rejuvenation in individual organs are idiosyncratic. Specialized cells in some organs are responsible for consistent repair and rejuvenation (blood rejuvenates through human adulthood, albeit at diminished rates), but other organs lack such cells (nerve cells rarely rejuvenate). The balance between injury/decay and repair/rejuvenation ultimately results in the integrity or degeneration of an organ.

8. Beyond understanding cells in isolation, deciphering the internal laws of cellular citizenship — tolerance, communication, specialization, diversity, boundary-formation, cooperation, niches, ecological relationships — will ultimately result in the birth of a new kind of cellular medicine.

9. The capacity to build new humans out of our building blocks — i.e., cells — lies very much within the reach of medicine today; cellular reengineering can ameliorate, or even reverse, cellular pathology.

10. Cellular engineering has already allowed us to rebuild parts of humans with reengineered cells. As our understanding of this arena grows, new medical and ethical conundrums will arise, intensifying and challenging the basic definition of who we are, and how much we wish to change ourselves.

These tenets continue to animate, drive — and even surprise — us today. As doctors, we learn these principles. As patients, we live them. As humans entering a new realm of medicine, we will have to learn how to embrace them, challenge them, and incorporate them into our cultures, societies, and selves.

EPILOGUE
"BETTER VERSIONS OF ME"

If we could be less human.
if we could stand out of the range
of the cataracts of the given
and not find our pockets
swollen with change
we haven't — but must have —
stolen, who wouldn't?

— Kay Ryan, "The Test We
Set Ourselves," 2010

But I too made things
That may one day be
Better versions of me.

— Walter Shrank, "Battle
Cries of Every Size," 2021

A few weeks before Paul Greengard died, we took another walk on the slippery marble stones at Rockefeller University. We walked past the building where George Palade had started his basement lab, and dissected parts and subparts of the cell using biochemistry

685

and electron microscopy. Part of the campus was cordoned off, and scaffolded; workers were building a new lab. I was interested in talking to Greengard about making new humans.

"You mean genetically?" he asked.

He was referring to new technologies, gene editing among them, that had allowed researchers such as He Jiankui to attempt deliberate alterations in the human genome.

But I didn't mean genetically — or at least not *just* genetically. Think of Emily Whitehead, whose immune system was rebuilt with T cells weaponized to kill her cancer. Louise Brown, the first baby born through IVF. Or Timothy Ray Brown, the patient with AIDS who had received a bone marrow transplant from a donor with cells resistant to HIV. He, too, had been rebuilt with new cells. Nancy Lowry, living with her sister's blood. Helen Mayberg's first patients, implanted with tiny electrical stimulators, with electrodes and bolts of energy running through the neurons in their brains.

Why not extend the building of human parts to other cellular systems? Rebuild the failing pancreas in a type 1 diabetic with insulin-producing cells, or replace an arthritic woman's grinding joints with new cartilage. I told him about Verve and how it was trying to create humans with liver cells that would permanently lower cholesterol.

Greengard nodded. He had just listened to a seminar on neural organoids — tiny clusters of neuronal cells that, cultured in a matrix-like solution in the lab, organize themselves into ball-like shapes. Researchers had started calling them "mini-brains" — an exaggeration, no doubt — but there was something undeniably creepy about watching tiny balls with human neurons firing and communicating with each other. Had a thought, no matter how garbled, ever fired within one such organelle? If we poked them, did they feel sensation?

One morning, Toghrul Jafarov, the postdoc in my lab, showed me a culture full of Gremlin-expressing cells he had harvested from a mouse. They glowed green because that fluorescent jellyfish protein — GFP — had been inserted into their genomes.

At first, nothing happened; the cells sat obdurately in the flask. But then, they began to divide, slowly at first, and then furiously. They formed tiny whirls of cartilage around themselves.

When the flask was full of millions of cells, Jafarov drew them up into a tiny needle, about as thick as two strands of human hair, and injected them into the knee joints of mice. He'd been working on the procedure for months and perfected it slowly: he had to enter the joint with the needle causing no injury, like a perfect diver slicing into water without causing a splash.

A few weeks later, he showed me the knee. The cells had formed a thin layer of cartilage at the joint. We had made a chimeric knee, with a jellyfish protein in its cells, glowing silently within the mouse. It was far from perfect — just a few cells had engrafted — but it was clearly the first step to build a new cellular joint.

In the strangest of Kazuo Ishiguro's novels, *Never Let Me Go,* we are cast into the future when human cloning has become legalized. We meet a group of schoolchildren. They live in a boarding school named Hailsham — perhaps a veiled reference to the sham school that houses them. Slowly, the students discover that their sole purpose is to act as donors for the adults that they've been cloned from. Bit by bit, organs are removed from them and "donated" to their older clones. Once the organs have been harvested, the children will inevitably die.

At one point in the novel, one of the children, Kathy, encounters some drawings made by Tom, a friend and eventual lover. "I was taken aback," she says, "at how densely detailed each one was. In fact, it took a moment to see they were animals at all. The first impression was like one you'd get if you took the back off a radio set: tiny canals, weaving tendons, miniature screws and wheels were all drawn with obsessive precision, and only when you held the page away could you see it was some kind of armadillo, say, or a bird [. . .] For all their busy, metallic features, there was something sweet, even vulnerable about each of them."

The "tiny canals, weaving tendons, miniature screws and wheels" are, perhaps, metaphors for anatomy — organs and cells — redrawn as movable fixtures that can be pulled out, reassembled, and transferred, like blocks, from one human to another. As the critic Louis Menand wrote in the *New Yorker*, "The shadowy backdrop in *Never Let Me Go* is genetic engineering and associated technologies." But that's not quite right. The backdrop is *cellular* engineering.

I read Ishiguro's novel while Jafarov harvested the cartilage cells from one mouse and transferred them into another. The first mouse had to be sacrificed. The experiment was not in vain: he was searching for a cure for human arthritis, a crippling, debilitating illness that leaves hundreds of thousands unable to move. But I cannot write this, and think of the experiment, without feeling a pang of remorse and the inescapable shiver of concern of what such a future might bring.

We've met "new humans" throughout this book. And we've encountered ideas about making newer humans, part by part, using cells. Some of these ideas lie in the far future, perhaps. But some of it is happening as I write. As I described earlier, a group of researchers including Jeff Karp and Doug Melton are making an "artificial pancreas," hoping to implant this neo-organ into type 1 diabetics. Two companies, Vertex and ViaCyte, are already enrolling patients to infuse them with pancreatic insulin-producing cells created by making stem cells turn into pancreatic cells. At the Mayo Clinic, scientists are making a bioartificial liver, building them out of liver cells. Hearts used to be harvested from cadavers — but one ambitious cell-engineering project involves mounting cardiac muscle cells, derived from stem cells, on a collagenous scaffold that resembles the heart to build a bioartificial heart out of cells.

Ishiguro's novel is described as science fiction. And fictional it is: I cannot imagine us cloning and sacrificing humans to act as organ donors. But what about cellular engineering as a means to human enhancement? One experiment that Toghrul Jafarov is trying to perform in the lab is to inject the bone-cartilage stem cells into the limbs and joints of very young mice. Would they become taller — with jackrabbit limbs, except in mouse bodies? "Mice-rabbits"? Again, this is not an experiment performed in vain. There are humans with extreme short stature, some of whom want to be taller. But not all: some with short stature argue that their lives are perfectly fine. Some are healthy and happy. Ascribing a "disability" to them, they argue, is to assign some unique "ability" (can height be construed as an *ability*?) to the rest of us.

But what if a "normal" human wanted to enhance his or her height through cellular therapy? That seems to be not science fictional; it could lie within our imagination of a shadowy future. Would we stop them — and if so, why?

The philosopher Michael Sandel has been pondering this question for a while. Years ago, I met him briefly in Aspen, Colorado, following a seminar he presented on genetic engineering and human cloning as a quest for perfection. It was a beautiful afternoon among the hills and quivering Aspen leaves. In his blue jacket and tie, Sandel looked manicured and professorial. (Then again, he is a professor in Harvard's philosophy department.) The talk was a provocation. Sandel challenged the human quest for enhancement, basing his argument, ultimately, on what the late theologian William May called "an openness to the unbidded."

The "unbidded" — the vagaries, or gifts, of chance — Sandel argues, is essential to human nature. Our children surprise us with their gifts, and those surprises, and our responses to them, would be extinguished if each one of us launched a quest for enhancement, for perfection. It would violate an essential part of the human spirit to do away with the "unbidded gifts." Better wrestle with these vagaries and make the best of them.

In 2004, Sandel consolidated his ideas in an essay, "The Case Against Perfection," which he soon expanded into a book. Reviewing the book in the *Times,* the ethicist William Saletan wrote: "[Sandel's] deeper worry is that some kinds of enhancement violate the norms embedded in human practices. Baseball, for example, is supposed to develop and celebrate an array of talents. Steroids warp the game. Parents are supposed to cultivate children through unconditional as well as conditional love. Selecting a baby's sex betrays that relationship."

To argue against human enhancement, Saletan continues, "Sandel needs something deeper: a common foundation for the various norms in sports, arts and parenting. He thinks he has found it in the idea of giftedness. To some degree, being a good parent, athlete or performer is about *accepting and cherishing the raw material you've been given to work with* [italics my own]. Strengthen your body, but respect it. Challenge your child, but love her. Celebrate nature. Don't try to control everything [. . .] Why should we accept our lot as a gift? Because the loss of such reverence would change our moral landscape."

I used to find Sandel's argument persuasive — but as the combined forces of genetics and cell engineering extend their reach to touch new depths of the human body and personhood, the "moral landscape" has changed radically: the borderlands between the emancipation from the ravages of disease (extreme short stature, or muscle-wasting cachexia) and the augmentation of human features (increasing height or bulking up muscle) are blurring. Augmentation *has become* the new emancipation. And the more the borderlands between ailment and enhancement blur, the more easily the "raw" material that Saletan describes is perceived as precisely that: "raw," and therefore awaiting to be molded into something else — a new kind of human, built anew. "Cooked," the opposite of "raw," carries a connotation of augmentation, but also of cheating. But is enhancement cheating? What if it was used to prevent a disease for an illness that may or may not occur. Should an aging knee be injected with cartilage-forming stem cells *before* it succumbs to osteoarthritis — i.e., in a state of *pre*-disease?

In Silicon Valley, not far from Stanford's hospital where children with leukemia await transplants to generate new blood, a start-up called Ambrosia is offering transfusions of matched young blood plasma "harvested from youths between sixteen and twenty-five years old" to supposedly rejuvenate the creaking, but very wealthy, shriveling bodies of aging billionaires. Rather than draining old blood from the dead, you infuse young blood into the aged — embalmment in reverse (I am tempted to draw an analogy to vampirism, but perhaps we will find a new euphemism for this chilling kind of attempted cellular rejuvenation — "rebalming," or "demummification"). One liter of "young blood" costs $8,000; two liters is a bargain at $12,000. In 2019, the FDA issued a stern cautionary note against the program, citing lack of benefit, although Ambrosia argues that the therapy works.

"Accepting and cherishing the raw material you've been given to work with." What raw material? Sandel's and Saletan's discussion focuses on genes — and indeed, gene therapy, gene editing, and genetic selection has preoccupied ethicists, doctors, and philosophers for the last decade. But genes are lifeless without cells. The real "raw material" of the human body is not information, but the way that information is enlivened, decoded, transformed, and integrated — i.e., by cells. "The genomic revolution has induced a kind of moral vertigo," Sandel writes. But it is the cellular revolution that will actualize this moral vertigo.

William K. was a young man with an ancient illness. I saw him when I was a fellow in hematology in Boston — first in the wards, and then in my clinic. He was twenty-one and had sickle cell anemia. He was admitted to the hospital about once a month with a "crisis" — a syndrome involving pain in his bones and chest so unfathomably intense that only a continuous intravenous morphine drip could quench it.

Sickle cell anemia is a disease that we understand at a cellular and molecular level. It is a disease of hemoglobin, the oxygen-carrying molecule found within red blood cells, which may be one of the most sophisticated molecular machines devised by evolution. Hemoglobin is a complex of four proteins; and it's shaped like a four-leafed clover. Two of the "leaves" are formed by a protein named Alpha-Globin, while two are another protein named Beta-Globin.

Nestled at the center of each of the proteins is another chemical: heme. And at the center of heme sits an atom of iron. It is a doll-within-a-doll-within-a-doll scheme. Red blood cells contain hemoglobin molecules that contain heme that, in turn, clasps iron atoms. It's the iron that binds and unbinds the oxygen.

The elaborate apparatus built around those four iron atoms in the hemoglobin molecule has a distinct molecular purpose. Red blood cells cannot simply bind oxygen and hold on to it; they have to release it. The red cells pick their payload — oxygen — from the capillaries of the lung and ferry it around. And when the cells reach oxygen-poor environments in the body — with the heart muscle pumping and pushing them around minute after minute — hemoglobin, literally, twists and unclasps the oxygen that the iron atoms have bound. Hemoglobin is blood's hidden secret — a complex of proteins so vital to our existence as organisms that we have evolved a cell whose principal job is to act as a suitcase to carry it around.

But this system of oxygen delivery fails if hemoglobin, the oxygen carrier, is malformed. In sickle cell anemia, you inherit a mutation in both copies of the Beta-Globin gene. The mutation is exquisitely subtle: it results in the change of *one* amino acid in Beta-Globin. But the effects are devastating: that single change creates a protein that — no longer a "glob" — clumps into fiber-like clusters in oxygen-poor environments. The fibrous clumps deform the shape of the red blood cell. Rather than a coin-shaped cell that floats easily through blood, the clumps of hemoglobin tug at the cell's membrane. The cell shrivels into a quarter-moon crescent, a sickle that cannot float easily in the blood; it aggregates and clogs the blood vessels, especially in tissues that have low oxygen content: the deep insides of the marrow, in the distal tips of the fingers and toes, or in the depths of the intestines. The pain that this clogging of capillaries produces is like a corkscrew driven into the bones

(William described each episode like being forced to enter a torture chamber. "And then all the doors around you get locked"). It's like a heart attack of the marrow, or of the intestines. The medical term for this syndrome is a "sickle cell crisis."

William K. had one such episode every month. He would be admitted to the hospital, writhing in agony. When the pain partially subsided, he would be discharged home with oral pain medicines. But the twin demons — of the possibility of addiction to opioids, and the anticipation of the next crisis — hung over him, as much as they hung over me. As a fellow assigned to his care, my job was to harness these demons, giving him just enough medicine to control his pain, but not overshooting the mark.

Between 2019 and 2021, multiple independent groups reported trials of gene therapy strategies to treat sickle cell anemia. One strategy involves harvesting a patient's blood stem cells, as with a standard transplant. A virus is then used to deliver a corrected copy of the Beta-Globin gene into the stem cells. These patient blood stem cells, now bearing a corrected copy of the gene, are transplanted back into the patient, and the blood that grows out of the stem cells now permanently bears the corrected gene. (Although several patients were treated and demonstrated benefits, the trial was halted because two patients developed a leukemia-like disease. Whether the leukemia was the consequence of the virus, or of the chemotherapy required for the transplant, remains unknown.)

Another strategy — ingenious in its take — exploits a twist in human physiology. Fetal blood cells, unlike adult red blood cells, express a different form of hemoglobin. Submerged in womb fluid, where oxygen levels are extremely low, the fetus needs to extract oxygen aggressively from its mother's blood cells that come through the umbilical cord (in later life, once its own lungs become functional, the fetus's cells switch over to adult hemoglobin). Fetal blood cells therefore carry a unique form of hemoglobin — fetal hemoglobin — that is especially designed to extricate oxygen in the fetal environment. Like adult hemoglobin, fetal hemoglobin also has four chains — two Alpha- and two Gamma-Globin chains. But since none of its chains is encoded by Beta hemoglobin (the gene mutated in sickle cell patients) there is no mutation to cause sickling; it is perfectly normal, doesn't have the property of distorting the blood cell, and, in fact, can function especially well in low-oxygen environments.

Stuart Orkin and David Williams, working with a team of researchers and a cellular therapy company, have found a way to permanently activate fetal hemoglobin in blood stem cells, thereby overriding the sickled form of adult hemoglobin. Blood stem cells are extracted from sickle cell patients, manipulated by gene editing to "reexpress" fetal hemoglobin in an adult, and then transplanted back into the patient. In essence, adult red blood cells turn into fetal cells, and are no longer susceptible to sickling. Old blood becomes young.

In a trial reported in 2021, a thirty-three-year-old woman with sickle cell anemia was treated with this strategy. The level of hemoglobin in her blood rose almost twofold over the next fifteen months. In the two years before treatment, she had experienced between seven to nine severe pain crises every year. In the year and a half after treatment, she has had none. Thus far, there have been no reports of leukemia in this study. It's also too early to tell if there are adverse effects that will emerge over time, but there is a chance that this woman has been cured of sickle cell anemia. At Stanford, another group, led by Matt Porteus, is using gene editing to rewrite and correct the culprit sickling mutation in hemoglobin beta (fetal hemoglobin is not activated, but rather the culprit mutation is gene edited). Porteus's strategy is also in trials, and early results have been promising.

I don't know if William K. will choose to be treated with any one of these novel therapies. I am no longer his doctor. But having known him intimately for a decade — and knowing his spirit of adventure, the terrifying frequency of his pain crisis and the haunting fear of opioid addiction — I would suspect that he may well be lining up for one of these trials.

When he does get transplanted, he, too, will cross a border. He will become a new human, built out of his own reengineered cells. He will be a new sum of new parts.

ACKNOWLEDGMENTS

There are innumerable people to thank for the genesis of this book. First, my many readers: Sarah Sze, Sujoy Bhattacharyya, Ranu Bhattacharyya, Nell Breyer, Leela Mukherjee-Sze, Aria Mukherjee-Sze, and Lisa Yuskavage.

An enormous amount of scientific input was added by Sean Morrison (stem cells); Cori Bargmann (development); Nick Lane and Martin Kemp (evolution); Marc Flajolet (the brain); Barry Coller (platelets); Laura Otis (history); Paul Nurse (cell cycle); Irving Weissman (immunology); Helen Mayberg (neurology); Tom Whitehead, Carl June, Bruce Levine, and Stephan Grupp (CAR-T therapy); Harold Varmus (cancer); Ron Levy (antibody therapy); and Fred Applebaum (transplantation). Conversations with Laura Otis, Paul Greengard, Enzo Cerundolo, and Francisco Marty were indispensable. The nurses at the Fred Hutchinson Cancer Research Center provided one of the most moving accounts of the early history of that procedure.

A note of immense gratitude to Nan Graham, my editor at Scribner; Stuart Williams, at Bodley Head; and Meru Gokhale at Penguin Random House. Rana Dasgupta and my agent, Sarah Chalfant, at the Wylie Agency provided crucial support. Jerry Marshall and Alexandra Truitt researched the wonderful photographic section.

Sabrina Pyun kept the production on a militant schedule, and Rachel Rojy performed a heroic job collating the notes and the bibliography. Philip Bashe performed such jujitsu with copyediting that no comma and no footnote went unnoticed.

And to Kiki Smith, who generously provided the most mesmerizing pictures of "cells" that grace this book: thank you, thank you, thank you.

NOTES

"In the sum of the parts, there are only the parts": Wallace Stevens, "On the Road Home," in *Selected Poems: A New Collection,* ed. John N. Serio (New York: Alfred A. Knopf, 2009), 119.

"[Life] is a continuing rhythmic movement": Friedrich Nietzsche, "Rhythmische Untersuchungen," in *Friedrich Nietzsche, Werke, Kristiche Gesamstaube,* vol. 2.3, ed. Fritz Bornmann and Mario Carpitella (Vorlesungsaufzeuchnungen [SS 1870-SS 1871]; Berlin: de Gruyter, 1993), 322.

Prelude: "The Elementary Particles of Organisms"

"'Elementary,' he said": Arthur Conan Doyle, *The Adventures of Sherlock Holmes* (Hertfordshire: Wordsworth, 1996), 378.

The conversation took place over dinner in October 1837: Schwann's memories of the dinner are recorded in a speech that he gave in 1878, and he further recorded the moment in Theodor Schwann, *Microscopical Researches into the Accordance in the Structure and Growth of Animals and Plants,* trans. Henry Smith (London: Sydenham Society, 1847), xiv. Laura Otis, *Müller's Lab* (New York: Oxford University Press, 2007), 62–64; Marcel Florkin, *Naissance et déviation de la théorie cellulaire dans l'oeuvre de Théodore Schwann* (Paris: Hermann, 1960), 62.

He had been "hay gathering": Ulrich Charpa, "Matthias Jakob Schleiden (1804–1881): The History of Jewish Interest in Science and the Methodology of Microscopic Botany," in *Aleph: Historical Studies in Science and Judaism,* vol. 3 (Bloomington: Indiana University Press, 2003), 213–45.

His collection was prized among botanists: Details of his collection can be found in Matthias Jakob Schleiden, "Beiträge zur Phytogenesis," *Archiv für Anatomie, Physiologie und Wissenschaftliche Medicin* (1838): 137–76.

"Each cell leads a double life": Matthias Jakob Schleiden, "Contributions to Our Knowledge of Phytogenesis," in *Scientific Memoirs, Selected from the Transactions of Foreign Academies of Science and Learned Societies and from Foreign Journals,* vol. 2, ed. Richard Taylor, trans. William Francis (London: Richard and John E. Taylor, 1841), 281.

The developing animal's microscopic structure:

Schwann's interest in the unity of cells as building blocks of animals and plants was also driven by the idea that if plants and animals were built of autonomous, independent living units, then there was no need to invoke a special "vital" fluid that was responsible for life or for the birth of cells — an idea that Johannes Müller stuck stubbornly to. Schleiden, a student of his, believed in vital fluids but had his own theory of cellular origin — a process that Schleiden thought was analogous to the formation of crystals — a theory proved later to be completely wrong. Ironically, then, the birth of cell theory is not the story of mistaken origins but of origins mistaken. The commonalities that Schleiden and Schwann saw in plant and animal tissues — such as that all living beings were composed of cells — was absolutely real, but Schleiden's theory (one that Schwann accepted, although with growing doubt) about how these cells were born, as we shall soon see, was to be proved wrong, most prominently by Rudolf Virchow.

It's hard to know whether Schleiden had already deduced that all plant tissues were built out of cellular units before his conversation with Schwann, or whether the conversation jolted him to examine (or reexamine) his specimens and observe the universality of their cellular structures in a new light.

I've thus used the phrase "returned to his botanical specimens" to signal some caution about how much Schleiden had concluded before his dinner with Schwann, and how much came immediately after. However, the date of the dinner (1837), Schleiden's publication of his paper soon after (in 1838), and the well-documented visit to Schwann's lab to observe the similarities between animal and plant cells suggest that the interaction with Schwann was an important catalyst in Schleiden's thinking about the fundamentals and universality of cell theory. Furthermore, the fact that both Schleiden and Schwann readily accepted each other's role as cofounder, not rivals, in the origin of modern cell theory also suggests that their interactions — over conversations at dinner, say — must have played at least some role in strengthening Schleiden's conviction that all plant tissues were built out of cells. Schwann, unlike Schleiden, is clearer about the importance of the evening conversation in 1837: it changed the fundamental direction of his research. In his 1878 speech mentioned above, he admits readily that Schleiden's observations on the development of plants were pivotal to his subsequent discovery that animal tissues were also built from cells.

"*common means of formation through cells*": Florkin, *Naissance et dêviation de la théorie cellulaire*, 45.

In 1838, Schleiden collected his observations: Schleiden, "Beiträge zur Phytogenesis," 137–76.

A year later, Schwann followed Schleiden's work on plants: Schwann, *Microscopical Researches,* 2.

"A bond of union" connects the different branches: Ibid., ix.

Schleiden left Berlin: Sara Parker, "Matthias Jacob Schleiden (1804–1881)," Embryo Project Encyclopedia, last modified May 29, 2017, https://embryo.asu.edu/pages/matthias-jacob-schleiden-1804-1881.

And in 1839, Schwann left, too: Otis, *Müller's Lab,* 65.

I wrote three articles for the New Yorker: Siddhartha Mukherjee, "The Promise and Price of Cellular Therapies," *New Yorker* online, last modified July 15, 2019; "Cancer's Invasion Equation," *New Yorker* online, last modified September 4, 2017; "How Does the Coronavirus Behave Inside a Patient?," *New Yorker* online, last modified March 26, 2020.

Roy Porter's The Greatest Benefit to Mankind: Roy Porter, *The Greatest Benefit to Mankind: A Medical History of Humanity from Antiquity to the Present* (London: HarperCollins, 1999).

Henry Harris's The Birth of the Cell: Henry Harris, *The Birth of the Cell* (New Haven, CT: Yale University Press, 2000).

Introduction: "We Shall Always Return to the Cell"

"No matter how we twist and turn": Rudolf Virchow, *Disease, Life and Man: Selected Essays*, trans. Lelland J. Rather (Stanford, CA: Stanford University Press, 1958), 81.

I watched my friend Sam P. die: Details regarding Sam P.'s case are from personal communication with Sam P. and his physician, 2016. Names and identifying details have been changed for anonymity.

Emily was Patient No. 7: Details regarding Emily Whitehead's case are from personal communication with Emily Whitehead, her parents, and her doctors, 2019; extracted from Mukherjee, "Promise and Price of Cellular Therapies."

"living atoms": Antonie van Leeuwenhoeck, "Observations, Communicated to the Publisher by Mr. Antony Van Leeuwenhoek, in a Dutch Letter of the 9th Octob. 1676. Here English'd: Concerning Little Animals by Him Observed in Rain-Well-Sea- and Snow Water; as Also in Water Wherein Pepper Had Lain Infused," *Philosophical Transactions of the Royal Society* 12, no. 133 (March 25, 1677): 821–32.

when Emily Whitehead received her infusion: "CAR T-cell Therapy," National Cancer Institute Dictionary online, accessed December 2021, https://www.cancer.gov/publications/dictionaries/cancer-terms/def/car-t-cell-therapy.

"Every theory, hypothesis, or point of view": Serhiy A. Tsokolov, "Why Is the Definition of Life So Elusive? Epistemological Considerations," *Astrobiology* 9, no. 4 (2009): 401–12.

Complex, multicellular living beings: To be clear, these "emergent" properties are not defining characteristics of life. They are, rather, properties that multicellular creatures have evolved out of systems of living cells.

these properties repose, ultimately, in the cells: Not all cells have all the properties. For instance, cellular specialization in complex organisms enables that storage of nutrients, for instance, reposes in certain cells, while dispensation of waste reposes in others. Single-cell organisms such as yeast and bacteria can have specialized subcellular structures that achieve these functions, but multicellular organisms, such as humans, have evolved specialized organs with specialized cells to achieve these functions.

Yale University virologist Akiko Iwasaki: Akiko Iwasaki, interview with the author, February 2020. See also "SARS-CoV-2 Variant Classifications and Definitions," Centers for Disease Control and Prevention online, last modified December 1, 2021, https://www.cdc.gov/coronavirus/2019-ncov/variants/variant-classifications.html. See also "Severe Acute Respiratory Syndrome (SARS)," World Health Organization online, accessed

December 2021, https://www.who.int
/health-topics/severe-acute-respiratory
-syndrome#tab=tab_1.

I was taken by this reclusive, progressive, soft-spoken German: Ibid. See also John Simmons, *The Scientific 100: A Ranking of the Most Influential Scientists, Past and Present* (New York: Kensington, 2000), 88–92. See also George A. Silver, "Virchow, The Heroic Model in Medicine: Health Policy by Accolade," *American Journal of Public Health* 77, no. 1 (1987): 82–88.

"cellular pathology," as he described it: Virchow, *Disease, Life and Man,* 81.

The Original Cell: An Invisible World

"True knowledge is to be aware of one's ignorance": Rudolf Virchow, "Letters of 1842," in *Letters to His Parents, 1839–1864,* ed. Marie Rable, trans. Lelland J. Rather (USA: Science History Publications, 1990), 28–29.

the softness of Rudolf Virchow's voice: Elliot Weisenberg, "Rudolf Virchow, Pathologist, Anthropologist, and Social Thinker," *Hektoen International* 1, no. 2 (Winter 2009): https://hekint.org/2017/01/29/rudolf
-virchow-pathologist-anthropologist-and
-6social-thinker/.

the Flemish scientist Andreas Vesalius: C. D. O'Malley, *Andreas Vesalius of Brussels 1514–1564* (Berkeley: University of California Press, 1964). See also David Schneider, *The*

710

Invention of Surgery: A History of Modern Medicine — from the Renaissance to the Implant Revolution (New York: Pegasus Books, 2020), 68–98.

The professors sat on "lofty chairs": Andreas Vesalius, *De Humani Corporis Fabrica (The Fabric of the Human Body), vol. 1, bk. 1, The Bones and Cartilages,* trans. William Frank Richardson and John Burd Carman (San Francisco: Norman, 1998), li–lii.

The intricate drawings that Vesalius produced: Andreas Vesalius, *The Illustrations from the Works of Andreas Vesalius of Brussels,* ed. Charles O'Malley and J. B. Saunders (New York: Dover, 2013).

De Humani Corporis Fabrica (The Fabric of the Human Body): Vesalius, *Fabric of the Human Body,* 7 vols.

published the same year that the Polish astronomer: Nicolaus Copernicus, *On the Revolutions of Heavenly Spheres,* trans. Charles Glenn Wallis (New York: Prometheus Books, 1995).

Why, for instance, did a woman who gave birth: Ignaz Semmelweis, *The Etiology, Concept, and Prophylaxis of Childbed Fever,* ed. and trans. K. Codell Carter (Madison: University of Wisconsin Press, 1983).

Virchow was a mere eighteen years old: Izet Masic, "The Most Influential Scientists in the Development of Public Health (2): Rudolf Ludwig Virchow (1821–1902)," *Materia*

Socio-medica 31, no. 2 (June 2019): 151–52, doi:10.5455/msm.2019.31.151-152.

At the Pépinière, the surgical institute: Rudolf Virchow, *Der Briefwechsel mit den Eltern 1839–1864: zum ersten Mal vollständig in historisch-kritischer Edition (The Correspondence with the Parents, 1839–1864: For the First Time Complete in a Historical-Critical Edition)* (Germany: Blackwell Wissenschafts, 2001), 32.

"It goes on like this every day": Ibid., 19.

"There is an urgent and far-reaching": Rudolf Virchow, *Der Briefwechsel mit den Eltern,* 246, letter of July 4, 1844.

"I am my own advisor": Manfred Stürzbecher, "Die Prosektur der Berliner Charité im Briefwechsel zwischen Robert Froriep und Rudolf Virchow," *Beiträge zur Berliner Medizingeschichte,* 186, letter of Virchow to Froriep, March 2, 1847.

The Visible Cell: "Fictitious Stories About the Little Animals"

Moravian monk Gregor Mendel discovered genes: Gregor Mendel, "Experiments in Plant Hybridization," trans. Daniel J. Fairbanks and Scott Abbott, *Genetics* 204, no. 2 (2016): 407–22.

The Russian geneticist Nikolai Vavilov: Nicolai Vavilov, "The Origin, Variation, Immunity and Breeding of Cultivated Plants," trans. K. Starr Chester, *Chronica Botanica* 13, no. 1/6 (1951).

Even the English naturalist Charles Darwin: Charles Darwin, *On the Origin of Species,* ed. Gillian Beer (Oxford, UK: Oxford University Press, 2008).

opticians, Hans and Zacharias Janssen: "Lens Crafters Circa 1590: Invention of the Microscope," This Month in Physics History, *APS Physics* 13, no. 3 (March 2004): 2, https://www.aps.org/publications/apsnews/200403/history.cfm.

Some historians have argued that Janssen's competitors: "Hans Lipperhey," in *Oxford Dictionary of Scientists* online, Oxford Reference, accessed December 2021, https://www.oxfordreference.com/view/10.1093/oi/authority.20110803100108176.

Seventeenth-century Netherlands was a booming: Donald J. Harreld, "The Dutch Economy in the Golden Age (16th–17th Centuries)," EH.Net Encyclopedia of Economic and Business History, ed. Robert Whaples, last modified August 12, 2004, http://eh.net/encyclopedia/the-dutch-economy-in-the-golden-age-16th-17th-centuries/. See also Charles Wilson, "Cloth Production and International Competition in the Seventeenth Century," *Economic History Review* 13, no. 2 (1960): 209–21.

Leeuwenhoek, then forty-two, gathered some: Leeuwenhoek, "Observations, Communicated to the Publisher by Mr. Antony Van Leeuwenhoek, in a Dutch Letter of the

9th Octob. 1676. Here English'd: Concerning Little Animals by Him Observed in Rain-Well-Sea- and Snow Water; as Also in Water Wherein Pepper Had Lain Infused," 821–31. See also J. R. Porter, "Antony van Leeuwenhoek: Tercentenary of His Discovery of Bacteria," *Bacteriological Reviews* 40, no. 2 (1976): 260–69.

"No greater pleasure has yet come to my eye": Leeuwenhoek, "Observations, Communicated to the Publisher . . . ," 821–31.

"the body of a Mouse Color, clear towards the oval point": Ibid.

"In the year 1675," he wrote: Ibid.

In 1677, Leeuwenhoek observed human spermatozoa: M. Karamanou et al., "Anton van Leeuwenhoek (1632–1723): Father of Micromorphology and Discoverer of Spermatozoa," *Revista Argentina de Microbiologia* 42, no. 4 (2010): 311–14. See also S. S. Howards, "Antonie van Leeuwenhoek and the Discovery of Sperm," *Fertility and Sterility* 67, no. 1 (1997): 16–17.

He found them "moving like a snake or an eel swimming in water": Lisa Yount, *Antoni van Leeuwenhoek: Genius Discoverer of Microscopic Life* (Berkeley, CA: Enslow, 2015), 62.

Henry Oldenburg, the secretary of the Royal Society: Nick Lane, "The Unseen World: Reflections on Leeuwenhoek (1677) 'Concerning Little Animals,'" *Philosophical*

Transactions of the Royal Society B 370, no. 1666 (April 19, 2015), https://doi.org/10.1098/rstb.2014.0344.

Leeuwenhoek was "neither a philosopher, a medical man, nor a gentleman": Steven Shapin, *A Social History of Truth: Civility and Science in the Seventeenth Century* (Chicago: University of Chicago Press, 2011), 307. See also Robert Hooke to Antoni van Leeuwenhoek, December 1, 1677, quoted in Antony van Leeuwenhoek, *Antony van Leeuwenhoek and His Little Animals: Being Some Account of the Father of Protozoology & Bacteriology and His Multifarious Discoveries in These Disciplines,* comp., ed., trans. Clifford Dobell (1932; New York: Russell and Russell, 1958), 183.

This was science by affidavit: Lane, "The Unseen World."

"My work, which I've done for a long time": Leeuwenhoek to unknown, June 12, 1763, quoted in Carl C. Gaither and Alma E. Cavazos-Gaither, eds., *Gaither's Dictionary of Scientific Quotations* (New York: Springer, 2008), 734.

Robert Hooke, an English scientist and polymath: Allan Chapman, *England's Leonardo: Robert Hooke and the Seventeenth-Century Scientific Revolution* (Bristol, UK: Institute of Physics, 2005).

aftermath of the Great Fire of London: Ben Johnson, "The Great Fire of London," Historic UK: The History and Heritage

Accommodation Guide, accessed December 2021, https://www.historic-uk.com /HistoryUK/HistoryofEngland/The-Great -Fire-of-London/.

"If . . . an Object, plac'd very near": Robert Hooke, preface, in *Microphagia: Or Some Physiological Descriptions of Minute Bodies Made by Magnifying Glasses with Observations and Inquiries Thereupon* (London: Royal Society, 1665).

"the most ingenious book I read in all my life": Samuel Pepys, *The Diary of Samuel Pepys,* ed. Henry B. Wheatley, trans. Mynors Bright (London: George Bell and Sons, 1893), available at Project Gutenberg, https://www .gutenberg.org/files/4200/4200-h/4200-h .htm.

Among the dozens of meticulous illustrations: Martin Kemp, "Hooke's Housefly," *Nature* 393 (June 25, 1998): 745, https://doi .org/10.1038/31608.

"The Eyes of a Fly . . . appear almost like a Lattice": Hooke, *Microphagia.*

Hooke got an ant drunk on brandy: Ibid., 204.

"I took a good clear piece of cork": Ibid., 110.

"a great many little boxes": Ibid.

"The first experiment there exhibited was the pepper-water": Thomas Birch, ed., *The History of the Royal Society of London, for Improving the Knowledge, from its First Rise* (London: A. Millar, 1757), 352.

"As it has often reached my ear": Antonie van

716

Leeuwenhoek, "To Robert Hooke." 12 November 1680. Letter 33 of *Alle de brieven: 1679–1683*. Vol. 3. De Digitale Bibliotheek voor de Nederlandse Letteren (DBNL), 333.

"Nay, we may yet carry it farther": Antonie van Leeuwenhoek, *The Select Works of Antony van Leeuwenhoek, Containing His Microscopal Discoveries in Many of the Works of Nature,* ed. and trans. Samuel Hoole (London: G. Sidney, 1800), iv.

"Hooke did not for a moment": Harris, *Birth of the Cell*, 2.

"the walls of a living cell in cork": Ibid., 7.

In 1687, Isaac Newton published: Isaac Newton, *The Principia: Mathematical Principles of Natural Philosophy,* trans. I. Bernard Cohen and Anne Whitman (Oakland: University of California Press, 1999).

Indeed, Hooke, and several other physicists: This was not the first time that Hooke and Newton had clashed. In the 1670s, Newton had presented the Royal Society with his experiment that white light, when passed through a prism, was broken up into a continuous, rainbowlike spectrum of individual colors. Recombine these with another prism, and you reconstitute white light. Hooke, then the society's curator, disagreed with Newton and wrote a scathing review of the paper, sending Newton, already paranoid about disclosing his work, into a fit of self-righteous rage. The two geniuses of seventeenth-century England, each with a planet-sized ego, would continue to argue over the next decades — culminating in Hooke's insistence on credit for the law of universal gravitation.

No definitive likeness, or portrait: In 2019, Dr. Larry Griffing, a biology professor in Texas, examined a painting of an unidentified scientist, painted by Mary Beale around 1680. Griffing believes that the painting is a portrait of Hooke: "Portraits," RobertHooke .org, accessed December 2021, http://robert hooke.org.uk/?page_id=227.

The Universal Cell: "The Smallest Particle of This Little World"

"I could exceedingly plainly perceive it": Hooke, *Microphagia,* 111.

"As soon as the microscope was applied": Schwann, *Microscopical Researches,* x.

"one of the strangest silences in the history of science": Leslie Clarence Dunn, *A Short History of Genetics: The Development of Some of the Main Lines of Thought, 1864–1939* (Ames: Iowa State University Press, 1991), 15.

"wonderfully minute living creatures": Leonard Fabian Hirst, *The Conquest of Plague: A Study of the Evolution of Epidemiology* (Oxford, UK: Clarendon Press, 1953), 82.

"living contagion": Ibid., 81.

Bichat, in particular, distinguished twenty-one (!)

forms: Xavier Bichat, *Traité Des Membranes en Général et De Diverses Membranes en Particulier* (Paris: Chez Richard, Caille et Ravier, 1816). See also Harris, *Birth of the Cell,* 18.

Raspail believed in doing: Dora B. Weiner, *Raspail: Scientist and Reformer* (New York: Columbia University Press, 1968).

"The Court is today confronted with an eminent scientist": Pierre Eloi Fouquier and Matthieu Joseph Bonaventure Orfila, *Procès et défense de F. V. Raspail poursuivi le 19 mai 1846, en exercice illégal de la medicine* (Paris: Schneider et Langrand, 1846), 21.

Characteristically, Raspail refused: By the mid-1840s, Raspail had changed his intellectual pursuits and decided to dedicate himself to antisepsis, sanitation, and social medicine, especially among the imprisoned and the poor. He was convinced that parasites and worms caused most diseases, although he never gravitated toward bacteria as causes of contagion. In 1843, he published *Histoire naturelle de la santé et de la maladie* and *Manuel annuaire de la santé.* The two books, both enormous successes, addressed personal sanitation and hygiene, including recommendations for diet, exercise, mental activity, and the benefits of fresh air. In later life, Raspail turned to politics and was elected to the Chamber of Deputies, where he continued to campaign for medical reform for the imprisoned and the poor, and for increased sanitation in cities, mirroring the work of crusading physician John Snow in London. Perhaps the lasting image of this man, who almost disappeared from medical literature, can be found in Vincent van Gogh's painting *Still Life with a Plate of Onions,* which depicts a copy of Raspail's *Manuel* lying on a table next to a plate of onions. Van Gogh, a hypochondriac, probably bought the book off the streets, but the lasting work of an acerbic man placed next to a plate of tear-inducing vegetables seems somehow appropriate. (François-Vincent Raspail, *Histoire naturelle de la santé et de la maladie chez les végétaux et chez les animaux en général, et en particulier chez l'homme* [Paris: Elibron Classics, 2006], and *Manuel-annuaire de la santé pour 1864, ou médecine et pharmacie domestiques* [Paris: Simon Bacon, 1854].)

"Each cell selects": Weiner, *Raspail.* For more details, see also Dora Weiner, "François-Vincent Raspail: Doctor and Champion of the Poor," *French Historical Studies* 1, no. 2 (1959): 149–71.

Tucked away as an epigraph: Detailed in Harris, *Birth of the Cell,* 33.

"And what if all of animated nature": Samuel Taylor Coleridge, "The Eolian Harp," in *The Poetical Works of Samuel Taylor Coleridge,* ed. William B. Scott (London: George Routledge and Sons, 1873), 132.

In 1694, the Dutch microscopist Nicolaas

Hartsoeker: Matthias Jakob Schleiden, "Contributions to Our Knowledge of Phytogenesis," trans. William Francis, in *Scientific Memoirs, Selected from the Transactions of Foreign Academies of Science and Learned Societies and from Foreign Journals,* vol. 2, ed. Richard Taylor (London: Richard and John E. Taylor, 1841), 281. This is also detailed in Raphaële Andrault, "Nicolas Hartsoeker, Essai de dioptrique, 1694," in Raphaële Andrault et al., eds., *Médecine et philosophie de la nature humaine de l'âge classique aux Lumières: Anthologie* (Paris: Classiques Garnier, 2014).

"an aggregate of fully individualized": Schleiden, "Beiträge zur Phytogenesis," 137–76.

"A great portion of animal tissues": Schwann, *Microscopical Researches,* 6.

"The extraordinary diversity in figure [of organs and tissues]": Ibid., 1.

A "conflicted, enigmatic, transitional figure": Laura Otis, her parents, and her doctors, interview with the author, 2022.

"We have, indeed, compared the growth of organisms with crystallization": Schwann, *Microscopical Researches,* 212.

"but [crystallization] involves very much that is uncertain and paradoxical": Ibid., 215.

"It must at any rate, however": J. Müller, *Elements of Physiology,* ed. John Bell, trans. W. M. Baly (Philadelphia: Lea and Blanchard, 1843), 15.

"The main outcome is that a common principle": Harris, *Birth of the Cell,* 102.

Virchow called it leukocythemia *and then simply* leukemia: Rudolf Virchow, "Weisses Blut, 1845," in *Gesammelte Abhandlungen zur Wissenschaftlichen Medicin,* ed. Rudolf Virchow (Frankfurt: Meidinger Sohn, 1856), 149–54; Virchow, "Die Leukämie," in ibid., 190–212.

"He is of dark complexion," Bennett wrote: John Hughes Bennett, "Case of Hypertrophy of the Spleen and Liver, Which Death Took Place from Suppuration of the Blood," *Edinburgh Medical and Surgical Journal* 64 (1845): 413–23.

"The following case seems to me particularly valuable": John Hughes Bennett, "On the Discovery of Leucocythemia," *Monthly Journal of Medical Science* 10, no. 58 (1854): 374–81.

In 1848, this restlessness acquired a political dimension: Byron A. Boyd, *Rudolf Virchow: The Scientist as Citizen* (New York: Garland, 1991).

Virchow wrote a furious article: Rudolf Virchow, "Erinnerungsblätter," in *Archiv für Pathologische Anatomie und Physiologie und für Klinische Medicin* 4, no. 4 (1852): 541–48. See also Theodore M. Brown and Elizabeth Fee, "Rudolf Carl Virchow: Medical Scientist, Social Reformer, Role Model," *American Journal of Public Health* 96, no.

12 (December 2006): 2104–5, doi:10.2105 /AJPH.2005.078436.

The cause of the disease: Kurd Schulz, *Rudolf Virchow und die Oberschlesische Typhusepidemie von 1848. Jahrbuch der Schlesischen Friedrich-Wilhelms-Universität zu Breslau.* Vol. 19. Ed. (Göttingen Working Group, 1978).

"The body is a cell state in which every cell is a citizen": Rudolf Virchow, quoted in Weisenberg, "Rudolf Virchow, Pathologist, Anthropologist, and Social Thinker."

Raspail's phrase had become Virchow's central tenet: François Raspail, "Classification Generalé des Graminées," in *Annales des Sciences Naturelles,* vol. 6, comp. Jean Victor Audouin, A. D. Brongniart, and Jean-Baptiste Dumas (Paris: Libraire de L'Académie Royale de Médicine, 1825), 287–92. See also Silver, "Virchow, the Heroic Model in Medicine," 82–88.

"There is no life," Virchow wrote, "except through direct succession": Quoted in Lelland J. Rather, *A Commentary on the Medical Writings of Rudolf Virchow: Based on Schwalbe's Virchow-Bibliographie, 1843–1901* (San Francisco: Norman, 1990), 53.

Cellular Pathology *detonated through the world of medicine:* Rudolf Virchow, *Cellular Pathology: As Based upon Physiological and Pathological Histology: Twenty Lectures Delivered in the Pathological Institute of Berlin During the Months of February, March, and April, 1858*

(London: John Churchill, 1858).

"Admission to a hospital must be open": Quoted in Rather, *Commentary on the Medical Writings of Rudolf Virchow,* 19.

He published the data: For details regarding Virchow's response to racism, see Rudolf Virchow, "Descendenz und Pathologie," *Archiv für Pathologische Anatomie und Physiologie und für Klinische Medicin* 103, no. 3 (1886): 413–36.

"Life is, in general, cell activity.": Quoted in Rather, *Commentary on the Medical Writings of Rudolf Virchow,* 4.

"Every disease depends on": Quoted in ibid., 101. See also "Eine Antwort an Herrn Spiess," *Virch. Arch. XIII,* 481. A Reply to Mr. Spiess. VA 13 (1858): 481–90.

The patient, M.K., a young man: Details regarding M.K.'s case are from my personal interactions with M.K., 2002. Names and identifying details have been changed for anonymity.

severe combined immunodeficiency (acronymed SCID): "Severe Combined Immunodeficiency (SCID)," National Institute of Allergy and Infectious Diseases (NIAID) online, last modified April 4, 2019, https://www.niaid.nih.gov/diseases-conditions/severe-combined-immunodeficiency-scid#:~:text=Severe%20combined%20immunodeficiency%20(SCID)%20is,highly%20susceptible%20to%20severe%20infections.

"every animal presents itself": Rudolf Virchow,

"Lecture I," *Cellular Pathology as Based upon Physiological and Pathological Histology: Twenty Lectures Delivered in the Pathological Institute of Berlin During the Months of February, March, and April, 1858,* trans. Frank Chance (London: John Churchill, 1860), 1–23.

The Pathogenic Cell: Microbes, Infections, and the Antibiotic Revolution

"Like hermits, microbes need only be concerned with feeding themselves": Elizabeth Pennisi, "The Power of Many," *Science* 360, no. 6396 (June 29, 2018): 1388–91, doi:10.1126/science.360.6396.1388.

In 1668, Francesco Redi published: Francesco Redi, *Experiments on the Generation of Insects,* trans. Mab Bigelow (Chicago: Open Court, 1909).

Redi concluded that maggots: Ibid. See also Paul Nurse, "The Incredible Life and Times of Biological Cells," *Science* 289, no. 5485 (September 8, 2000): 1711–16, doi:10.1126/science.289.5485.1711.

In Paris, in 1859, Louis Pasteur took Redi's: René Vallery-Radot, *The Life of Pasteur,* vol. 1., trans. R. L. Devonshire (New York: Doubleday, Page, 1920), 141.

In Wollstein, Germany, Robert Koch: Thomas D. Brock, *Robert Koch: A Life in Medicine and Bacteriology* (Madison, WI: Science Tech, 1988), 32.

725

In early 1876, he learned to isolate anthrax bacteria: Robert Koch, "The Etiology of Anthrax, Founded on the Course of Development of Bacillus Anthracis" (1876), in *Essays of Robert Koch.,* ed. and trans. K. Codell Carter (New York: Greenwood Press, 1987), 1–18.

"In view of this fact," he wrote in his notes: Quoted in Thomas Goetz, *The Remedy: Robert Koch, Arthur Conan Doyle, and the Quest to Cure Tuberculosis* (New York: Gotham Books, 2014), 74. See also Steve M. Blevins and Michael S. Bronze, "Robert Koch and the 'Golden Age' of Bacteriology," *International Journal of Infectious Diseases* 14, no. 9 (September 2010): e744–e51.

"Bacillus anthracis of the Germans": Quoted in Robert Koch, "Über die Milzbrandimpfung. Eine Entgegung auf den von Pasteur in Genf gehaltenen Vortrag," in *Gesammelte Werke von Robert Koch,* ed. J. Schwalbe, G. Gaffky, and E. Pfuhl (Leipzig, Ger.: Verlag von Georg Thieme, 1912), 207–31.

"Up to now, Pasteur's work on anthrax has led to nothing": Ibid. See also Robert Koch, "On the Anthrax Inoculation," in *Essays of Robert Koch,* 97–107.

Pasteur's use of the term was: Agnes Ullmann, "Pasteur-Koch: Distinctive Ways of Thinking About Infectious Diseases," *Microbe* 2, no. 8 (August 2007): 383–87, http:// www.antimicrobe.org/h04c.files/history

/Microbe%202007%20Pasteur-Koch.pdf.
See also Richard M. Swiderski, *Anthrax: A History* (Jefferson, NC: McFarland, 2004), 60.

The first hint of a potential link came from a Hungarian obstetrician: Semmelweis, *Childbed Fever.*

"What protected those who delivered outside the clinic from": Ibid., 81.

Semmelweis could hardly help but notice that Kolletschka's symptoms: Ibid., 19.

an English physician named John Snow: John Snow, *On the Mode of Communication of Cholera* (London: John Churchill, 1849).

"I found that nearly all the deaths had taken place": John Snow, "The Cholera Near Golden-Square, and at Deptford," *Medical Times and Gazette* 9 (September 23, 1854): 321–22.

"For the morbid matter of cholera": Snow, *Mode of Communication of Cholera,* 15.

Yet in Lister's time, surgeons paid: Dennis Pitt and Jean-Michel Aubin, "Joseph Lister: Father of Modern Surgery," *Canadian Journal of Surgery* 55, no. 5 (October 2012): e8–e9, doi:10.1503/cjs.007112.

arsphenamine, was discovered by Drs. Paul Ehrlich and Sahachiro Hata: Felix Bosch and Laia Rosich, "The Contributions of Paul Ehrlich to Pharmacology: A Tribute on the Occasion of the Centenary of His Nobel Prize," *Pharmacology* 82, no. 3 (October

2008): 171–79, doi:10.1159/000149583.

among them penicillin . . . discovered in molding plates by Alexander Fleming in 1928: Siang Yong Tan and Yvonne Tatsumura, "Alexander Fleming (1881–1955): Discoverer of Penicillin," *Singapore Medical Journal* 56, no. 7 (2015): 366–67, doi:10.11622/smedj.2015105.

and the anti-TB drug streptomycin, isolated: H. Boyd Woodruff, "Selman A. Waksman, Winner of the 1952 Nobel Prize for Physiology or Medicine," *Applied and Environmental Microbiology* 80, no. 1 (January 2014): 2–8, doi:10.1128/AEM.01143-13.

Ed Yong's seminal book: Ed Yong, *I Contain Multitudes: The Microbes Within Us and a Grander View of Life* (New York: Ecco, 2016).

An infectious disease specialist once told me: Francisco Marty, interview with the author, February 2018.

we had misclassified not just some arcane microbe: Carl R. Woese and G. E. Fox. "Phylogenetic Structure of the Prokaryotic Domain: The Primary Kingdoms," *Proceedings of the National Academy of Sciences of the United States of America* 74, no. 11 (November 1977): 5088–90, https://doi.org/10.1073/pnas.74.11.5088.

Archaea, Woese argued, were not "almost like": Carl R. Woese, O. Kandler, and M. L. Wheelis, "Towards a Natural System of Organisms: Proposal for the Domains Archaea,

Bacteria, and Eucarya," *Proceedings of the National Academy of Sciences of the United States of America* 87, no. 12 (June 1990): 4576–79, doi:10.1073/pnas.87.12.4576.

In 1998, Ernst Mayr: Ernst Mayr, "Two Empires or Three?," *Proceedings of the National Academy of Sciences of the United States of America* 95, no. 17 (August 18, 1998): 9720–23, https://doi.org/10.1073/pnas.95.17.9720.

The journal Science *described Woese as a "scarred revolutionary":* Virginia Morell, "Microbiology's Scarred Revolutionary," *Science* 276, no. 5313 (May 2, 1997): 699–702, doi:10.1126/science.276.5313.699.

As Nick Lane, the evolutionary biologist: Nick Lane, *The Vital Question: Energy, Evolution, and the Origins of Complex Life* (New York: W. W. Norton, 2015), 8.

These three components: Jack Szostak, David Bartel, and P. Luigi Luisi, "Synthesizing Life," *Nature* 409 (January 2001): 387–90, https://doi.org/10.1038/35053176.

At some point, the biologists believe: Ting F. Zhu and Jack W. Szostak, "Coupled Growth and Division of Model Protocell Membranes," *Journal of the American Chemical Society* 131, no. 15 (April 2009): 5705–13.

"This ancestor," as Lane puts it: Lane, *The Vital Question*, 2.

"modern" eukaryotic cell arose within archaea: James T. Staley and Gustavo Caetano-Anollés, "Archaea-First and

the Co-Evolutionary Diversification of Domains of Life," *BioEssays* 40, no. 8 (August 2018): e1800036, doi:10.1002/bies.201800036. See also "BioEsssays: Archaea-First and the Co-Evolutionary Diversification of the Domains of Life," YouTube, 8:52, WBLifeSciences, https://www.youtube.com/watch?v=9yVWn_Q9faY&ab_channel=CrashCourse.

"unexplained void . . . the black hole at the heart of biology": Lane, *The Vital Question,* 1.

The Organized Cell: The Interior Anatomy of the Cell

"Give me an organic vesicle": François-Vincent Raspail, quoted in Lewis Wolpert, *How We Live and Why We Die: The Secret Lives of Cells* (New York: W. W. Norton, 2009), 14.

"Cell biology finally makes possible a century-old dream": George Palade, banquet speech at the Nobel Banquet, December 10, 1974, Nobel Prize online, http://nobelprize.org/nobel_prizes/medicine/laureates/1974/palade-speech.html.

"The cell," Rudolf Virchow proposed in 1852: Rather, *Commentary on the Medical Writings of Rudolf Virchow,* 38.

The cell membrane must be an oily layer: Ernest Overton, *Über die osmotischen Eigenschaften der lebenden Pflanzen-und Tierzelle* (Zurich: Fäsi & Beer, 1895), 159–84. See also Overton, *Über die allgemeinen*

osmotischen Eigenschaften der Zelle, ihre ver-
mutlichen Ursachen und ihre Bedeutung für
die Physiologie (Zurich: Fäsi & Beer, 1899).
See also Overton, "The Probable Origin
and Physiological Significance of Cellular
Osmotic Properties," in Papers on Biological Membrane Structure, ed. Daniel Branton
and Roderic B. Park (Boston: Little, Brown,
1968), 45–52. See also Jonathan Lombard,
"Once upon a Time the Cell Membranes:
175 Years of Cell Boundary Research," Biology Direct 9, no. 32 (December 19, 2014),
https://doi.org/10.1186/s13062-014-0032-7.

In the 1920s, Evert Gorter and François Grendel:
Evert Gorter and François Grendel, "On
Bimolecular Layers of Lipoids on the Chro-
mocytes of the Blood," Journal of Experimental Medicine 41, no. 4 (March 31, 1925):
439–43, doi:10.1084/jem.41.4.439.

Two biochemists, Garth Nicolson and Seymour
Singer: Seymour Singer and Garth Nicolson,
"The Fluid Mosaic Model of the Structure
of Cell Membranes," Science 175, no. 4023
(February 18, 1972): 720–31, doi:10.1126
/science.175.4023.720.

When a white blood cell creeps toward a microbe: Orion D. Weiner et al., "Spatial
Control of Actin Polymerization During
Neutrophil Chemotaxis," Nature Cell Biology 1, no. 2 (June 1999): 75–81, https://doi
.org/10.1038/10042.

Romanian American cell biologist George

Palade: James D. Jamieson, "A Tribute to George E. Palade," *Journal of Clinical Investigation* 118, no. 11 (November 3, 2008): 3517–18, doi:10.1172/JCI37749.

you're likely to see is a kidney-shaped organelle: Richard Altmann, *Die Elementarorganismen und ihre Beziehungen zu den Zellen* (Leipzig, Ger.: Verlag von Veit, 1890), 125.

In 1967, evolutionary biologist Lynn Margulis: Lynn Sagan, "On the Origin of Mitosing Cells," *Journal of Theoretical Biology* 14, no. 3 (March 1967): 225–74, doi:10.1016/0022-5193(67)90079-3.

As Nick Lane explains in The Vital Question, *Margulis argued:* Lane, *Vital Question,* 5.

"Should all the billions of gently burning little fires cease to burn": Eugene I. Rabinowitch, "Photosynthesis — Historical Development of Scientific Interpretation and Significance of the Process," in *The Physical and Economic Foundation of Natural Resources: I. Photosynthesis — Basic Features of the Process* (Washington, DC: Interior and Insular Affairs Committee, House of Representatives, United States Congress, 1952), 7–10.

"[It] finally makes possible a century-old dream": George Palade, quoted in Andrew Pollack, "George Palade, Nobel Winner for Work Inspiring Modern Cell Biology, Dies at 95," *New York Times,* October 8, 2008, B19.

"He thought of himself like a scientific version": Paul Greengard, personal interaction with

the author, February 2019.

The dungeon, however unbecoming: Ibid. See also George Palade, "Intracellular Aspects of the Process of Protein Secretion" (Nobel Lecture, Stockholm, December 12, 1974).

"The new field had virtually no tradition": G. E. Palade, "Keith Roberts Porter and the Development of Contemporary Cell Biology," *Journal of Cell Biology* 75, no. 1 (November 1977): D3–D10, https://doi.org/10.1083/jcb.75.1.D1.

Palade launched crucial collaborations with Porter and Claude: Unfortunately, Claude would leave the Rockefeller Institute in 1949 to return to his native Belgium. In 1974, he shared the Nobel Prize with Palade and another cell biologist, Christian de Duve. Palade, "Keith Roberts Porter and the Development of Contemporary Cell Biology," D3–D18.

"[S]tructure — as traditionally envisaged": Palade, "Intracellular Aspects of the Process of Protein Secretion," Nobel Lecture.

From there, the labeled proteins traveled: George E. Palade, "Intracellular Aspects of the Process of Protein Synthesis," *Science* 189, no. 4200 (August 1, 1975): 347–58, doi:10.1126/science.1096303.

The Belgian biologist Christian de Duve: David D. Sabatini and Milton Adesnik, "Christian de Duve: Explorer of the Cell Who Discovered New Organelles by Using a

Centrifuge," *Proceedings of the National Academy of Sciences of the United States of America* 110, no. 33 (August 13, 2013): 13234–35, doi:10.1073/pnas.1312084110.

String together all the DNA in every human being: Barry Starr, "A Long and Winding DNA," *KQED* online, last modified on February 2, 2009, https://www.kqed.org /quest/1219/a-long-and-winding-dna.

"We can only hope that what the geneticist J. B. S. Haldane": Thoru Pederson, "The Nucleus Introduced," *Cold Spring Harbor Perspectives in Biology* 3, no. 5 (May 1, 2011): a000521, doi:10.1101/cshperspect.a000521.

"La fixité du milieu intérieur est la condition de la vie libre, indépendante": Claude Bernard, *Lectures on the Phenomena of Life Common to Animals and Plants,* trans. Hebbel E. Hoff, Roger Guillemin, and Lucienne Guillemin (Springfield, IL: Charles C. Thomas, 1974).

eleven-year-old hockey player named Jared: Valerie Byrne Rudisill, *Born with a Bomb: Suddenly Blind from Leber's Hereditary Optic Neuropathy,* ed. Margie Sabol and Leslie Byrne (Bloomington, IN: AuthorHouse, 2012).

For further details regarding Leber hereditary optic neuropathy (LHON): "Leber Hereditary Optic Neuropathy (Sudden Vision Loss)," Cleveland Clinic online, last modified February 26, 2021.

The culprit gene was found and mapped in

1988: D. C. Wallace et al., "Mitochondrial DNA Mutation Associated with Leber's Hereditary Optic Neuropathy," *Science* 242, no. 4884 (December 9, 1988): 1427–30, doi:10.1126/science.3201231.

"I wish this had been the combination": Jared, quoted in Rudisill, *Born with a Bomb.*

"So here I am at the Musicians Institute": Ibid.

In 2011, a group of ophthalmologists: Byron Lam et al., "Trial End Points and Natural History in Patients with G11778A Leber Hereditary Optic Neuropathy," *JAMA Ophthalmology* 132, no. 4 (April 1, 2014): 428–36, doi:10.1001/jamaophthalmol.2013.7971.

In 2011, Chinese physicians: Shuo Yang et al., "Long-term Outcomes of Gene Therapy for the Treatment of Leber's Hereditary Optic Neuropathy," *eBioMedicine* (August 10, 2016): 258–68, doi:10.1016/j.ebiom.2016.07.002.

RESCUE trial, in which gene therapy was used: Nancy J. Newman et al., "Efficacy and Safety of Intravitreal Gene Therapy for Leber Hereditary Optic Neuropathy Treated Within 6 Months of Disease Onset," *Ophthalmology* 128, no. 5 (May 2021): 649–60, doi: 10.1016/j.ophtha.2020.12.012.

The Dividing Cell: Cellular Reproduction and the Birth of IVF

"There is no such thing as reproduction": Andrew Solomon, *Far from the Tree: Parents,*

Children and the Search for Identity (New York: Scribner, 2013), 1.

As the French biologist François Jacob once put it: Quoted in Jacques Monod, *Chance and Necessity: An Essay on the Natural Philosophy of Modern Biology* (New York: Alfred A. Knopf, 1971), 20.

Walther Flemming, the son of a psychiatrist: Neidhard Paweletz, "Walther Flemming: Pioneer of Mitosis Research," *Nature Reviews Molecular Cell Biology* 2, no. 1 (January 1, 2001): 72–75, https://doi.org/10.1038/35048077.

"Do the shifts in the positions of the visible": Walther Flemming, "Contributions to the Knowledge of the Cell and Its Vital Processes: Part 2," *Journal of Cell Biology* 25, no. 1 (April 1, 1965): 1–69, https://www.ncbi.nlm.nih.gov/pmc/articles/PMC2106612/.

"nuclear figures began to organize themselves": Ibid., 1–9.

Theodor Boveri and Walter Sutton would make: Walter Sutton, "The Chromosomes in Heredity," *Biological Bulletin* 4, no. 5 (April 1903): 231–51, https://doi.org/1535741; Theodor Boveri, *Ergebnisse über die Konstitution der chromatischen Substanz des Zellkerns* (Jena, Ger.: Verlag von Gustav Fischer, 1904).

Guardians of the Genome — among them the p53: "The p53 Tumor Suppressor Protein," in *Genes and Disease* (Bethesda, MD:

National Center for Biotechnology Information, last modified January 31, 2021), 215–16, available online at https://www.ncbi.nlm.nih.gov/books/NBK22268/.

"My dad was a blue-collar worker": Paul Nurse, interviewed by the author, March 2017. "Sir Paul Nurse: I Looked at My Birth Certificate. That Was Not My Mother's Name," *Guardian* (International edition) online, last modified August 9, 2014, https://www.theguardian.com/culture/2014/aug/09/paul-nurse-birth-certificate-not-mothers-name.

"By 1982," he wrote, "work on the control of protein": Tim Hunt, "Biographical," Nobel Prize online, accessed February 20, 2022, https://www.nobelprize.org/prizes/medicine/2001/hunt/biographical/.

"tubes and tips and gel plates, and even a peristaltic pump": Tim Hunt, "Protein Synthesis, Proteolysis, and Cell Cycle Transitions" (Nobel Lecture, Stockholm, December 9, 2021).

"We were just looking at the same thing": Nurse, interviewed by the author, March 2017.

In the mid-1950s, an unorthodox, secretive professor: Stuart Lavietes, "Dr. L. B. Shettles, 93, Pioneer in Human Fertility," *New York Times,* February 16, 2003, 1041.

For details regarding Landrum Shettles's experiment: Tabitha M. Powledge, "A Report from the Del Zio Trial," *Hastings Center Report* 8, no. 5 (October 1978): 15–17, https://

www.jstor.org/stable/3561442.

in the words of science historian Marga-ret Marsh, "careful mavericks": Quoted in "Test Tube Babies: Landrum Shettles," PBS *American Experience* online, accessed March 14, 2022, https://www.pbs.org/wgbh/americanexperience/features/babies-bio-shettles/.

For details regarding Robert Edwards and Patrick Steptoe's work: Martin H. Johnson, "Robert Edwards: The Path to IVF," *Reproductive Biomedicine Online* 23, no. 2 (August 23, 2011): 245–62, doi:10.1016/j.rbmo.2011.04.010. See also James Le Fanu, *The Rise and Fall of Modern Medicine* (New York: Carroll & Graf, 2000), 157–76.

"My grants were spent, and I was in debt": Robert Geoffrey Edwards and Patrick Christopher Steptoe, *A Matter of Life: The Story of a Medical Breakthrough* (New York: William Morrow, 1980), 17.

Harvard scientists John Rock and Miriam Menkin: John Rock and Miriam F. Menkin, "In Vitro Fertilization and Cleavage of Human Ovarian Eggs," *Science* 100, no. 2588 (August 4, 1944): 105–7, doi:10.1126/science.100.2588.105.

In 1951, Min Chueh Chang: M. C. Chang, "Fertilizing Capacity of Spermatozoa Deposited into the Fallopian Tubes," *Nature* 168, no. 4277 (October 20, 1951): 697–98, doi:10.1038/168697b0.

[T]hree, six, nine, and twelve hours": Edwards and Steptoe, *A Matter of Life,* 43.

"ripening programme in the eggs of primates": Ibid., 44.

"After 18 hours exactly": Ibid., 45.

"Excitement beyond belief": Ibid.

"Laparoscopy is of no use whatsoever": Ibid., 62.

One afternoon, late in the winter of 1968, Jean Purdy: Quoted in "Recipient of the 2019 IETS Pioneer Award: Dr. Barry Bavister," *Reproduction, Fertility and Development* 31, no. 3 (2019): vii–viii, https://doi.org/10.1071 /RDv31n3_PA.

"A spermatozoon was just passing into the first egg": Jean Purdy, quoted in ibid.

The paper by Edwards, Steptoe, and Bavister: Robert G. Edwards, Barry D. Bavister, and Patrick C. Steptoe, "Early Stages of Fertilization *In Vitro* of Human Oocytes Matured *In Vitro*," *Nature* 221, no. 5181 (February 15, 1969): 632–35, https://doi .org/10.1038/221632a0.

"It is perhaps difficult now to comprehend": Johnson, "Robert Edwards: The Path to IVF," 245–62.

"contraceptive development research increased": Martin H. Johnson et al., "Why the Medical Research Council Refused Robert Edwards and Patrick Steptoe Support for Research on Human Conception in 1971," *Human Reproduction* 25, no. 9

(September 2010): 2157–74, doi: 10.1093/humrep/deq155.

On November 10, 1977, a tiny cluster of living embryonic cells: Robin Marantz Henig, *Pandora's Baby: How the First Test Tube Babies Sparked the Reproductive Revolution* (Boston: Houghton Mifflin, 2004).

"[The baby] didn't have to be resuscitated": Martin Hutchinson, "I Helped Deliver Louise," BBC News online, last modified July 24, 2003, http://news.bbc.co.uk/2/hi/health/3077913.stm.

"I felt quite whacked, really": Ibid.

test tubes were hardly used: Victoria Derbyshire, "First IVF Birth: 'It Makes Me Feel Really Special,'" BBC News Two online, last modified July 23, 2015, https://www.bbc.co.uk/programmes/p02xv7jc.

"[T]he Browns have . . . degraded": Quoted in Ciara Nugent, "What It Was Like to Grow Up as the World's First 'Test-Tube Baby,'" *Time* online, last modified July 25, 2018, https://time.com/5344145/louise-brown-test-tube-baby/.

The July 31 cover of Time *borrowed:* Cover image, *Time,* July 31, 1978, available online at http://content.time.com/time/magazine/0,9263,7601780731,00.html.

"born in a slightly different way than everybody else": Derbyshire, "First IVF Birth." See also Elaine Woo and *Los Angeles Times,* "Lesley Brown, British Mother of First In Vitro

Baby, Dies at 64," Health & Science, *Washington Post* online, June 25, 2012, https://www.washingtonpost.com/national/health-science/lesley-brown-british-mother-of-first-in-vitro-baby-dies-at-64/2012/06/25/gJQAkavb2V_story.html.

a 1962 scientific paper of his: Robert G. Edwards, "Meiosis in Ovarian Oocytes of Adult Mammals," *Nature* 196 (November 3, 1962): 446–50, https://doi.org/10.1038/196446a0.

Release these molecules and activate them: Deepak Adhikari et al., "Inhibitory Phosphorylation of Cdk1 Mediates Prolonged Prophase I Arrest in Female Germ Cells and Is Essential for Female Reproductive Lifespan," *Cell Research* 26 (2016): 1212–25, https://doi.org/10.1038/cr.2016.119.

The Stanford group took 242 human embryos: Krysta Conger, "Earlier, More Accurate Prediction of Embryo Survival Enabled by Research," Stanford Medicine News Center, last modified October 2, 2010, https://med.stanford.edu/news/all-news/2010/10/earlier-more-accurate-prediction-of-embryo-survival-enabled-by-research.html.

The fact that only about one-third: Ibid.

The Tampered Cell: Lulu, Nana, and the Transgressions of Trust

"Deem and I were chatting about something else": Jon Cohen, "The Untold Story of the 'Circle of Trust' Behind the World's First

741

Gene-Edited Baby," Asia/Pacific News, *Science* online, last modified August 1, 2019, https://www.science.org/content/article/untold-story-circle-trust-behind-world-s-first-gene-edited-babies.

fable of JK — of the seduction to alter human babies: Ibid.

They could, in effect, select the "right" embryos to implant: Richard Gardner and Robert Edwards, "Control of the Sex Ratio at Full Term in the Rabbit by Transferring Sexed Blastocysts," *Nature* 218 (April 27, 1968): 346–48, https://doi.org/10.1038/218346a0.

"Numerous attempts have been made to control": Ibid.

This list is necessarily abbreviated: Https://www.broadinstitute.org/what-broad/areas-focus/project-spotlight/crispr-timeline.

Previous studies had shown that humans: L. Meyer et al., "Early Protective Effect of CCR-5 Delta 32 Heterozygosity on HIV-1 Disease Progression: Relationship with Viral Load. The SEROCO Study Group," *AIDS* 11, no. 11 (September 1997): F73–F78, doi:10.1097/00002030-199711000-00001.

Not long afterward, he must have biopsied: "28 Nov 2018 — International Summit on Human Genome Editing — He Jiankui Presentation and Q&A," YouTube, 1:04.28, WCSethics, https://www.youtube.com/watch?v=tLZufCrjrN0.

It read: "Good news!": Pam Belluck,

"Gene-Edited Babies: What a Chinese Scientist Told an American Mentor," *New York Times,* April 14, 2019, A1.

"I spent the next half hour, forty-five minutes": Cohen, "Untold Story of the 'Circle of Trust.'"

"FYI, this is probably the first human germ line editing": Ibid.

"Just to remind everyone here": Robin Lovell-Badge, introduction, "28 Nov 2018 — International Summit on Human Genome Editing — He Jiankui Presentation and Q&A," YouTube.

Cells biopsied from one blastocyst, he said: David Cyranoski, "First CRISPR Babies: Six Questions That Remain," News, *Nature* online, last modified November 30, 2018, https://www.nature.com/articles/d41586 -018-07607-3.

"Successfully is iffy here": Mark Terry, "Reviewers of Chinese CRISPR Research: 'Ludicrous' and 'Dubious at Best,'" BioSpace, last modified December 5, 2019, https:// www.biospace.com/article/peer-review -of-china-crispr-scandal-research-shows -deep-flaws-and-questionable-results/.

"I don't think it has been a transparent process": Badge, introduction, "28 Nov 2018 — International Summit on Human Genome Editing — He Jiankui Presentation and Q&A," YouTube. See also US National Academy of Sciences and US National Academy of Medicine, the Royal Society of the United

Kingdom, and the Academy of Sciences of Hong Kong, *Second International Summit on Human Genome Editing: Continuing the Global Discussion, November, 27–29, University of Hong Kong, China* (Washington, DC: National Academies Press, 2018).

"Honestly, I thought, This is fake, right?": Cohen, "Untold Story of the 'Circle of Trust.'"

"Having listened to Dr. He": David Cyranoski, "CRISPR-baby Scientist Fails to Satisfy Critics," News, *Nature* online, last modified November 30, 2018, https://www.nature.com/articles/d41586-018-07573-w.

But careful or not, he wants to be a maverick: David Cyranoski, "Russian 'CRISPR-baby' Scientist Has Started Editing Genes in Human Eggs with Goal of Altering Deaf Gene," News, *Nature* online, last modified October 18, 2019, https://www.nature.com/articles/d41586-019-03018-0.

Which begs another question: Nick Lane, interview with the author, January 2022.

multicellularity has "been viewed as a major transition": László Nagy, quoted in Pennisi, "The Power of Many," 1388–91.

But perhaps the most astonishing feature: Richard K. Grosberg and Richard R. Strathmann, "The Evolution of Multicellularity: A Minor Major Transition?," *Annual Review of Ecology, Evolution, and Systematics* 38 (December 2007): 621–54, doi/10.1146/annurev.ecolsys.36.102403.114735.

The transformation from single cells into multi-cellularity: Ibid.

In one of the most intriguing attempts: William C. Ratcliff et al., "Experimental Evolution of Multicellularity," *Proceedings of the National Academy of Sciences of the United States of America* 109, no. 5 (2012): 1595–600, https://doi.org/10.1073/pnas.1115323109.

Skinny and boundlessly enthusiastic: William Ratcliff, interview with the author, December 2021.

It was a simulation: Ibid.

Evolutionary scientists have performed variations: Elizabeth Pennisi, "Evolutionary Time Travel," *Science* 334, no. 6058 (November 18, 2011): 893–95, doi:10.1126/science.334.6058.893.

Evolution raced toward collective existence: Enrico Sandro Colizzi, Renske M. A. Vroomans, and Roeland M. H. Merks, "Evolution of Multicellularity by Collective Integration of Spatial Information," *eLife* 9 (October 16, 2020): e56349, doi:10.7554/eLife.56349. See also Matthew D. Herron et al., "*De Novo* Origins of Multicellularity in Response to Predation," *Scientific Reports* 9 (February 20, 2019), https://doi.org/10.1038/s41598-019-39558-8.

The Developing Cell: A Cell Becomes an Organism

"Life not so much 'is' but 'becomes'": Ignaz

Döllinger, quoted in Janina Wellmann, *The Form of Becoming: Embryology and the Epistemology of Rhythm, 1760–1830,* trans. Kate Sturge (New York: Zone Books, 2017), 13.

Wolff, wrote a doctoral dissertation titled: Caspar Friedrich Wolff, "Theoria Generationis" (dissertation, U Halle, 1759. Halle: U H, 1759).

"It is becoming aware of the form": Johann Wolfgang von Goethe, "Letter to Frau von Stein," *The Metamorphosis of Plants* (Cambridge, MA: MIT Press, 2009), 15.

The observations by Albertus Magnus and, later, Caspar Wolff: Joseph Needham, *History of Embryology* (Cambridge, UK: University of Cambridge Press, 1934).

For a thorough review of the development of the trophoblast: Martin Knöfler et al., "Human Placenta and Trophoblast Development: Key Molecular Mechanisms and Model Systems," *Cellular and Molecular Life Sciences* 76, no. 18 (September 2019): 3479–96, doi: 10.1007/s00018-019-03104-6.

"at a certain stage there emerges a single cell": Lewis Thomas, *The Medusa and the Snail: More Notes of a Biology Watcher* (New York: Penguin Books, 1995), 131.

Spemann and Mangold knew that this cluster of cells: Edward M. De Robertis, "Spemann's Organizer and Self-Regulation in Amphibian Embryos," *Nature Reviews Molecular*

Cell Biology 7, no. 4 (April 2006): 296–302, doi:10.1038/nrm1855.

The tissue extracted from the second tadpole embryo: Scott F. Gilbert, Development Biology, vol. 2 (Sunderland, UK: Sinauer Associates, 2010), 241–86. See also Richard Harland, "Induction into the Hall of Fame: Tracing the Lineage of Spemann's Organizer," Development 135, no. 20 (October 15, 2008): 3321–23, fig. 1, https://doi.org/10.1242/dev.021196. See also Robert C. King, William D. Stansfield, and Pamela K. Mulligan, "Heteroplastic Transplantation," in A Dictionary of Genetics, 7th ed. (New York: Oxford University Press, 2007), 205. See also "Hans Spemann, the Nobel Prize in Physiology or Medicine 1935," the Nobel Prize online, accessed February 4, 2022, https://www.nobelprize.org/prizes/medicine/1935/spemann/facts/. See also Samuel Philbrick and Erica O'Neil, "Spemann-Mangold Organizer," The Embryo Project Encyclopedia, last modified January 12, 2012, http://embryo.asu.edu/pages/spemann-mangold-organizer. See also Hans Spemann and Hilde Mangold, "Induction of Embryonic Primordia by Implantation of Organizers from a Different Species," International Journal of Developmental Biology 45, no. 1 (2001): 13–38.

In 1957, a German company called Chemie

Grünenthal: Katie Thomas, "The Story of Thalidomide in the U.S., Told Through Documents," *New York Times,* March 23, 2020. See also James H. Kim and Anthony R. Scialli, "Thalidomide: The Tragedy of Birth Defects and the Effective Treatment of Disease," *Toxicological Sciences* 122, no. (2011): 1–6.

"The burden of proof that the drug is safe": Interagency Coordination in Drug Research and Regulations: Hearings Before the Subcommittee on Reorganization and International Organizations of the Committee on Government Operations, US Senate, 87th Congress. 93 (1961) (letter from Frances O. Kelsey).

"In this connection, we are much concerned": Ibid.

"In view of the great public interest": Thomas, "Story of Thalidomide in the U.S."

"physicians of the highest professional standing": Ibid.

A multitude of cells are affected: Tomoko Asatsuma-Okumura, Takumi Ito, and Hiroshi Handa, "Molecular Mechanisms of the Teratogenic Effects of Thalidomide," *Pharmaceuticals* 13, no. 5 (2020): 95.

In 1962, she was awarded the Presidential Medal of Honor: Robert D. McFadden, "Frances Oldham Kelsey, Who Saved U.S. Babies from Thalidomide, Dies at 101," *New York Times,* August 7, 2015.

The Restless Cell: Circles of Blood

"The cell . . . is a nexus": Maureen A. O'Malley and Staffan Müller-Wille, "The Cell as Nexus: Connections Between the History, Philosophy and Science of Cell Biology," *Studies in History and Philosophy of Science Part C: Studies in History and Philosophy of Biological and Biomedical Sciences* 41, no. 3 (September 2010): 169–71, doi:10.1016/j.shpsc.2010.07.005.

"There is much that is irresolute and restless about me": Rudolf Virchow, "Letters of 1842," 26 January 1843, *Letters to his Parents, 1839 to 1864,* ed. Marie Rable, trans. Lelland J. Rather (United States of America: Science History, 1990), 29.

Around AD 150, Galen of Pergamon: Rachel Hajar, "The Air of History: Early Medicine to Galen (Part 1)," *Heart Views* 13, no. 3 (July–September 2012): 120–28, doi:10.4103/1995-705X.102164.

Harvey upended this theory: William Harvey, *On the Motion of the Heart and Blood in Animals,* ed. Alexander Bowie, trans. Robert Willis (London: George Bell and Sons, 1889).

"I began privately to think": Ibid., 48.

"[Blood] flows through the lungs and heart": William Harvey, "An Anatomical Study on the Motion of the Heart and the Blood in Animals," in *Medicine and Western Civilisation,* ed. David J. Rothman, Steven Marcus,

749

and Stephanie A. Kiceluk (New Brunswick, NJ: Rutgers University Press, 1995), 68–78.

"[T]hose sanguineous globules [red blood cells]": Antonie van Leeuwenhoek, "Mr. H. Oldenburg." 14 August 1675. Letter 18 of *Alle de brieven: 1673–1676.* De Digitale Bibliotheek voor de Nederlandse Letteren (DBNL). 301.

Marcello Malpighi, the seventeenth-century Italian anatomist: Marcello Malpighi, "De Polypo Cordis Dissertatio," Italy, 1666.

anatomist and physiologist named William Hewson: William Hewson, "On the Figure and Composition of the Red Particles of the Blood, Commonly Called Red Globules," *Philosophical Transactions of the Royal Society of London* 63 (1773): 303–23.

In 1840, Friedrich Hünefeld, a German physiologist: Friedrich Hünefeld, *Der Chemismus in der thierischen Organisation: Physiologisch-chemische Untersuchungen der materiellen Veränderungen oder des Bildungslebens im thierischen Organismus, insbesondere des Blutbildungsprocesses, der Natur der Blut körperchenund und ihrer Kenrchen: Ein Beitrag zur Physiologie und Heilmittellehre* (Leipzig, Ger.: Brockhaus, 1840).

Later that year, Denys tried to transfuse: Peter Sahlins, "The Beast Within: Animals in the First Xenotransfusion Experiments in France, ca. 1667–68," *Representations* 129,

750

no. 1 (2015): 25–55, https://doi.org/10.1525/rep.2015.129.1.25.

Landsteiner mixed blood from one individual: Karl Landsteiner, "On Individual Differences in Human Blood" (Nobel Lecture, Stockholm, December 11, 1930).

"[T]he result was exactly the same as if the blood cells": Ibid.

"Accidents following transfusion": Reuben Ottenberg and David J. Kaliski, "Accidents in Transfusion: Their Prevention by Preliminary Blood Examination — Based on an Experience of One Hundred Twenty-eight Transfusions," *Journal of the American Medical Association (JAMA)* 61, no. 24 (December 13, 1913): 2138–40, doi:10.1001/jama.1913.04350250024007.

"This great stride forward in the technique": Geoffrey Keynes, *Blood Transfusion* (Oxford, UK: Oxford Medical, 1922), 17.

In 1923, there were 123 transfusions: Ennio C. Rossi and Toby L. Simon, "Transfusions in the New Millennium," in *Rossi's Principles of Transfusion Medicine,* ed. Toby L. Simon et al. (Oxford, UK: Wiley Blackwell, 2016), 8.

"On the 13th June": A. C. Taylor to Bruce Robertson, letter, August 14, 1917, L. Bruce Robertson Fonds, Archives of Ontario, Toronto.

By the end of the war, the Red Cross had: "History of Blood Transfusion," American

Red Cross Blood Services online, accessed March 15, 2022, https://www.redcross blood.org/donate-blood/blood-donation -process/what-happens-to-donated-blood /blood-transfusions/history-blood-transfusion .html.

"War has never lavished gifts on humanity": "Blood Program in World War II," *Annals of Internal Medicine* 62, no. 5 (May 1, 1965): 1102, https:// doi.org/10.7326/0003-4819-62-5-1102_1.

The Healing Cell: Platelets, Clots, and a "Modern Epidemic"

"Imperious Caesar, dead and turned to": William Shakespeare, *Hamlet,* ed. David Bevington (New York: Bantam Books, 1980), 5.1: 213–16.

In 1881, the Italian pathologist and microscopist Giulio Bizzozero: Douglas B. Brewer, "Max Schultze (1865), G. Bizzozero (1882) and the Discovery of the Platelet," *British Journal of Haematology* 133, no. 3 (May 2006): 251–58, https://doi.org /10.1111/j.1365-2141.2006.06036.x.

in 1865, a German microscopist and anatomist: Max Schultze, "Ein heizbarer Objecttisch und seine Verwendung bei Untersuchungen des Blutes," Archiv für mikroskopische Anatomie 1 (December 1865): 1–14, https:// doi.org/10.1007/BF02961404.

"those who are concerned with the in-depth study": Ibid.

"The existence of a constant blood particle": Giulio Bizzozero, "Su di un nuovo elemento morfologico del sangue dei mammiferi e sulla sua importanza nella trombosi e nella coagulazione," *Osservatore Gazetta delle Cliniche* 17 (1881): 785–87.

"Blood platelets, swept along": Ibid.

In 1924, a Finnish hematologist, Erik von Willebrand: I. M. Nilsson, "The History of von Willebrand Disease," *Haemophilia* 5, supp. no. 2 (May 2002): 7–11, doi: 10.1046/j.1365-2516.1999.0050s2007.x.

In 1886, William Osler, one of the founders: William Osler, *The Principles and Practice of Medicine* (New York: D. Appleton, 1899). See also William Osler, "Lecture III: Abstracts of the Cartwright Lectures: On Certain Problems in the Physiology of the Blood Corpuscles" (lecture, Association of the Alumni of the College of Physicians and Surgeons, New York, March 23, 1886), 917–19.

The unraveling of the mechanisms of cholesterol metabolism: Joseph L. Goldstein et al., "Heterozygous Familial Hypercholesterolemia: Failure of Normal Allele to Compensate for Mutant Allele at a Regulated Genetic Locus," *Cell* 9, no. 2 (October 1, 1976): 195–203, https://doi.org/10.1016/0092-8674(76)90110-0.

"The modern epidemic of heart disease": James Le Fanu, *The Rise and Fall of Modern*

Medicine (London: Abacus, 2000), 322.

In 1897, a young chemist named Felix Hoffman: G. Tsoucalas, M. Karamanou, and G. Androutsos, "Travelling Through Time with Aspirin, a Healing Companion," *European Journal of Inflammation* 9, no. 1 (January 1, 2011): 13–16, https://doi.org/10.1177/17217 27X1100900102.

In the 1940s and 1950s, Lawrence Craven: Lawrence L. Craven, "Coronary Thrombosis Can Be Prevented," *Journal of Insurance Medicine* 5, no. 4 (1950): 47–48.

There are clot-dissolving drugs: Marc S. Sabatine and Eugene Braunwald, "Thrombolysis in Myocardial Infarction (TIMI) Study Group: JACC Focus Seminar 2/8," *Journal of the American Journal of Cardiology* 77, no. 22 (2021): 2822–45, doi: 10.1016/j.jacc.2021.01.060. See also X. R. Xu et al., "The Impact of Different Doses of Atorvastatin on Plasma Endothelin and Platelet Function in Acute ST-segment Elevation Myocardial Infarction After Emergency Percutaneous Coronary Intervention," *Zhonghua nei ke za zhi* 55, no. 12 (2016): 932–36, doi: 10.3760/cma.j .issn.0578-1426.2016.12.005.

The Guardian Cell: Neutrophils and Their *Kampf* Against Pathogens

"In 1736 I lost one of my sons": Benjamin Franklin, *Autobiography of Benjamin Franklin*

(New York: John B. Alden, 1892), 96.

In the 1840s, a French pathologist in Paris, Gabriel Andral: Gabriel Andral, *Essai D'Hematologie Pathologique* (Paris: Fortin, Masson et Cie Libraires, 1843).

In 1843, an English doctor named William Addison: William Addison, *Experimental and Practical Researches on Inflammation and on the Origin and Nature of Tubercles of the Lung* (London: J. Churchill, 1843), 10.

"A fine young man, aged 20": Ibid., 62.

"tubercles, in considerable numbers": Ibid., 57.

"filled with granules": Ibid., 61.

professor of zoology, Elie (or Ilya) Metchnikoff: Siddhartha Mukherjee, "Before Virus, After Virus: A Reckoning," *Cell* 183 (October 15, 2020): 308–14, doi: 10.1016/j .cell.2020.09.042.

"thick cushion layer": Ilya Mechnikov, "On the Present State of the Question of Immunity in Infectious Diseases" (Nobel Lecture, Stockholm, December 11, 1908).

"[T]he accumulation of mobile cells": Ibid.

He called the phenomenon phagocytosis: Elias Metchnikoff, "Über eine Sprosspilzkrankheit der Daphnien: Beitrag zur Lehre über den Kampf der Phagocyten gegen Krankheitserreger," *Archiv für Pathologische Anatomie und Physiologie und für Klinische Medicin* 96 (1884): 177–95.

encapsulate the relationship between an organism and its invaders: Mechnikov, "Present

State of the Question of Immunity."
monocytes, and neutrophils — are among the very first cells: Katia D. Filippo and Sara M. Rankin, "The Secretive Life of Neutrophils Revealed by Intravital Microscopy," *Frontiers in Cell and Developmental Biology* 8, no. 1236 (November 10, 2020), https://doi.org/10.3389/fcell.2020.603230. See also Pei Xiong Liew and Paul Kubes, "The Neutrophil's Role During Health and Disease," *Physiological Reviews* 99, no. 2 (February 2019): 1223–48, doi:10.1152/physrev.00012.2018.

Ehrlich, calling this idea specific affinity: Paul R. Ehrlich, *The Collected Papers of Paul Ehrlich,* ed. F. Himmelweit, Henry Hallett Dale, and Martha Marquardt (London: Elsevier Science & Technology, 1956), 3.

"The place where the punctures were made": Quoted in O. P. Jaggi, *Medicine in India* (Oxford, UK: Oxford University Press, 2000), 138.

A prior bout with an illness somehow protected the body: Arthur Boylston, "The Origins of Inoculation," *Journal of the Royal Society of Medicine* 105, no. 7 (July 2012): 309–13, doi:10.1258/jrsm.2012.12k044.

Chinese doctors harvested a smallpox scab: Wee Kek Koon, "Powdered Pus up the Nose and Other Chinese Precursors to Vaccinations," Opinion, *South China Morning Post* online, April 6, 2020, https://www

.scmp.com/magazines/post-magazine/short
-reads/article/3078436/powdered-pus-nose
-and-other-chinese-precursors.

In the 1760s, traditional healers in Sudan: Ahmed
Bayoumi, "The History and Traditional
Treatment of Smallpox in the Sudan,"
*Journal of Eastern African Research & Devel-
opment* 6, no. 1 (1976): 1–10, https://www
.jstor.org/stable/43661421.

"There is a set of old women": Lady Mary
Wortley Montagu, *Letters of the Right Hon-
ourable Lady M——y W——y M——u: Written
During Her Travels in Europe, Asia, and Africa,
to Persons of Distinction, Men of Letters, &c. in
Different Parts of Europe* (London: S. Payne,
A. Cook, and H. Hill, 1767), 137–40.

*In 1775, a Dutch diplomat who dabbled in medi-
cine:* Anne Marie Moulin, *Le dernier langage
de la médecine: Histoire de l'immunologie de
Pasteur au Sida* (Paris: Presses universitaires
de France, 1991), 23.

*apothecary's apprentice named Edward Jenner
heard:* Stefan Riedel, "Edward Jenner and
the History of Smallpox and Vaccination,"
Baylor University Medical Center Proceedings
18, no. 1 (2005): 21–25, https://doi.org/10
.1080/08998280.2005.11928028. See also
Susan Brink, "What's the Real Story About
the Milkmaid and the Smallpox Vaccine?,"
History, National Public Radio (NPR) online, February 1, 2018.

"the Disease makes its progress": Edward

Jenner, "An Inquiry into the Causes and Effects of the Variole Vaccine, or Cow-pox, 1798," in *The Three Original Publications on Vaccination Against Smallpox by Edward Jenner,* Louisiana State University, Law Center, https://biotech.law.lsu.edu/cphl/history /articles/jenner.htm#top.

in 1774, Benjamin Jesty, a hefty, prosperous farmer: James F. Hammarsten, William Tattersall, and James E. Hammarsten, "Who Discovered Smallpox Vaccination? Edward Jenner or Benjamin Jesty?," *Transactions of the American Clinical and Climatological Association* 90 (1979): 44–55, https://www.ncbi .nlm.nih.gov/pmc/articles/PMC2279376 /pdf/tacca00099-0087.pdf.

In mice, genetic inactivation of innate immunity: Mar Naranjo-Gomez et al., "Neutrophils Are Essential for Induction of Vaccine-like Effects by Antiviral Monoclonal Antibody Immunotherapies," *JCI Insight* 3, no. 9 (May 3, 2018): e97339, published online May 3, 2018, doi:10.1172/jci.insight.97339. See also Jean Louis Palgen et al., "Prime and Boost Vaccination Elicit a Distinct Innate Myeloid Cell Immune Response," *Scientific Reports* 8, no. 3087 (2018): https://doi.org/10.1038 /s41598-018-21222-2.

The Defending Cell: When a Body Meets a Body

"Gin a body meet a body": Robert Burns,

"Comin Thro' the Rye" (1782), in James Johnson, ed., *The Scottish Musical Museum; Consisting of Upwards of Six Hundred Songs, with Proper Basses for the Pianoforte,* vol. 5 (Edinburgh: William Blackwood and Sons, 1839), 430–31.

He was sent to convalesce in Egypt: Cay-Rüdiger Prüll, "Part of a Scientific Master Plan? Paul Ehrlich and the Origins of his Receptor Concept," *Medical History* 47, no. 3 (July 2003): 332–56, https://www.ncbi.nlm.nih.gov/pmc/articles/PMC1044632/.

experiment that implicitly reminded Ehrlich of the Egyptian: Paul Ehrlich, "Ehrlich, P. (1891), Experimentelle Untersuchungen über Immunität. I. Über Ricin," *DMW — Deutsche Medizinische Wochenschrift* 17, no. 32 (1891): 976–79.

Kitasato and von Behring demonstrated that the serum: Emil von Behring and Shibasaburo Kitasato, "Über das Zustandekommen der Diphtherie-Immunität und der Tetanus-Immunität bei Thieren," *Deutschen Medicinischen Wochenschrift* 49 (1890): 1113–14, https://doi.org/10.17192/eb2013.0164.

In a rather desultory footnote to the diphtheria paper: J. Lindenmann, "Origin of the Terms 'Antibody' and 'Antigen,'" *Scandinavian Journal of Immunology* 19, no. 4 (April 1984): 281–85, doi:10.1111/j.1365-3083.1984.tb00931.x.

What was this antitoxisch: Emil von Behring,

"Untersuchungen über das Zustandekommen der Diphtherie-Immunität bei Thieren," *Deutschen Medicinischen Wochenschrift* 50 (1890): 1145–48. See also William Bulloch, *The History of Bacteriology* (London: Oxford University Press, 1938). See also L. Brieger, S. Kitasato, and A. Wassermann, "Über Immunität und Giftfestigung," *Zeitschrift für Hygiene und Infektionskrankheiten* 12 (1892): 254–55. See also L. Deutsch, "Contribution à l'étude de l'origine des anticorps typhiques," *Annales de l'Institut Pasteur* 13 (1899), 689–727. See also Paul Ehrlich, "Experimentelle Untersuchungen über Immunität. II. Ueber Abrin," *Deutsche Medizinische Wochenschrift* 17 (1891): 1218–19; and "Über Immunität durch Vererbung und Säugung," *Zeitschrift für Hygiene und Infektionskrankheiten, medizinische Mikrobiologie, Immunologie und Virologie* 12 (1892): 183–203.

"the two words were destined": Lindenmann, "Origin of the Terms 'Antibody' and 'Antigen,'" 281–85.

The true molecular "shape" of an antibody: Rodney R. Porter, "Structural Studies of Immunoglobulins" (Nobel Lecture, Stockholm, December 12, 1972).

between 1959 and 1962, Gerald Edelman: Gerald M. Edelman, "Antibody Structure and Molecular Immunology" (Nobel Lecture, Stockholm, December 12, 1972).

In 1940, the fabled chemist: Linus Pauling, "A Theory of the Structure and Process of Formation of Antibodies," *Journal of the American Chemical Society* 62, no. 10 (1940): 2643–57.

Joshua Lederberg challenged Pauling's ideas: Joshua Lederberg, "Genes and Antibodies," *Science* 129, no. 3364 (1959): 1649–53.

Burnet extended the analogy to B cells: Frank Macfarlane Burnet, "A Modification of Jerne's Theory of Antibody Production Using the Concept of Clonal Selection," *CA: A Cancer Journal for Clinicians* 26, no. 2 (March–April 1976): 119–21. See also Burnet, "Immunological Recognition of Self" (Nobel Lecture, Stockholm, December 12, 1960).

"When the connection is made": Lewis Thomas, *The Lives of a Cell: Notes of a Biology Watcher* (New York: Penguin Books, 1978), 91–102.

In the 1980s, a series: Susumu Tonegawa, "Somatic Generation of Antibody Diversity," *Nature* 302 (1983): 575–81.

Milstein and Köhler's paper was published in Nature: Georges Köhler and Cesar Milstein, "Continuous Cultures of Fused Cells Secreting Antibody of Predefined Specificity," *Nature* 256 (August 7, 1975): 495–97, https://doi.org/10.1038/256495a0.

In August 1975, N.B., a fifty-three-year-old: Lee Nadler et al., "Serotherapy of a Patient with a Monoclonal Antibody Directed Against a

Human Lymphoma-Associated Antigen," *Cancer Research* 40, no. 9 (September 1980): 3147–54, PMID: 7427932.

"We'd isolate single plasma cells": Ron Levy, interview with the author, December 2021.

The Discerning Cell: The Subtle Intelligence of the T Cell

"For centuries, the thymus": Jacques Miller, "Revisiting Thymus Function," *Frontiers in Immunology* 5 (August 28, 2014): 411, https://doi.org/10.3389/fimmu.2014.00411.

In 1961, a thirty-year-old PhD student in London: Jacques F. Miller, "Discovering the Origins of Immunological Competence," *Annual Review of Immunology* 17 (1999): 1–17, doi:10.1146/annurev.immunol.17.1.1.

it remained alive and intact, growing "luxuriant hair": Ibid.

Each of these viruses: Margo H. Furman and Hidde L. Ploegh, "Lessons from Viral Manipulation of Protein Disposal Pathways," *Journal of Clinical Investigation* 110, no. 7 (2002): 875–79, https://doi.org/10.1172/JCI16831.

Townsend wrote about Enzo in the journal Nature Immunology: Alain Townsend, "Vincenzo Cerundolo 1959–2020," *Nature Immunology* 21, no. 3 (March 2020): 243, doi: 10.1038/s41590-020-0617-5.

In the 1970s, Rolf Zinkernagel and Peter Doherty: Rolf M. Zinkernagel and Peter C. Doherty,

"Immunological Surveillance Against Altered Self Components by Sensitised T Lymphocytes in Lymphocytes Choriomeningitis," *Nature* 251, no. 5475 (October 11, 1974): 547–48, doi: 10.1038/251547a0.

"That protein, NP, never makes it to the cell surface": Alain Townsend, interview with the author, 2019.

solved by the crystallographer Pam Bjorkman: Pam Bjorkman and P. Parham, "Structure, Function, and Diversity of Class I Major Histocompatibility Complex Molecules," *Annual Review of Biochemistry* 59 (1990): 253–88, doi:10.1146/annurev.bi.59.070190.001345.

"[E]very immunologist's pulse will race": Alain Townsend and Andrew McMichael, "MHC Protein Structure: Those Images That Yet Fresh Images Beget," *Nature* 329, no. 6139 (October 8–14, 1987): 482–83, doi:10.1038/329482a0.

"Those images that yet / Fresh images beget": William Butler Yeats, "Byzantium," in *The Collected Poems of W. B. Yeats* (Hertfordshire, UK: Wordsworth Editions, 1994), 210–11.

Mark Davis's at Stanford, Tak Mak's in Toronto: James Allison, B. W. McIntyre, and D. Bloch, "Tumor-Specific Antigen of Murine T-Lymphoma Defined with Monoclonal Antibody," *Journal of Immunology* 129, no. 5 (November 1982): 2293–300,

PMID: 6181166. See also Yusuke Yanagi et al., "A Human T cell–Specific cDNA Clone Encodes a Protein Having Extensive Homology to Immunoglobulin Chains," *Nature* 308 (March 8, 1984): 145–49, https://doi.org/10.1038/308145a0. See also Stephen M. Hedrick et al., "Isolation of cDNA Clones Encoding T cell–Specific Membrane-Associated Proteins," *Nature* 308 (March 8, 1984): 149–53, https://doi.org/10.1038/308149a0.

In the 1990s, Emil Unanue: Javier A. Carrero and Emil R. Unanue, "Lymphocyte Apoptosis as an Immune Subversion Strategy of Microbial Pathogens," *Trends in Immunology* 27, no. 11 (November 2006): 497–503, https://doi.org/10.1016/j.it.2006.09.005.

These are detected by a second class of T cells, called CD4 T cells: Charles A. Janeway et al., *Immunobiology: The Immune System in Health and Disease,* 5th ed. (New York: Garland Science, 2001): 114–30, https://www.ncbi.nlm.nih.gov/books/NBK27098/.

"Lymphocytes, like wasps": Lewis Thomas, *A Long Line of Cells: Collected Essays* (New York: Book of the Month Club, 1990), 71.

The article is called: Philip D. Greenberg, "Ralph M. Steinman: A Man, a Microscope, a Cell, and So Much More," *Proceedings of the National Academy of Sciences of the United States of America* 108, no. 52 (December 8, 2011): 20871–72, https://doi

.org/10.1073/pnas.1119293109.

"mononucleosis-like syndrome, marked by a hectic fever": Mirko D. Grmek, *History of AIDS: Emergence and Origin of a Modern Pandemic,* trans. Russell C. Maulitz and Jacalyn Duffin (Princeton, NJ: Princeton University Press, 1993), 3.

Nick came down with an odd, wasting disease: Ibid., 5.

In April 1981 a pharmacist: Robert D. McFadden, "Frances Oldham Kelsey, Who Saved U.S. Babies from Thalidomide, Dies at 101," *New York Times,* August 8, 2015, A1.

a landmark date, the Morbidity and Mortality Weekly Report: "*Pneumocystis* Pneumonia — Los Angeles," US Centers for Disease Control *Morbidity and Mortality Weekly Report (MMWR)* 30, no. 21 (June 5, 1981): 1–3, https://stacks.cdc.gov/view/cdc/1261.

"The occurrence of pneumocystosis in these 5": Ibid.

"[A]ll 3 patients tested had abnormal cellular-immune function": Ibid.

In March 1981, the Lancet *published:* Kenneth B. Hymes et al., "Kaposi's Sarcoma in Homosexual Men — A Report of Eight Cases," *Lancet* 318, no. 8247 (September 19, 1981): 598–600, doi:10.1016/s0140-6736(81)92740-9.

In a 1981 issue of the Lancet, *a letter to the editor:* Robert O. Brennan and David T. Durack, "Gay Compromise Syndrome,"

Letters to the Editor, *Lancet* 318, no. 8259 (December 12, 1981): 1338–39, https://doi.org/10.1016/S0140-6736(81)91352-0.

Some called it "gay-related immune deficiency": Grmek, *History of AIDS,* 6–12.

In July 1982, as doctors still searched: "Acquired Immuno-Deficiency Syndrome — AIDS," US Centers for Disease Control *Morbidity and Mortality Weekly Report (MMWR),* 31, no. 37 (September 24, 1982): 507, 513–14, available at https://stacks.cdc.gov/view/cdc/35049.

As early as 1981, three independent groups: M. S. Gottlieb et al., "Pneumocystis Carinii Pneumonia and Mucosal Candidiasis in Previously Healthy Homosexual Men: Evidence of a New Acquired Cellular Immunodeficiency," *New England Journal of Medicine* 305, no. 24 (December 10, 1981): 1425–31, doi:10.1056/NEJM198112103052401. See also H. Masur et al., "An Outbreak of Community-Acquired Pneumocystis Carinii Pneumonia: Initial Manifestation of Cellular Immune Dysfunction," *New England Journal of Medicine* 305, no. 24 (December 10, 1981): 1431–38, doi:10.1056/NEJM198112103052402. See also F. P. Siegal et al., "Severe Acquired Immunodeficiency in Male Homosexuals, Manifested by Chronic Perianal Ulcerative Herpes Simplex Lesions," *New England Journal of Medicine* 305, no. 24 (December 10, 1981): 1439–44,

doi:10.1056/NEJM198112103052403.
"the first human disease to be characterized": Jonathan M. Kagan et al., "A Brief Chronicle of CD4 as a Biomarker for HIV/AIDS: A Tribute to the Memory of John L. Fahey," *Forum on Immunopathological Diseases and Therapeutics* 6, no. 1/2 (2015): 55–64, doi:10.1615/ForumImmunDisTher.2016014169.
AIDS-causing virus was finally revealed on March 20: Françoise Barré-Sinoussi et al., "Isolation of a T-Lymphotropic Retrovirus from a Patient at Risk for Acquired Immune Deficiency Syndrome (AIDS)," *Science* 220, no. 4599 (May 20, 1983): 868–71, doi:10.1126/science.6189183.
the team published four papers in the journal Science: J. Schüpbach et al., "Serological Analysis of a Subgroup of Human T-Lymphotropic Retroviruses (HTLV-III) Associated with AIDS," *Science* 224, no. 4648 (May 4, 1984): 503–5, doi:10.1126/science.6200937; Robert C. Gallo et al., "Frequent Detection and Isolation of Cytopathic Retroviruses (HTLV-III) from Patients with AIDS and at Risk for AIDS," *Science* 224, no. 4648 (May 4, 1984): 500–503, doi: 10.1126/science.6200936; M. G. Sarngadharan et al., "Antibodies Reactive with Human T-Lymphotropic Retroviruses (HTLV-III) in the Serum of Patients with AIDS," *Science* 224, no. 4648 (May 4, 1984): 506–8, doi:10.1126/science.6324345;

and M. Popovic et al., "Detection, Isolation, and Continuous Production of Cytopathic Retroviruses (HTLV-III) from Patients with AIDS and Pre-AIDS," *Science* 224, no. 4648 (May 4, 1984): 497–500, doi: 10.1126/science.6200935.

It was christened human immunodeficiency virus, or HIV: Robert C. Gallo, "The Early Years of HIV/AIDS," *Science* 298, no. 5599 (November 29, 2002): 1728–30, doi: 10.1126/science.1078050.

For a full collection of crucial papers in this area, see: Ruth Kulstad, ed., *AIDS: Papers from Science, 1982–1985* (Washington DC: American Association for the Advancement of Science, 1986).

In his 1981 novel Midnight's Children: Salman Rushdie, *Midnight's Children* (Toronto: Alfred A. Knopf, 2010).

In one study, a two-dose: L. Gyuay et al., "Intrapartum and Neonatal Single-Dose Nevirapine Compared with Zidovudine for Prevention of Mother-to-Child Transmission of HIV-1 in Kampala, Uganda: HIVNET 012 Randomised Trial," *Lancet* 354, no. 9181 (September 4, 1999): 795–802, https://doi.org/10.1016/S0140-6736(99)80008-7 (https://www.sciencedirect.com/science/article/pii/S0140673699800087).

On February 7, 2007, Timothy Ray Brown: Timothy Ray Brown, "I Am the Berlin Patient: A Personal Reflection," *AIDS Research and*

Human Retroviruses 31, no. 1 (January 1, 2015): 2–3, doi:10.1089/aid.2014.0224. See also Sabin Russell, "Timothy Ray Brown, Who Inspired Millions Living with HIV, Dies of Leukemia," Hutch News Stories, Fred Hutchinson Cancer Research Center online, last modified September 30, 2020, https://www.fredhutch.org/en/news/center-news/2020/09/timothy-ray-brown-obit.html.

"I became delirious, nearly went blind": Brown, "I Am the Berlin Patient," 2–3.

The Tolerant Cell: The Self, Horror Autotoxicus, and Immunotherapy

"And what I assume you shall assume": Walt Whitman, "Song of Myself," in *Leaves of Grass: Comprising All the Poems Written by Walt Whitman* (New York: Modern Library, 1892), 24.

As the Caterpillar in Alice in Wonderland *asked Alice:* Lewis Carroll, *Alice in Wonderland* (Auckland, NZ: Floating Press, 2009), 35.

"A clear nonconfluent margin separates different species": Elda Gaino, Giorgio Bavestrello, and Giuseppe Magnino, "Self/Non-self recognition in Sponges," *Italian Journal of Zoology* 66, no. 4 (1999): 299–315, doi:10.1080/11250009909356270.

Aristotle imagined the self as the core of being: Aristotle, *De Anima,* trans. R. D. Hicks (New York: Cosimo Classics, 2008).

some *Vedic philosophers in India:* Brian Black, *The Character of the Self in Ancient India: Priests, Kings, and Women in the Early Upanishads* (Albany: State University of New York Press, 2007).

Surgeons in India, particularly Sushruta: Marios Loukas et al., "Anatomy in Ancient India: A Focus on Susruta Samhita," *Journal of Anatomy* 217, no. 6 (December 2010): 646–50, doi:10.1111/j.1469-7580.2010.01294.x.

In 1942, a twenty-two-year-old woman: James F. George and Laura J. Pinderski, "Peter Medawar and the Science of Transplantation: A Parable," *Journal of Heart and Lung Transplantation* 29, no. 9 (September 1, 2001), 927, https//:doi.org/10.1016/S1053 -2498)01)00345-X.

The nonself, Medawar argued, was recognized: Ibid.

Snell used these animals to dissect the genetics: George D. Snell, "Studies in Histocompatibility" (Nobel Lecture, Stockholm, December 8, 1980).

Owen inverted the question: Ray D. Owen, "Immunogenetic Consequences of Vascular Anastomoses Between Bovine Twins," *Science* 102, no. 2651 (October 19, 1945): 400– 401, doi: 10.1126/science.102.2651.400.

"Recognition that an antigenic determinant": Macfarlane Burnet, *Self and Not-Self* (London: Cambridge University Press, 1969), 25.

Philippa Marrack and John Kappler, an im-munologist couple: J. W. Kappler, M. Roehm, and P. Marrack, "T Cell Tolerance by Clonal Elimination in the Thymus," *Cell* 49, no. 2 (April 24, 1987): 273–80, doi:10.1016/0092-8674(87)90568-x.

Beyond central tolerance, there is a phenom-enon: Carolin Daniel, Jens Nolting, and Harald von Boehmer, "Mechanisms of Self-Nonself Discrimination and Possible Clinical Relevance," *Immunotherapy* 1, no. 4 (July 2009): 631–44, doi:10.2217/imt.09.29.

horror autotoxicus — the body poisoning itself: Paul Ehrlich, *Collected Studies on Immunity* (New York: John Wiley & Sons, 1906), 388.

My mind returned to an image: William Shakespeare, "When Icicles Hang by the Wall," *Love's Labour's Lost, London Sunday Times* online, last modified December 30, 2012, https://www.thetimes.co.uk/article/when -icicles-hang-by-the-wall-by-william-shake speare-1564-1616-5kgxk93bnwc.

William Coley, had tried to treat cancer pa-tients: William B. Coley, "The Treatment of Inoperable Sarcoma with the Mixed Toxins of Erysipelas and Bacillus Pro-digiosus: Immediate and Final Results in One Hundred Forty Cases," *Journal of the American Medical Association (JAMA)* 31, no. 9 (August 27, 1898): 456–65, doi:10.1001 /jama.1898.92450090022001g; William B. Coley "The Treatment of Malignant Tumors by Repeated Inoculation of Erysipelas," *Journal of the American Medical Association (JAMA)* 20, no. 22 (June 3, 1893): 615–16, doi:10.1001/jama.1893.02420490019007; and William B. Coley "II. Contribution to the Knowledge of Sarcoma," *Annals of Surgery* 14, no. 3 (September 1891): 199–200, doi:10.1097/00000658-189112000-00015.

Rosenberg's team had grown these tumor-infiltrating lymphocytes: Steven A. Rosenberg and Nicholas P. Restifo, "Adoptive Cell Transfer as Personalized Immunotherapy for Human Cancer," *Science* 348, no. 6230 (April 2015): 62–68, doi:10.1126/science.aaa4967.

Allison injected some of the mice with antibodies: James P. Allison, "Immune Checkpoint Blockade in Cancer Therapy" (Nobel Lecture, Stockholm, December 7, 2018).

But Honjo's team slowly converged on the function of PD-1: Tasuku Honjo, "Serendipities of Acquired Immunity" (Nobel Lecture, Stockholm, December 7, 2018).

A new class of drugs grew out of this work: Julie R. Brahmer et al., "Safety and Activity of anti-PD-L1 Antibody in Patients with Advanced Cancer," *New England Journal of Medicine* 366, no. 26 (June 28, 2012): 2455–65, doi:10.1056/NEJMoa1200694. See also Omid Hamid et al., "Safety and Tumor Responses with Lambrolizumab (anti-PD-1) in Melanoma," *New England Journal of*

Medicine 369, no. 2 (July 11, 2013): 134–44, doi:10.1056/NEJMoa1305133.

The Pandemic

"Into the notable city of Florence": Giovanni Boccaccio, *The Decameron of Giovanni Boccaccio,* trans. John Payne (Frankfurt, Ger.: Outlook Verlag, 2020), 5.

To read that first case report: Mechelle L. Holshue et al., "First Case of 2019 Novel Coronavirus in the United States," *New England Journal of Medicine* 382, no. 10 (2020): 929–36, doi: 10.1056/NEJMoa2001191.

A second wave that moved through India in April 2021: The Wire and Murad Banaji, "As Delta Tore Through India, Deaths Skyrocketed in Eastern UP, Analysis Finds," *The Wire,* February 11, 2022, https://science .thewire.in/health/covid-19-excess-deaths -eastern-uttar-pradesh-cjp-investigation/.

A sixty-five-year-old journalist: Aggarwal, Mayank Aggarwal, "Indian Journalist Live-Tweeting Wait for Hospital Bed Dies from Covid," *Asia, India. Independent,* April 21, 2021, https://www.independent.co.uk/asia /india/india-journalist-tweet-covid-death -b1834362.html.

Akiko Iwasaki, a virologist: Akiko Iwasaki, interview with the author, April 2020.

a thirty-three-year-old man from Munich: Camilla Rothe et al., "Transmission of 2019-nCoV Infection from an Asymptomatic

Contact in Germany," *New England Journal of Medicine* 328 (2020): 970–71, doi: 10.1056/NEJMc2001468.

In 2020, a group of Dutch: Caspar I. van der Made et al., "Presence of Genetic Variants Among Young Men with Severe COVID-19," *Journal of the American Medical Association (JAMA)* 324, no. 7 (2020): 663–73, doi: 10.1001/jama.2020.13719.

At Ben tenOever's lab in New York: Daniel Blanco-Melo et al., "Imbalanced Host Response to SARS-CoV-2 Drives Development of COVID-19," *Cell* 181, no. 5 (2020): 1036–45, doi: 10.1016/j.cell.2020.04.026.

"It's almost as if the virus hijacked the cell": Ben tenOever, interview with the author, January 2020.

At Rockefeller University in New York, Jean-Laurent Casanova: Qian Zhang et al., "Inborn Errors of Type I IFN Immunity in Patients with Life-Threatening COVID-19," *Science* 370, no. 6515 (2020): eabd4570, doi: 10.1126/science.abd4570. See also Paul Bastard et al., "Autoantibodies Against Type I IFNs in Patients with Life-Threatening COVID-19," *Science* 370, no. 6515 (2020): eabd4585, doi: 10.1126/science.abd4585.

"a raider that had come into an unlocked house": James Somers, "How the Coronavirus Hacks the Immune System," *New Yorker* (November 2, 2020), https://www.newyorker.com/magazine/2020/11/09

/how-the-coronavirus-hacks-the-immune-system.

"There appears to be a fork in the road": Akiko Iwasaki, interview with the author, April 2020.

There's a haunting image in one of Zadie Smith's essays: Zadie Smith, "Fascinated to Presume: In Defense of Fiction," *New York Review of Books,* October 24, 2019, https://www.nybooks.com/articles/2019/10/24/zadie-smith-in-defense-of-fiction/.

The Citizen Cell: The Benefits of Belonging

"The crowd, suddenly there where there was nothing before": Elias Canetti, *Crowds and Power,* trans. Carol Stewart (New York: Continuum, Farrar, Straus and Giroux, 1981), 16.

"For the concept of a circuit": William Harvey, *The Circulation of the Blood: Two Anatomical Essays,* trans. Kenneth J. Franklin (Oxford, UK: Blackwell Scientific Publications, 1958), 12.

Medicine, I wrote, isn't a doctor: Siddhartha Mukherjee, "What the Coronavirus Crisis Reveals about American Medicine," *New Yorker* (April 27, 2020), https://www.newyorker.com/magazine/2020/05/04/what-the-coronavirus-crisis-reveals-about-american-medicine.

Aristotle, for one, thought: Aristotle, *On the Soul, Parva Naturalia, On Breath,* trans. W. S.

Hett (London: William Heinemann, 1964).
"the heart is, as it were": Galen, *On the Usefulness of the Parts of the Body,* trans. Margaret Tallmadge May (New York: Cornell University Press, 1968), 292.

The medieval physiologist Ibn Sina: Izet Masic, "Thousand-Year Anniversary of the Historical Book: "Kitab al-Qanun fit-Tibb" — The Canon of Medicine, Written by Abdullah ibn Sina," *Journal of Research in Medical Sciences* 17, no. 11 (2012): 993–1000, https://www.ncbi.nlm.nih.gov/pmc/articles/PMC3702097/.

English physiologist William Harvey, working: D'Arcy Power, *William Harvey: Masters of Medicine* (London: T. Fisher Unwin, 1897). See also W. C. Aird, "Discovery of the Cardiovascular System: From Galen to William Harvey," *Journal of Thrombosis and Hemostasis* 9, no. 1 (2011): 118–29, doi: 10.1111/j.1538-7836.2011.04312.x.

"eyes small, round": Edgar F. Mauer, "Harvey in London," *Bulletin of the History of Medicine* 33, no. 1 (1959): 21–36, https://www.jstor.org/stable/44450586.

"pleased some more, others less": William Harvey, *On the Motion of the Heart and Blood in Animals,* trans. Robert Willis, ed. Jarrett A. Carty (Eugene, OR: Resource Publications, 2016), 36.

On January 17, 1912, Alexis Carrel: Hannah Landecker, *Culturing Life: How Cells Became*

Technologies (Cambridge: Harvard University Press, 2007), 75.

"[T]he fragment pulsated regularly": Alexis Carrel, "On the Permanent Life of Tissue Outside of the Organism," *Journal of Experimental Medicine* 15, no. 5 (1912): 516–30, https://www.ncbi.nlm.nih.gov/pmc/articles /PMC2124948/pdf/516.pdf.

That same year, W. T. Porter: W. T. Porter, "Coordination of Heart Muscle Without Nerve Cells," *Journal of the Boston Society of Medical Sciences* 3, no. 2 (1898), https:// pubmed.ncbi.nlm.nih.gov/19971205/.

In the 1880s, German biologist Friedrich Bidder had noted: Carl J. Wiggers, "Some Significant Advances in Cardiac Physiology During the Nineteenth Century," *Bulletin of the History of Medicine* 34, no. 1 (1960): 1–15, https://www.jstor.org/stable/44446654.

a Hungarian-born physiologist, Albert Szent-Györgyi: Beáta Bugyi and Miklós Kellermayer, "The Discovery of Actin: 'To See What Everyone Else Has Seen, and to Think What Nobody Has Thought,'" *Journal of Muscle Research and Cell Motility* 41 (2020): 3–9, https://doi.org/10.1007/ s10974-019-09515-z. See also Andrzej Grzybowski and Krzysztof Pietrzak, "Albert Szent Györrgi (1893–1986): The Scientist who Discovered Vitamin C," *Clinics in Dermatology* 31 (2013): 327–31, https://www .cidjournal.com/action/showPdf?pii=S0738

-081X%2812%2900171-X. See also Albert Szent-Györgyi, "Contraction in the Heart Muscle Fibre," *Bulletin of the New York Academy of Medicine* 28, no. 1 (1952): 3–10, https://www.ncbi.nlm.nih.gov/pmc/articles /PMC1877124/pdf/bullnyacadmed00430 -0012.pdf.

"In order to make a system which can shorten": Ibid.

The Contemplating Cell: The Many-Minded Neuron

"The Brain — is wider than the Sky —": Emily Dickinson, "The Brain Is Wider than the Sky," 1862, *The Complete Poems of Emily Dickinson,* ed. Thomas H. Johnson (Boston: Little, Brown, 1960), 312–13.

In 1873 Camillo Golgi, the Italian biologist: Camillo Golgi, "The Neuron Doctrine — Theory and Facts," Nobel Lecture. Sweden (December 11, 1906), https://www.nobel prize.org/uploads/2018/06/golgi-lecture.pdf.

"inextricable tangle" of interconnected: Ennio Pannese, "The Golgi Stain: Invention, Diffusion and Impact on Neurosciences," *Journal of the History of the Neurosciences* 8, no. 2 (1999): 132–40, doi: 10.1076/jhin .8.2.132.1847.

"shy, unsociable, secretive, brusque": Larry W. Swanson, Eric Newman, Alfonso Araque, and Janet M. Dubinsky, *The Beautiful Brain: The Drawings of Santiago Ramon y Cajal*

(New York: Abrams, 2017), 12.

Santiago Ramón y Cajal was the son of an anatomy teacher: Marina Bentivoglio, "Life and Discoveries of Santiago Ramón y Cajal," *Nobel Prize* (April 20, 1998), https://www.nobelprize.org/prizes/medicine/1906/cajal/article/. See also Luis Ramón y Cajal, "Cajal, as Seen by His Son," *Cajal Club* (1984), https://cajalclub.org/wp-content/uploads/sites/9568/2019/08/Cajal-As-Seen-By-His-Son-by-Luis-Ram%C3%B3n-y-Cajal-p.-73.pdf, and Santiago Ramón y Cajal, "The Structure and Connections of Neurons," Nobel Lecture, Sweden (December 12, 1906), https://www.nobelprize.org/uploads/2018/06/cajal-lecture.pdf.

He would later call this drawing habit his "irresistible mania": Santiago Ramón y Cajal, *Recollections of My Life,* trans. E. Horne Craigie, and Juan Cano (Cambridge: MIT Press, 1996), 36.

In 1906, Cajal and Golgi were jointly awarded: "The Nobel Prize in Physiology or Medicine 1906," Nobel Prize, https://www.nobelprize.org/prizes/medicine/1906/summary/.

To see his drawings of neurons: Pablo Garcia-Lopez, Virginia Garcia-Marin, and Miguel Freire, "The Histological Slides and Drawings of Cajal," *Frontiers in Neuroanatomy* 4, no. 9 (2010), doi: 10.3389/neuro.05.009.2010.

The story, likely apocryphal: Henry Schmidt,

"Frogs and Animal Electricity," *Explore Whipple Collections, Whipple Museum of the History of Science* (University of Cambridge), https://www.whipplemuseum.cam.ac.uk/explore-whipple-collections/frogs/frogs-and-animal-electricity.

In 1939, Alan Hodgkin, an undergraduate: Christof J. Schwiening, "A Brief Historical Perspective: Hodgkin and Huxley," *Journal of Physiology* 590, no. 11 (2012): 2571–75, doi: 10.1113/jphysiol.2012.230458.

they rushed their paper off to Nature: Alan Hodgkin and Andrew Huxley, "Action Potentials Recorded from Inside a Nerve Fibre," *Nature* 144, no. 3651 (1939): 710–11, doi: 10.1038/144710a0.

"It takes a courageous person," the poet Kay Ryan: Kay Ryan, "Leaving Spaces," *The Best of It: New and Selected Poems* (New York: Grove Press, 2010), 38.

"The idea of a chemical mediator released": J. F. Fulton, *Physiology of the Nervous System* (New York: Oxford University Press, 1949).

In the 1920s and 1930s, for the English neurophysiologist Henry Dale: Henry Dale, "Some Recent Extensions of the Chemical Transmission of the Effects of Nerve Impulses," Nobel Lecture (December 12, 1936), https://www.nobelprize.org/prizes/medicine/1936/dale/lecture/.

Unusual for his times, Dale: Report of the Wellcome Research Laboratories at the Gordon

Memorial College, Khartoum, vol. 3 (Khartoum: Wellcome Research Laboratories, 1908), 138.

In Graz, Austria, another neurophysiologist, Otto Loewi: Otto Loewi, "The Chemical Transmission of Nerve Action," Nobel Lecture (December 12, 1936), https://www.nobelprize.org/prizes/medicine/1936/loewi/lecture/. See also Alli N. McCoy and Yong Siang Tan, "Otto Loewi (1873–1961): Dreamer and Nobel Laureate," Singapore Medical Journal 55, no. 1 (2014): 3–4, doi: 10.11622/smedj.2014002.

"I awoke" he wrote: Otto Loewi, "An Autobiographical Sketch," Perspectives in Biology and Medicine 4, no. 1 (1960): 3–25, https://muse.jhu.edu/article/404651/pdf.

"the conversion of Saul on the road to Damascus": Don Todman, "Henry Dale and the Discovery of Chemical Synaptic Transmission," European Neurology 60 (2008): 162–64, https://doi.org/10.1159/000145336.

A small number of neurons in animals: Stephen G. Rayport and Eric R. Kandel, "Epileptogenic Agents Enhance Transmission at an Identified Weak Electrical Synapse in Aplysia," Science 213, no. 4506 (1981): 462–64, https://www.jstor.org/stable/1686531.

Those complex circuits, you might surmise: Annapurna Uppala et al., "Impact of Neurotransmitters on Health through Emotions," International Journal of Recent

781

Scientific Research 6, no. 10 (2015): 6632–36, doi: 10.1126/science.1089662.

"If a subject [. . .] has a glamorous aura": Edward O. Wilson, *Letters to a Young Scientist* (New York: Liveright, 2013), 46.

Glial cells are present all over the nervous system: Christopher S. von Bartheld, Jami Bahney, and Suzana Herculano-Houzel, "The Search for True Numbers of Neurons and Glial Cells in the Human Brain: A Review of 150 Years of Cell Counting," *Journal of Comparative Neurology* 524, no. 18 (2016): 3865–95, doi: 10.1002/cne.24040.

Unlike neurons, they don't generate electrical impulses: Sarah Jäkel and Leda Dimou, "Glial Cells and Their Function in the Adult Brain: A Journey through the History of Their Ablation," *Frontiers in Cellular Neuroscience* 11 (2017), https://doi.org/10.3389/fncel.2017.00024.

Neural connections between the eyes and the brain: Dorothy P. Schafer et al., "Microglia Sculpt Postnatal Neural Circuits in an Activity and Complement-Dependent Manner," *Neuron* 74, no. 4 (2012): 691–705, doi: 10.1016/j.neuron.2012.03.026.

"Cells that fire together, wire together": Carla J. Shatz, "The Developing Brain," *Scientific American* 267, no. 3 (1992): 60–67, https://www.jstor.org/stable/24939213.

"It reinforces an old intuition": Hans Agrawal, interview with the author, October 2015.

In 2007, they announced a startling discovery: Beth Stevens et al., "The Classical Complement Cascade Mediates CNS Synapse Elimination," *Cell* 131, no. 6 (2007): 1164–78, https://doi.org/10.1016/j.cell.2007.10.036.

"The questions we took on in the new lab": Beth Stevens, interview with the author, February 2016.

They are the brain's "constant gardeners": Virginia Hughes, "Microglia: The Constant Gardeners," *Nature* 485 (2012): 570–72, https://doi.org/10.1038/485570a.

Recent experiments suggest that dysfunctions: Andrea Dietz, Steven A. Goldman, and Maiken Nedergaard, "Glial Cells in Schizophrenia: A Unified Hypothesis," *Lancet Psychiatry* 7, no. 3 (2019): 272–81, doi: 10.1016/S2215-0366(19)30302-5.

reciting the lines of Kenneth Koch's poem: Kenneth Koch, "One Train May Hide Another," *One Train* (New York: Alfred A. Knopf, 1994).

All I could feel was the "dank joylessness": William Styron, *Darkness Visible: A Memoir of Madness* (New York: Open Road, 2010), 10.

"Oh, he'll eventually come around": Paul Greengard, interview with the author, January 2019.

"Depression is a slow brain problem": Ibid. See also Jung-Hyuck Ahn et al., "The B"/PR72 Subunit Mediates Ca2+-dependent Dephosphorylation of DARPP-32 by Protein

Phosphatase 2A," *Proceedings of the National Academy of Sciences* 104, no. 23 (2007): 9876–81, doi: 10.1073/pnas.0703589104.

I was reminded of the Carl Sandburg poem: Carl Sandburg, "Fog," *Chicago Poems* (New York: Henry Holt, 1916), 71.

Andrew Solomon, the writer, once described: Andrew Solomon, *The Noonday Demon: An Atlas of Depression* (New York: Scribner, 2001), 33.

In the autumn of 1951, doctors treating: Robert A. Maxwell and Shohreh B. Eckhardt, *Drug Discovery: A Casebook and Analysis* (New York: Springer Science+Business Media, 1990), 143–54. See also Siddhartha Mukherjee, "Post-Prozac Nation," *New York Times Magazine* (April 19, 2012), https://www.nytimes.com/2012/04/22/magazine/the-science-and-history-of-treating-depression.html., and Alexis Wnuk, "Rethinking Serotonin's Role in Depression," *BrainFacts* (March 8, 2019), https://www.sfn.org/sitecore/content/home/brainfacts2/diseases-and-disorders/mental-health/2019/rethinking-serotonins-role-in-depression-030819.

When Life *magazine sent a photographer:* "TB Milestone: Two New Drugs Give Real Hope of Defeating the Dread Disease," *Life* 32, no. 9 (1952): 20–21.

In the 1970s, Arvid Carlsson, a biochemist: Arvid Carlsson, "A Half-Century of

Neurotransmitter Research: Impact on Neurology and Psychiatry," Nobel Lecture, Sweden (December 8, 2000), https://www.nobelprize.org/uploads/2018/06/carlsson-lecture.pdf.

In her bestselling 1994 memoir Prozac Nation: Elizabeth Wurtzel, *Prozac Nation* (New York: Houghton Mifflin, 1994), 203.

"One morning I woke up and really did want to live": Ibid., 454–55.

he had shown that one such factor, called DARPP-32: Per Svenningsson et al., "P11 and Its Role in Depression and Therapeutic Responses to Antidepressants," *Nature Reviews Neuroscience* 14 (2013): 673–80, doi: 10.1038/nrn3564. For Greengard's classic paper on dopamine signaling, see John W. Kebabian, Gary L. Petzold, and Paul Greengard, "Dopamine-Sensitive Adenylate Cyclase in Caudate Nucleus of Rat Brain, and Its Similarity to the 'Dopamine Receptor,'" *Proceedings of the National Academy of Science* 69, no. 8 (August 1972): 2145–49. doi:10.1073/pnas.69.8.2145.

In the early 2000s, taking a radical: Helen S. Mayberg, "Targeted Electrode-Based Modulation of Neural Circuits for Depression," *Journal of Clinical Investigation* 119, no. 4 (2009): 717–25, doi: 10.1172/JCI38454.

"In a pair of pale-pink curves of neural flesh": David Dobbs, "Why a 'Lifesaving' Depression Treatment Didn't Pass Clinical Trials,"

Atlantic (April 17, 2018), https://www.the atlantic.com/science/archive/2018/04/zapping-peoples-brains-didnt-cure-their-depression-until-it-did/558032/.

"I remember each one of those patients": Helen Mayberg, interview with the author, November 2021.

"[a]ll patients spontaneously reported acute effects": Helen S. Mayberg et al., "Deep Brain Stimulation for Treatment-Resistant Depression," *Neuron* 45 (2005): 651–60, doi: 10.1016/j.neuron.2005.02.014. See also H. Johansen-Berg et al., "Anatomical Connectivity of the Subgenual Cingulate Region Targeted with Deep Brain Stimulation for Treatment-Resistant Depression," *Cerebral Cortex* 18, no. 6 (2008): 1374–83, doi: 10.1093/cercor/bhm167.

Significantly, a pivotal study (called BROADEN): Dobbs, "Why a 'Lifesaving' Depression Treatment Didn't Pass Clinical Trials."

"The experience, which was searing,": Peter Tarr, "'A Cloud Has Been Lifted': What Deep-Brain Stimulation Tells Us About Depression and Depression Treatments," *Brain and Behavior Research Foundation* (September 17, 2018), https://www.bbrfoundation.org/content/cloud-has-been-lifted-what-deep-brain-stimulation-tells-us-about-depression-and-depression.

"Electroceuticals are in": "BROADEN Trial of DBS for Treatment-Resistant Depression

Halted by the FDA," *The Neurocritic* (January 18, 2014), https://neurocritic.blogspot.com/2014/01/broaden-trial-of-dbs-for-treatment.html.

In a paper published in Lancet Psychiatry *in 2017:* Paul E. Holtzheimer et al., "Subcallosal Cingulate Deep Brain Stimulation for Treatment-Resistant Depression: A Multisite, Randomised, Sham-Controlled Trial," *Lancet Psychiatry* 4, no. 11 (2017): 839–49, doi: 10.1016/S2215-0366(17)30371-1.

The Orchestrating Cell: Homeostasis, Fixity, and Balance

"Every cell has its own special action": Rudolf Virchow, "Lecture I: Cells and the Cellular Theory," trans. Frank Chance, *Cellular Pathology as Based Upon Physiological and Pathological Histology: Twenty Lectures Delivered in the Pathological Institute of Berlin* (London: John Churchill, 1860), 1–23.

"Now we will count to twelve": Pablo Neruda, "Keeping Still," trans. Dan Bellum, *Literary Imagination* 8, no. 3 (2016): 512.

"mysterious, hidden": Salvador Navarro, "A Brief History of the Anatomy and Physiology of a Mysterious and Hidden Gland Called the Pancreas," *Gastroenterología y hepatología* 37, no. 9 (2014): 527–34, doi: 10.1016/j.gastrohep.2014.06.007.

The Alexandrian anatomist Herophilus: John M. Howard and Walter Hess, *History of the*

787

Pancreas: Mysteries of a Hidden Organ (New York: Springer Science+Business Media, 2002).

"As the vein, the artery and the nerve": Quoted in ibid., 6.

"big glandular body": Ibid., 12.

"[I]t would be totally useless to animals": Ibid., 15.

As Wirsung extracted the pancreas: Ibid., 16.

he was walking in an alley outside his house: Sanjay A. Pai, "Death and the Doctor," *Canadian Medical Association Journal* 167, no. 12 (2002): 1377–78, https://www.ncbi.nlm.nih.gov/pmc/articles/PMC138651/.

In 1856 Bernard published Mémoire sur le Pancréas: Claude Bernard, "Sur L'usage du suc pancréatique," *Bulletin de la Société Philomatique* (1848): 34–36. See also Claude Bernard, *Mémoire sur le pancréas, et sur le role du suc pancréatique dans les phénomènes digestifs; particulièrement dans la digestion des matières grasses neutres* (Paris: Kessinger Publishing, 2010).

In July 1920, Frederick Banting was a surgeon: Michael Bliss, *Banting: A Biography* (Toronto: University of Toronto Press, 1992).

Late one evening in October 1920, he read an article: Lars Rydén and Jan Lindsten, "The History of the Nobel Prize for the Discovery of Insulin," *Diabetes Research and Clinical Practice* 175 (2021), https://doi.org/10.1016/j.diabres.2021.108819.

The initial meeting, on November 8, 1920: Ian Whitford, Sana Qureshi, and Alessandra L. Szulc, "The Discovery of Insulin: Is There Glory Enough for All?" *Einstein Journal of Biology and Medicine* 28, no. 1 (2016): 12–17, https://einsteinmed.edu/uploadedFiles/Pulications/EJBM/28.1_12-17_Whitford.pdf.

In some dogs: Siang Yong Tan and Jason Merchant, "Frederick Banting (1891–1941): Discoverer of Insulin," *Singapore Medical Journal* 58, no. 1 (2017): 2–3, doi: 10.11622/smedj.2017002.

The first attempts were failures: "Banting & Best: Progress and Uncertainty in the Lab," *Insulin100: The Discovery and Development, DefiningMomentsCanada* (n.d.), https://definingmomentscanada.ca/insulin100/timeline/banting-best-progress-and-uncertainty-in-the-lab/.

Late that summer, with the temperature still rising: Michael Bliss, *The Discovery of Insulin* (Toronto: McClelland & Stewart, 2021), 67–72.

is a disease in which immune cells attack: Justin M. Gregory, Daniel Jensen Moore, and Jill H. Simmons, "Type 1 Diabetes Mellitus," *Pediatrics in Review* 34, no. 5 (2013): 203–15, doi: 10.1542/pir.34-5-203.

In 2014, a team led by Doug Melton: Douglas Melton, "The Promise of Stem Cell-Derived Islet Replacement Therapy," *Diabetologia*

64 (2021): 1030–36, https://doi.org/10.1007/s00125-020-05367-2.

Then both of Melton's children developed type 1 diabetes: David Ewing Duncan, "Doug Melton: Crossing Boundaries," *Discover* (June 5, 2005), https://www.discovermagazine.com/health/doug-melton-crossing-boundaries.

For a while, as Melton told a journalist: Karen Weintraub, "The Quest to Cure Diabetes: From Insulin to the Body's Own Cells," *The Price of Health,* WBUR (June 27, 2019), https://www.wbur.org/news/2019/06/27/future-innovation-diabetes-drugs.

One evening in 2014, a postdoctoral researcher: Gina Kolata, "A Cure for Type 1 Diabetes? For One Man, It Seems to Have Worked," *New York Times* (November 27, 2021), https://www.nytimes.com/2021/11/27/health/diabetes-cure-stem-cells.html.

"express markers found in mature beta cells": Felicia W. Pagliuca et al., "Generation of Functional Human Pancreatic β Cells in Vitro," *Cell* 159, no. 2 (2014): 428–39, doi: 10.1016/j.cell.2014.09.040.

One of the first patient to receive the infusion: Kolata, "A Cure for Type 1 Diabetes?"

Liver cells, in contrast, have evolved: John Y. L. Chiang, "Liver Physiology: Metabolism and Detoxification," *Pathobiology of Human Disease,* ed. Linda M. McManus and Richard N. Mitchell (San Diego:

Elsevier, 2014), 1770–82, doi: 10.1016 /B978-0-12-386456-7.04202-7.

Their miraculous fixity in midair: Carl Zimmer, *Life's Edge: The Search for What It Means to Be Alive* (New York: Penguin Random House, 2021), 128–37.

Part Six: Rebirth

"Old age is a massacre": Philip Roth, *Everyman* (London: Penguin Random House, 2016), 133.

The Renewing Cell: Stem Cells and the Birth of Transplantation

"He not busy being born is busy dying": Rachel Kushner, *The Hard Crowd* (New York: Scribner, 2021), 229.

"Stem cells don't simply transform themselves": Joe Sornberger, *Dreams and Due Diligence: Till and McCulloch's Stem Cell Discovery and Legacy* (Toronto: University of Toronto Press, 2011), 30–31.

On August 6, 1945, at about eight fifteen: Jessie Kratz, "Little Boy: The First Atomic Bomb," *Pieces of History, National Archives* (August 6, 2020), https://prologue.blogs.archives. gov/2020/08/06/little-boy-the-first-atomic -bomb/. See also Katie Serena, "See the Eerie Shadows of Hiroshima That Were Burned into the Ground by the Atomic Bomb," *All That's Interesting* (March 19, 2018), https:// allthatsinteresting.com/hiroshima-shadows.

"I was trying to describe the mushroom [cloud]": George R. Caron and Charlotte E. Meares, *Fire of a Thousand Suns: The George R. "Bob" Caron Story: Tail Gunner of the Enola Gay* (Littleton, CO: Web Publishing, 1995).

"Survivors began to notice": Robert Jay Lifton, "On Death and Death Symbolism," *American Scholar* 34, no. 2 (1965): 257–72, https://www.jstor.org/stable/41209276.

As the scientists Irving Weissman and Judith Shizuru put it: Irving L. Weissman and Judith A. Shizuru, "The Origins of the Identification and Isolation of Hematopoietic Stem Cells, and Their Capability to Induce Donor-Specific Transplantation Tolerance and Treat Autoimmune Diseases," *Blood* 112, no. 9 (2008): 3543–53, doi: 10.1182/blood-2008-08-078220.

Cynthia Ozick, the essayist, once wrote: Cynthia Ozick, *Metaphor and Memory* (London: Atlantic Books, 2017), 109.

In 1868, the German embryologist Ernst Haeckel: Ernst Haeckel, *Natürliche Schöpfungsgeschichte Gemeinverständliche wissenschaftliche Vorträge über die Entwickelungslehre im Allgemeinen und diejenige von Darwin, Göthe und Lamarck im Besonderen, über die Anwendung derselben auf den Ursprung des Menschen und andern damit zusammenhängende Gründfragen der Natur-Wissenschaft. Mit Tafeln, Holzschnitten, systematischen und genealogischen Tabellen* (Berlin: Berlag von

Georg Reimer, 1868). See also Miguel Ramalho-Santos and Holger Willenbring, "On the Origin of the Term 'Stem Cell,'" *Cell* 1, no. 1 (2007): 35–38, https://doi.org/10.1016/j.stem.2007.05.013.

In 1892, the zoologist Valentin Hacker: Valentin Hacker, "Die Kerntheilungsvorgänge bei der Mesoderm-und Entodermbildung von Cyclops," *Archiv für mikroskopische Anatomie* (1892): 556–81, https://www.biodiversitylibrary.org/item/49530#page/7/mode/1up.

The cytologist Artur Pappenheim, studying: Artur Pappenheim, "Ueber Entwickelung und Ausbildung der Erythroblasten," *Archiv für mikroskopische Anatomie* (1896): 587–643, https://doi.org/10.1007/BF0196990.

In 1896, the biologist Edmund Wilson used the phrase "stem cell": Edmund Wilson, *The Cell in Development and Inheritance* (New York: Macmillan, 1897).

As the idea of a "stem cell" gathered popularity: Wojciech Zakrzewski et al., "Stem Cells: Past, Present and Future," *Stem Cell Research and Therapy* 10, no. 68 (2019), https://doi.org/10.1186/s13287-019-1165-5.

About Ernest McCulloch and James Til's lives and experiments: Lawrence K. Altman, "Ernest McCulloch, Crucial Figure in Stem Cell Research, Dies at 84," *New York Times* (February 1, 2011), https://www.nytimes.com/2011/02/01/health/research/01mcculloch.html.

"Old Money Toronto" family: Joe Sornberger, *Dreams and Due Diligence: Till and McCulloch's Stem Cell Discovery and Legacy* (Toronto: University of Toronto Press, 2011). See also Edward Shorter, *Partnership for Excellence: Medicine at the University of Toronto and Academic Hospitals* (Toronto: University of Toronto Press, 2013), 107–14.

Till and McCulloch published their data: James E. Till Ernest McCulloch, "A Direct Measurement of the Radiation Sensitivity of Normal Mouse Bone Marrow Cells," *Radiation Research* 14, no. 2 (1961): 213–22, https://tspace.library.utoronto.ca/retrieve/4606/RadRes_1961_14_213.pdf.

"You have to remember that it was a fairly small group": Sornberger, *Dreams and Due Diligence,* 33.

"The paper represented an entirely new way": Ibid.

"The real discovery," he said later: Ibid., 38.

"all arose from the same stem cell": Irving Weissman, interview with the author, 2019.

Inspired by the Till and McCulloch experiments, Weissman began: Gerald J. Spangrude, Shelly Heimfeld, and Irving L. Weissman, "Purification and Characterization of Mouse Hematopoietic Stem Cells," *Science* 241, no. 4861 (1988): 58–62, doi: 10.1126/science.2898810. See also Hideo Ema et al., "Quantification of Self-Renewal Capacity in Single Hematopoietic Stem Cells from

Normal and Lnk-Deficient Mice," *Developmental Cell* 8, no. 6 (2006): 907–14, https://doi.org/10.1016/j.devcel.2005.03.019.

Weissman went through dozens of permutations: Spangrude, Heimfeld, and Weissman, "Purification and Characterization of Mouse Hematopoietic Stem Cells," 58–62, doi: 10.1126/science.2898810. See also C. M. Baum et al., "Isolation of a Candidate Human Hematopoietic Stem-Cell Population," *Proceedings of the National Academy of Sciences of the United States of America* 89, no. 7 (1992): 2804–08, doi: 10.1073/pnas.89.7.2804, and B. Péault, Irving Weissman, and C. Baum, "Analysis of Candidate Human Blood Stem Cells in "Humanized" Immune-Deficiency SCID Mice," *Leukemia* 7, suppl. 2 (1993): S98–101, https://pubmed.ncbi.nlm.nih.gov/7689676/.

in the 1960s, the Australian researcher Donald Metcalf: W. Robinson, Donald Metcalf, and T. R. Bradley, "Stimulation by Normal and Leukemic Mouse Sera of Colony Formation *in Vitro* by Mouse Bone Marrow Cells," *Journal of Cellular Therapy* 69, no. 1 (1967): 83–91, https://doi.org/10.1002/jcp.1040690111. See also E. R. Stanley and Donald Metcalf, "Partial Purification and Some Properties of the Factor in Normal and Leukaemic Human Urine Stimulating Mouse Bone Marrow Colony Growth in Vitro," *Australian Journal of Experimental*

Biology and Medical Science 47, no. 4 (1969): 467–83, doi: 10.1038/icb.1969.51.

In the spring of 1960, a six-year-old girl: Carrie Madren, "First Successful Bone Marrow Transplant Patient Surviving and Thriving at 60," *American Association for the Advancement of Science* (October 2, 2014), https://www.aaas.org/first-success ful-bone-marrow-transplant-patient-sur viving-and-thriving-60. See also Siddhartha Mukherjee, "The Promise and Price of Cellular Therapies," Annals of Medicine, *New Yorker* (July 15, 2019), https:// www.newyorker.com/magazine/2019/07/22 /the-promise-and-price-of-cellular-therapies.

One of them knew of a physician-scientist named E. Donnall "Don" Thomas: Frederick R. Appelbaum, "Edward Donnall Thomas (1920–2012)," *The Hematologist* 10, no. 1 (January 1, 2013), https://doi.org/10.1182 /hem.V10.1.1088.

First, he would try to eradicate: Israel Henig and Tsila Zuckerman, "Hematopoietic Stem Cell Transplantation — 50 Years of Evolution and Future Perspectives," *Rambam Maimonides Medical Journal* 5, no. 4 (2014), doi: 10.5041/RMMJ.10162.

In 1958, the French pioneer of bone marrow transplantation: Geoff Watts, "Georges Mathé," *Lancet* 376, no. 9753 (2010): 1640, https:// doi.org/10.1016/S0140-6736(10)62088-0. See also Douglas Martin, "Dr. Georges

Mathé, Transplant Pioneer, Dies at 88," *New York Times* (October 20, 2010), https://www.nytimes.com/2010/10/21/health/research/21mathe.html.

Over the next few years, Thomas assembled: Sandi Doughton, "Dr. Alex Fefer, 72, Whose Research Led to First Cancer Vaccine, Dies," *Seattle Times* (October 29, 2010), https://www.seattletimes.com/seattle-news/obituaries/dr-alex-fefer-72-whose-research-led-to-first-cancer-vaccine-dies/. See also Gabriel Campanario, "At 79, Noted Scientist Still Rows to Work and for Play," *Seattle Times* (August 15, 2014), https://www.seattletimes.com/seattle-news/at-79-noted-scientist-still-rows-to-work-and-for-play/, and Susan Keown, "Inspiring a New Generation of Researchers: Beverly Torok-Storb, Transplant Biologist and Mentor," *Spotlight on Beverly Torok-Storb, Fred Hutch,* Fred Hutchinson Cancer Research Center (July 7, 2014), https://www.fredhutch.org/en/faculty-lab-directory/torok-storb-beverly/torok-storb-spotlight.html?&link=btn.

the phenomenon called graft-versus-host: Marco Mielcarek et al., "CD34 Cell Dose and Chronic Graft-Versus-Host Disease after Human Leukocyte Antigen-Matched Sibling Hematopoietic Stem Cell Transplantation," *Leukemia & Lymphoma* 45, no. 1 (2004): 27–34, doi: 10.1080/1042819031000151103.

But as Fred Applebaum: Frederick R.

Appelbaum, "Haematopoietic Cell Transplantation as Immunotherapy," *Nature* 411 (2001): 385–89, doi: https://doi.org/10.1038/35077251.

When I reminded Applebaum of those early transplants: Frederick Appelbaum, interview with the author, June 2019.

In 1986, when the nuclear reactor blew up: "Anatoly Grishchenko, Pilot at Chernobyl, 53," *New York Times* (July 4, 1990), https://www.nytimes.com/1990/07/04/obituaries/anatoly-grishchenko-pilot-at-chernobyl-53.html. See also Tim Klass, "Chernobyl Helicopter Pilot Getting Bone-Marrow Transplant in Seattle," *AP News* (April 13, 1990), https://apnews.com/article/5b6c22bda9eba11ec767dffa5bbb665b.

Downstairs, in the lobby of the Hutch: Avichai Shimoni et al., "Long-Term Survival and Late Events after Allogeneic Stem Cell Transplantation from HLA-Matched Siblings for Acute Myeloid Leukemia with Myeloablative Compared to Reduced-Intensity Conditioning: A Report on Behalf of the Acute Leukemia Working Party of European Group for Blood and Marrow Transplantation," *Journal of Hematology & Oncology* 9 (2016), https://doi.org/10.1186/s13045-016-0347-1. See also "Acute Myeloid Leukemia (AML) — Adult," *Transplant Indications and Outcomes, Disease-Specific Indications and Outcomes. Be the Match.* National Marrow

Donor Program, https://bethematchclinical
.org/transplant-indications-and-outcomes
/disease-specific-indications-and-outcomes
/aml---adult/.

In 1998, James Thomson, an embryologist:
Gina Kolata, "Man Who Helped Start Stem
Cell War May End It," *New York Times*
(November 22, 2007), https://www.nytimes
.com/2007/11/22/science/22stem.html.

*More recent work has shown that under some
culture conditions:* Sophie M. Morgani et al.,
"Totipotent Embryonic Stem Cells Arise
in Ground-State Culture Conditions," *Cell
Reports* 3, no. 6 (2013): 1945–57, doi:
10.1016/j.celrep.2013.04.034.

Thomson's paper, published in Science *in 1998:*
James A. Thomson et al., "Embryonic Stem
Cell Lines Derived from Human Blastocysts," *Science* 282, no. 5391 (1998): 1145–
47, doi: 10.1126/science.282.5391.1145.

But critics, mostly from the religious right: David
Cyranoski, "How Human Embryonic Stem
Cells Sparked a Revolution," *Nature* (March
20, 2018), https://www.nature.com/articles
/d41586-018-03268-4.

*In 2001 President George W. Bush, pres-
sured:* Varnee Murugan, "Embryonic Stem
Cell Research: A Decade of Debate from
Bush to Obama," *Yale Journal of Biology
and Medicine* 82, no. 3 (2009): 101–3,
https://www.ncbi.nlm.nih.gov/pmc/articles
/PMC2744932/#:~:text=On%20August

%209%2C%202001%2C%20U.S.,still%
20be%20eligible%20for%20funding.

In 2006, working in Kyoto, Japan: Kazutoshi Takahashi and Shinya Yamanaka, "Induction of Pluripotent Stem Cells from Mouse Embryonic and Adult Fibroblast Cultures by Defined Factors," *Cell* 126, no. 4 (2006): 663–76, doi: 10.1016/j.cell.2006.07.024. See also Shinya Yamanaka, "The Winding Road to Pluripotency," Nobel Lecture, Sweden, (December 7, 2012), https://www.nobelprize.org/uploads/2018/06/yamanaka-lecture.pdf.

"We have colonies," the postdoc shouted out: Megan Scudellari, "A Decade of iPS Cells," *Nature* 534 (2016): 310–12, doi: 10.1038/534310a.

All of this derived from a skin fibroblast: M. J. Evans and M. H. Kaufman, "Establishment in Culture of Pluripotential Cells from Mouse Embryos," *Nature* 292 (1981): 154–56, https://doi.org/10.1038/292154a0.

In 2007, Yamanaka used this technique: Kazutoshi Takahashi et al., "Induction of Pluripotent Stem Cells from Adult Human Fibroblasts by Defined Factors," *Cell* 131, no. 5 (2007): 861–72, https://doi.org/10.1016/j.cell.2007.11.019.

The Repairing Cell: Injury, Decay, and Constancy

"Tenderness and rot share a border": Ryan,

800

"Tenderness and Rot," *The Best of It,* 232.

"Me job's to risk me life and limb": Robert Service, "Bonehead Bill," *Canadian Poets, Best Poems Encyclopedia,* https://www.best -poems.net/robert_w_service/bonehead_ bill.html.

Early experiments show that one such protein: Sarah C. Moser and Bram C. J. van der Eerden, "Osteocalcin — A Versatile Bone-Derived Hormone," *Frontiers in Endocrinology* 9 (January 2019): 794, https:// doi.org/10.3389/fendo.2018.00794. See also Cassandra R. Diegel et al., "An Osteocalcin-Deficient Mouse Strain Without Endocrine Abnormalities," *PLoS Genetics* 16, no. 5 (2020): e1008361, https://doi.org/10.1371 /journal.pgen.1008361, and T. Moriishi et al., "Osteocalcin Is Necessary for the Alignment of Apatite Crystallites, but Not Glucose Metabolism, Testosterone Synthesis, or Muscle Mass," *PLoS Genetics* 16, no. 5 (2020): e1008586, https://doi.org/10.1371 /journal.pgen.1008586.

There is a much vaster catalogue of cells: Li Ding et al., "Clonal Evolution in Relapsed Acute Myeloid Leukaemia Revealed by Whole-Genome Sequencing," *Nature* 481 (2012): 506–10, https://doi.org/10.1038 /nature10738. See also Lei Ding and Sean J. Morrison, "Haematopoietic Stem Cells and Early Lymphoid Progenitors Occupy Distinct Bone Marrow Niches," *Nature*

495, no. 7440 (2013): 231–35, doi: 10.1038 /nature11885, and L. M. Calvi et al., "Osteoblastic Cells Regulate the Haematopoietic Stem Cell Niche," *Nature* 425, no. 6960 (2003): 841–46, doi: 10.1038/nature02040.

Dan published his paper with me and Tim Wang: Daniel L. Worthley et al., "Gremlin 1 Identifies a Skeletal Stem Cell with Bone, Cartilage, and Reticular Stromal Potential," *Cell* 160, no. 1–2 (2015): 269–84, doi: 10.1016/j .cell.2014.11.042.

At the very same time, Chuck Chan: Charles K. F. Chan et al., "Identification of the Human Skeletal Stem Cell," *Cell* 175, no. 1 (2018): 43–56.e21, doi: 10.1016/j.cell.2018.07.029.

Unlike the Gremlin-labeled cells, Morrison's cells: Bo O. Zhou et al., "Leptin-Receptor-Expressing Mesenchymal Stromal Cells Represent the Main Source of Bone Formed by Adult Bone Marrow," *Cell Stem Cell* 15, no. 2 (August 2014): 154–68, doi: 10.1016/j .stem.2014.06.008.

"If the Michelson-Morley experiment": Albrecht Fölsing, *Albert Einstein: A Biography,* trans. Ewald Osers (New York: Penguin Books, 1998), 219.

We proposed a radically new hypothesis about osteoarthritis: Ng Jia, Toghrul Jafarov, and Siddhartha Mukherjee unpublished data.

A recent paper by Henry Kronenberg and colleagues suggests: Koji Mizuhashi et al., "Resting Zone of the Growth Plate Houses a

Unique Class of Skeletal Stem Cells," *Nature* 563 (2018): 254–58, https://doi.org/10.1038 /s41586-018-0662-5.

"At death you break up": Philip Larkin, "The Old Fools," *High Windows* (London: Faber & Faber, 2012).

The Selfish Cell: The Ecological Equation and Cancer

"Those who have not trained in chemistry": William H. Woglom, "General Review of Cancer Therapy," *Approaches to Tumor Chemotherapy,* ed. F. R. Moulton (Washington, DC: American Association for the Advancement of Sciences, 1947), 1–10.

For a general review of cancer: Vincent DeVita, Samuel Hellman, and Steven Rosenberg, *Cancer: Principles & Practice of Oncology,* 2nd ed., ed. Ramaswamy Govindan (Philadelphia: Lippincott Williams & Wilkins, 2012). See also Siddhartha Mukherjee, *The Emperor of All Maladies: A Biography of Cancer* (London: Harper Collins, 2011).

For a review of "driver" and "passenger" cell mutations: K. Anderson et al., "Genetic Variegation of Clonal Architecture and Propagating Cells in Leukaemia," *Nature* 469 (2011): 356–61, https://doi.org/10.1038 /nature09650. See also Noemi Andor et al., "Pan-Cancer Analysis of the Extent and Consequences of Intratumor Heterogeneity," *Nature Medicine* 22 (2016): 105–13, https://doi.org/10.1038/nm.3984, and Fabio Vandin, "Computational Methods for Characterizing Cancer Mutational Heterogeneity," *Frontiers in Genetics* 8, no. 83 (2017), doi: 10.3389/fgene.2017.00083.

That single gene encodes a many-fingered protein: Andrei V. Krivstov et al., "Transformation from Committed Progenitor to Leukaemia Stem Cell Initiated by MLL-AF9," *Nature* 442, no. 7104 (2006): 818–22, doi: 10.1038/nature04980.

In a study performed in 2002: Robert M. Bachoo et al., "Epidermal Growth Factor Receptor and Ink4a/Arf: Convergent Mechanisms Governing Terminal Differentiation and Transformation Along the Neural Stem Cell to Astrocyte Axis," *Cancer Cell* 1, no. 3 (2002): 269–77, doi: 10.1016/s1535-6108(02)00046-6. See also E. C. Holland, "Gliomagenesis: Genetic Alterations and Mouse Models," *Nature Reviews Genetics* 2, no. 2 (2001): 120–29, doi: 10.1038/35052535.

Dick has termed these "leukemia stem cells": John E. Dick and Tsvee Lapidot, "Biology of Normal and Acute Myeloid Leukemia Stem Cells," *International Journal of Hematology* 82, no. 5 (2005): 389–96, doi: 10.1532/IJH97.05144.

Sean Morrison has argued: Elsa Quintana et al., "Efficient Tumor Formation by Single Human Melanoma Cells," *Nature* 456

(2008): 593–98, doi: https://doi.org/10.1038
/nature07567.

Herceptin, for instance, for Her-2 positive breast cancer: Ian Collins and Paul Workman, "New Approaches to Molecular Cancer Therapeutics," *Nature Chemical Biology* 2 (2006): 689–700, doi: https://doi.org/10.1038/nchembio840.

And so researchers ran two kinds: Jay J. H. Park et al., "An Overview of Precision Oncology Basket and Umbrella Trials for Clinicians," *CA: A Cancer Journal for Clinicians* 70, no. 2 (2020): 125–37, https://doi.org/10.3322/caac.21600.

In one landmark study published in 2015: David M. Hyman et al., "Vemurafenib in Multiple Nonmelanoma Cancers with BRAF V600 Mutations," *New England Journal of Medicine* 373 (2015): 726–36, doi: 10.1056/NEJMoa1502309.

A major trial, called BATTLE-2: Chul Kim and Giuseppe Giaccone, "Lessons Learned from BATTLE-2 in the War on Cancer: The Use of Bayesian Method in Clinical Trial Design," *Annals of Translational Medicine* 4, no. 23 (2016): 466, doi: 10.21037/atm.2016.11.48.

"Ultimately," one reviewer commented dejectedly: Sawsan Rashdan and David E. Gerber, "Going into BATTLE: Umbrella and Basket Clinical Trials to Accelerate the Study of Biomarker-Based Therapies," *Annals of*

Translational Medicine 4, no. 24 (2016): 529, doi: 10.21037/atm.2016.12.57.

"We biomedical scientists are addicted to data": Michael B. Yaffe, "The Scientific Drunk and the Lamppost: Massive Sequencing Efforts in Cancer Discovery and Treatment," *Science Signaling* 6, no. 269 (2013): pe13, doi: 10.1126/scisignal.2003684.

"Cancer is no more a disease of cells": D. W. Smithers and M. D. Cantab, "Cancer: An Attack on Cytologism," *Lancet* 279, no. 7228 (1962): 493–99, https://doi.org/10.1016/S0140-6736(62)91475-7.

In the 1920s, the German physiologist Otto Warburg: Otto Warburg, K. Posener, and E. Negelein, "The Metabolism of Cancer Cells," *Biochemische Zeitschrift* 152 (1924): 319–44.

Ralph DeBerardinis, has shown that the Warburg effect: Ralph J. DeBerardinis and Navdeep S. Chandel, "We Need to Talk About the Warburg Effect," *Nature Metabolism* 2, no. 2 (2020): 127–29, doi: 10.1038/s42255-020-0172-2.

The Songs of the Cell

"I do not know which to prefer": Wallace Stevens, "Thirteen Ways of Looking at a Blackbird," *The Collected Poems of Wallace Stevens* (New York: Alfred A. Knopf, 1971), 92–95.

"nods and replies with downcast eyes": Amitav

Ghosh, *The Nutmeg's Curse: Parables for a Planet in Crisis* (Chicago: University of Chicago Press, 2021), 96.

"sensitive organ of the cell": Barbara McClintock, "The Significance of Responses of the Genome to Challenge," Nobel Lecture, Sweden (December 8, 1983), https://www.nobelprize.org/uploads/2018/06/mcclintock-lecture.pdf.

"a little, yellow-skinned, owl-faced": Carl Ludwig Schleich, *Those Were Good Days: Reminiscences,* trans. Bernard Miall (London: George Allen & Unwin, 1935), 151.

Epilogue: "Better Versions of Me"

"If we could be less human": Ryan, "The Test We Set Ourselves," *The Best of It,* 66.

But I too made things: Walter Shrank, *Battle Cries of Every Size* (Blurb, 2021), 45.

"You mean genetically?": Paul Greengard, interview with the author, February 2019.

In the strangest of Kazuo Ishiguro's novels: Kazuo Ishiguro, *Never Let Me Go* (London: Faber & Faber, 2009).

"I was taken aback": Ibid., 171–72.

"tiny canals, weaving tendons,": Ibid., 171.

"The shadowy backdrop in Never Let Me Go": Louis Menand, "Something About Kathy," *New Yorker* (March 28, 2005).

At the Mayo Clinic, scientists are making a bio-artificial liver: Doris A. Taylor et al., "Building a Total Bioartificial Heart: Harnessing

Nature to Overcome the Current Hurdles," *Artificial Organs* 42, no. 10 (2018): 970–82, doi: 10.1111/aor.13336.

The philosopher Michael Sandel has been pondering: Michael J. Sandel, "The Case Against Perfection," *Atlantic* (April 2004), https://www.theatlantic.com/magazine/archive/2004/04/the-case-against-perfection/302927/.

"an openness to the unbidded": Quoted in ibid.

"[Sandel's] deeper worry": William Saletan, "Tinkering with Humans," *New York Times* (July 8, 2007), https://www.nytimes.com/2007/07/08/books/review/Saletan.html.

"harvested from youths": Luke Darby, "Silicon Valley Doofs Are Spending $8,000 to Inject Themselves with the Blood of Young People," *GQ* (February 20, 2019), https://www.gq.com/story/silicon-valley-young-blood.

"The genomic revolution has induced": Sandel, "The Case Against Perfection."

Between 2019 and 2021, multiple independent groups reported: Ornob Alam, "Sickle-Cell Anemia Gene Therapy," *Nature Genetics* 53, no. 8 (2021): 1119, doi: 10.1038/s41588-021-00918-8. See also Arthur Bank, "On the Road to Gene Therapy for Beta-Thalassemia and Sickle Cell Anemia," *Pediatric Hematology and Oncology* 25, no. 1 (2008): 1–4, doi: 10.1080/08880010701773829. G. Lucarelli et al., "Allogeneic Cellular Gene Therapy in Hemoglobinopathies — Evaluation of

Hematopoietic SCT in Sickle Cell Anemia," *Bone Marrow Transplantation* 47, no. 2 (2012): 227–30, doi: 10.1038/bmt.2011.79. R. Alami et al., "Anti-Beta S-Ribozyme Reduces Beta S mRNA Levels in Transgenic Mice: Potential Application to the Gene Therapy of Sickle Cell Anemia," *Blood Cells, Molecules and Diseases* 25, no. 2 (1999): 110–19, doi: 10.1006/bcmd.1999.0235. A. Larochelle et al., "Engraftment of Immune-Deficient Mice with Primitive Hematopoietic Cells from Beta-Thalassemia and Sickle Cell Anemia Patients: Implications for Evaluating Human Gene Therapy Protocols," *Human Molecular Genetics* 4, no. 2 (1995): 163–72, doi: 10.1093/hmg/4.2.163. W. Misaki, "Bone Marrow Transplantation (BMT) and Gene Replacement Therapy (GRT) in Sickle Cell Anemia," *Nigerian Journal of Medicine* 17, no. 3 (2008): 251–56, doi: 10.4314/njm.v17i3.37390. Also see Julie Kanter et al., "Biologic and Clinical Efficacy of LentiGlobin for Sickle Cell Disease," *New England Journal of Medicine* 10, no. 1056 (2021), https://www.nejm.org/doi/full/10.1056/NEJMoa2117175.

Whether the leukemia was the consequence: Sunita Goyal et al., "Acute Myeloid Leukemia Case after Gene Therapy for Sickle Cell Disease," *New England Journal of Medicine* (2022), https://www.nejm.org/doi/full/10.1056/NEJMoa2109167. See also

Nick Paul Taylor, "Bluebird Stops Gene Therapy Trials after 2 Sickle Cell Patients Develop Cancer," *Fierce Biotech* (February 16, 2021), https://www.fiercebiotech.com /biotech/bluebird-stops-gene-therapy-trials -after-2-sickle-cell-patients-develop-cancer.

Stuart Orkin and David Williams, working with a team: Christian Brendel et al., "Lineage-Specific BCL11A Knockdown Circumvents Toxicities and Reverses Sickle Phenotype," *Journal of Clinical Investigation* 126, no. 10 (2016): 3868–78, doi: 10.1172/JCI87885.

In a trial reported in 2021, a thirty-three-year-old woman: Erica B. Esrick et al., "Post-Transcriptional Genetic Silencing of BCL11A to Treat Sickle Cell Disease," *New England Journal of Medicine* 384 (2021): 205–15, doi: 10.1056/NEJMoa2029392.

At Stanford, another group, led by Matt Porteus: Adam C. Wilkinson et al., "Cas9-AAV6 Gene Correction of Beta-Globin in Autologous HSCs Improves Sickle Cell Disease Erythropoiesis in Mice," *Nature Communications* 12, no. 1 (2021): 686, doi: 10.1038 /s41467-021-20909-x.

Porteus's strategy is also in trials: Michael Eisenstein, "Graphite Bio: Gene Editing Blood Stem Cells for Sickle Cell Disease," *Nature* (July 7, 2021), https://www.nature .com/articles/d41587-021-00010-w.

BIBLIOGRAPHY

Ackerknecht, Erwin Heinz. *Rudolf Virchow: Doctor, Statesman, Anthropologist.* Madison: University of Wisconsin Press, 1953.

Ackerman, Margaret E., and Falk Nimmerjahn. *Antibody Fc: Linking Adaptive and Innate Immunity.* Amsterdam: Elsevier, 2014.

Addison, William. *Experimental and Practical Researches on Inflammation and on the Origin and Nature of Tubercles of the Lung.* London: J. Churchill, 1843.

Aktipis, Athena. *The Cheating Cell: How Evolution Helps Us Understand and Treat Cancer.* Princeton, NJ: Princeton University Press, 2020.

Alberts, B., A. Johnson, J. Lewis, M. Raff, and K. Roberts. *Molecular Biology of the Cell.* 5th ed. New York: Garland Science, 2002.

Alberts, B., D. Bray, K. Hopkin, A. D. Johnson, J. Lewis, M. Raff, K. Roberts, and P. Walter. *Essential Cell Biology.* 4th ed. New York: Garland Science, 2013.

Appelbaum, Frederick R. *E. Donnall Thomas, 1920–2012.* Biographical Memoirs. National Academy of Sciences online, 2021, http://www.nasonline.org/publications/biographical-memoirs/memoir-pdfs/thomas-e-donnall.pdf.

Aristotle. *De Anima.* Translated by R. D. Hicks. New York: Cosimo Classics, 2008.

———. *On the Soul, Parva Naturalia, On Breath.* Translated by W. S. Hett. London: William Heinemann, 1964. First published 1691.

Aubrey, John. *Aubrey's Brief Lives.* London: Penguin Random House UK, 2016.

Barton, Hazel B., and Rachel J. Whitaker, eds. *Women in Microbiology.* Washington, DC: American Society for Microbiology Press, 2018.

Bazell, Robert. *Her-2: The Making of Herceptin, a Revolutionary Treatment for Breast Cancer.* New York: Random House, 1998.

Biss, Eula. *On Immunity: An Inoculation.* Minneapolis: Graywolf Press, 2014.

Black, Brian. *The Character of the Self in Ancient India: Priests, Kings, and Women in the Early Upanishads.* Albany: State University of New York Press, 2007.

Bliss, Michael. *Banting: A Biography.* Toronto: University of Toronto Press, 1992.

———. *The Discovery of Insulin.* Toronto: McClelland & Stewart, 2021.

Boccaccio, Giovanni. *The Decameron of Giovanni Boccaccio.* Translated by John

Payne. Frankfurt, Ger.: Outlook Verlag GmbH, 2020.

Boyd, Byron A. *Rudolf Virchow: The Scientist as Citizen.* New York: Garland, 1991.

Bradbury, S. *The Evolution of the Microscope.* Oxford, UK: Pergamon Press, 1967.

Brasier, Martin. *Secret Chambers: The Inside Story of Cells and Complex Life.* Oxford, UK: Oxford University Press, 2012.

Brivanlou, Ali H., ed. *Human Embryonic Stem Cells in Development.* Cambridge, MA: Academic Press, 2018.

Burnet, Macfarlane. *Self and Not-Self.* London: Cambridge University Press, 1969.

Cajal, Santiago Ramón y. *Recollections of My Life.* Translated by E. Horne Craigie and Juan Cano. Cambridge, MA: MIT Press, 1996.

Camara, Niels Olsen Saraiva, and Tárcio Teodoro Braga, eds. *Macrophages in the Human Body: A Tissue Level Approach.* London: Elsevier Science, 2022.

Campbell, Alisa M. *Monoclonal Antibody Technology: The Production and Characterization of Rodent and Human Hybridomas.* Amsterdam: Elsevier, 1984.

Canetti, Elias. *Crowds and Power.* Translated by Carol Stewart. New York: Continuum, Farrar, Straus and Giroux, 1981.

Carey, Nessa. *The Epigenetics Revolution: How Modern Biology Is Rewriting Our Understanding of Genetics, Disease and Inheritance.* London: Icon Books, 2011.

Caron, George R., and Charlotte E. Meares. *Fire of a Thousand Suns: The George R. "Bob" Caron Story: Tail Gunner of the* Enola Gay. Westminster, CO: Web, 1995.

Carroll, Lewis. *Alice in Wonderland.* London: Penguin Books, 1998.

Chapman, Allan. *England's Leonardo: Robert Hooke and the Seventeenth-Century Scientific Revolution.* Bristol, UK: Institute of Physics Publishing, 2005.

Conner, Clifford D. *A People's History of Science: Miners, Midwives, and "Low Mechanicks."* New York: Nation Books, 2005.

Copernicus, Nicolaus. *On the Revolutions of Heavenly Spheres.* Translated by Charles Glenn Wallis. New York: Prometheus Books, 1995.

Crawford, Dorothy H. *The Invisible Enemy: A Natural History of Viruses.* Oxford, UK: Oxford University Press, 2002.

Danquah, Michael K., and Ram I. Mahato, eds. *Emerging Trends in Cell and Gene Therapy.* New York: Springer, 2013.

Darwin, Charles. *On the Origin of Species.* Edited by Gillian Beer. Oxford, UK: Oxford University Press, 2008.

Davis, Daniel Michael. *The Compatibility Gene: How Our Bodies Fight Disease, Attract Others, and Define Our Selves.* Oxford, UK: Oxford University Press, 2014.

Dawkins, Richard. *The Selfish Gene.* Oxford, UK: Oxford University Press, 1989.

Dettmer, Philipp. *Immune: A Journey into the Mysterious System That Keeps You Alive.* New York: Random House, 2021.

DeVita, Vincent, Samuel Hellman, and Steven Rosenberg. *Cancer: Principles & Practice of Oncology.* 2nd ed. Edited by Ramaswamy Govindan. Philadelphia: Lippincott Williams & Wilkins, 1985.

Dickinson, Emily. *The Complete Poems of Emily Dickinson.* Edited by Thomas H. Johnson. Boston: Little, Brown, 1960.

Dobson, Mary. *The Story of Medicine: From Leeches to Gene Therapy.* New York: Quercus, 2013.

Döllinger, Ignaz. *Was ist Absonderung und wie geschieht sie?: Eine akademische Abhandlung von Dr. Ignaz Döllinger.* Würzburg, Ger.: Nitribitt, 1819.

Doyle, Arthur Conan. *The Adventures of Sherlock Holmes.* Hertfordshire, UK: Wordsworth, 1996.

Dunn, Leslie. *Rudolf Virchow: Four Lives in One.* Self-published, 2016.

Dunn, Leslie Clarence. *A Short History of Genetics: The Development of Some of the Main Lines of Thought, 1864–1939.* Ames: Iowa State University Press, 1991.

Dyer, Betsey Dexter, and Robert Allan Obar. *Tracing the History of Eukaryotic Cells: The Enigmatic Smile.* New York: Columbia University Press, 1994.

Edwards, Robert Geoffrey, and Patrick

Christopher Steptoe. *A Matter of Life: The Story of a Medical Breakthrough.* New York: William Morrow, 1980.

Ehrlich, Paul R. *The Collected Papers of Paul Ehrlich.* Edited by F. Himmelweit, Henry Hallett Dale, and Martha Marquardt. London: Elsevier Science & Technology, 1956.

———. *Collected Studies on Immunity.* New York: John Wiley & Sons, 1906.

Florkin, Marcel. *Papers About Theodor Schwann.* Paris: Liège, 1957.

Frank, Lone. *The Pleasure Shock: The Rise of Deep Brain Stimulation and Its Forgotten Inventor.* New York: Penguin Random House, 2018.

Friedman, Meyer, and Gerald W. Friedland. *Medicine's 10 Greatest Discoveries.* New Haven, CT: Yale University Press, 1998.

Galen. *On the Usefulness of the Parts of the Body.* Translated by Margaret Tallmadge May. Ithaca, NY: Cornell University Press, 1968.

Geison, Gerald L. *The Private Science of Louis Pasteur.* Princeton, NJ: Princeton University Press, 1995.

Ghosh, Amitav. *The Nutmeg's Curse: Parables for a Planet in Crisis.* Chicago: University of Chicago Press, 2021.

Glover, Jonathan. *Choosing Children: Genes, Disability, and Design.* Oxford, UK: Oxford University Press, 2006.

Godfrey, E. L. B. *Dr. Edward Jenner's Discovery*

of *Vaccination.* Philadelphia: Hoeflich & Senseman, 1881.

Goetz, Thomas. *The Remedy: Robert Koch, Arthur Conan Doyle, and the Quest to Cure Tuberculosis.* New York: Gotham Books, 2014.

Goodsell, David S. *The Machinery of Life.* New York: Springer, 2009.

Greely, Henry T. *CRISPR People: The Science and Ethics of Editing Humans.* Cambridge, MA: MIT Press, 2022.

Grmek, Mirko D. *History of AIDS: Emergence and Origin of a Modern Pandemic.* Translated by Russell C. Maulitz and Jacalyn Duffin. Princeton, NJ: Princeton University Press, 1993.

Gupta, Anil. *Understanding Insulin and Insulin Resistance.* Oxford, UK: Elsevier, 2022.

Hakim, Nadey S., and Vassilios E. Papalois, eds. *History of Organ and Cell Transplantation.* London: Imperial College Press, 2003.

Harold, Franklin M. *In Search of Cell History: The Evolution of Life's Building Blocks.* Chicago: University of Chicago Press, 2014.

Harris, Henry. *The Birth of the Cell.* New Haven, CT: Yale University Press, 2000.

Harvey, William. *On the Motion of the Heart and Blood in Animals.* Edited by Jarrett A. Carty. Translated by Robert Willis. Eugene, OR: Resource, 2016.

———. *The Circulation of the Blood: Two Anatomical Essays.* Translated by Kenneth J.

Franklin. Oxford, UK: Blackwell Scientific, 1958.

Henig, Robin Marantz. *Pandora's Baby: How the First Test Tube Babies Sparked the Reproductive Revolution.* Cold Spring Harbor, NY: Cold Spring Harbor Laboratory Press, 2006.

Hirst, Leonard Fabian. *The Conquest of Plague: A Study of the Evolution of Epidemiology.* Oxford, UK: Clarendon Press, 1953.

Ho, Anthony D., and Richard E. Champlin, eds. *Hematopoietic Stem Cell Transplantation.* New York: Marcel Dekker, 2000.

Ho, Mae-Wan. *The Rainbow and the Worm: The Physics of Organisms.* 3rd ed. Hackensack, NJ: World Scientific, 2008.

Hofer, Erhard, and Jürgen Hescheler, eds. *Adult and Pluripotent Stem Cells: Potential for Regenerative Medicine of the Cardiovascular System.* Dordrecht, Neth.: Springer, 2014.

Hooke, Robert. *Microphagia: Or Some Physiological Description of Minute Bodies Made by Magnifying Glasses with Observations and Inquiries Thereupon.* London: Royal Society, 1665.

Howard, John M., and Walter Hess. *History of the Pancreas: Mysteries of a Hidden Organ.* New York: Springer Science+Business Media, 2002.

Ishiguro, Kazuo. *Never Let Me Go.* London: Faber & Faber, 2009.

Jaggi, O. P. *Medicine in India: Modern Period.* Oxford, UK: Oxford University Press, 2000.

Janeway, Charles A., et al. *Immunobiology: The Immune System in Health and Disease.* 5th ed. New York: Garland Science, 2001.

Jauhar, Sandeep. *Heart: A History.* New York: Farrar, Straus and Giroux, 2018.

Jenner, Edward. *On the Origin of the Vaccine Inoculation.* London: G. Elsick, 1863.

Joffe, Stephen N. *Andreas Vesalius: The Making, the Madman, and the Myth.* Bloomington, IN: AuthorHouse, 2014.

Kaufmann, Stefan H. E., Barry T. Rouse, and David Lawrence Sacks, eds. *The Immune Response to Infection.* Washington, DC: ASM Press, 2011.

Kemp, Walter L., Dennis K. Burns, and Travis G. Brown. *The Big Picture: Pathology.* New York: McGraw-Hill, 2008.

Kenny, Anthony. *Ancient Philosophy.* Oxford, UK: Clarendon Press, 2006.

Kettenmann, Helmut, and Bruce R. Ransom, eds. *Neuroglia.* 3rd ed. Oxford, UK: Oxford University Press, 2013.

Kirksey, Eben. *The Mutant Project: Inside the Global Race to Genetically Modify Humans.* Bristol, UK: Bristol University Press, 2021.

Kitamura, Daisuke, ed. *How the Immune System Recognizes Self and Nonself: Immunoreceptors and Their Signaling.* Tokyo: Springer, 2008.

Kitta, Andrea. *Vaccinations and Public Concern in History: Legend, Rumor and Risk Perception.* New York: Routledge, 2012.

Koch, Kenneth. *One Train.* New York: Alfred A. Knopf, 1994.

Koch, Robert. *Essays of Robert Koch.* Edited and translated by Ed. K. Codell Carter. New York: Greenwood Press, 1987.

Kulstad, Ruth. *AIDS: Papers from Science, 1982–1985.* New York: Avalon Books, 1986.

Kushner, Rachel. *The Hard Crowd: Essays, 2000–2020.* New York: Scribner, 2021.

Lagerkvist, Ulf. *Pioneers of Microbiology and the Nobel Prize.* Singapore: World Scientific, 2003.

Lal, Pranay. *Invisible Empire: The Natural History of Viruses.* Haryana, Ind.: Penguin/Viking, 2021.

Landecker, Hannah. *Culturing Life: How Cells Became Technologies.* Cambridge, MA: Harvard University Press, 2007.

Lane, Nick. *Power, Sex, Suicide: Mitochondria and the Meaning of Life.* Oxford, UK: Oxford University Press, 2005.

———. *The Vital Question: Energy, Evolution, and the Origins of Complex Life.* New York: W. W. Norton, 2015.

Lee, Daniel W., and Nirali N. Shah, eds. *Chimeric Antigen Receptor T-Cell Therapies for Cancer.* Amsterdam: Elsevier, 2020.

Le Fanu, James. *The Rise and Fall of Modern Medicine.* London: Abacus, 2000.

Lewis, Jessica L., ed. *Gene Therapy and Cancer Research Progress.* New York: Nova Biomedical, 2008.

Lostroh, Phoebe. *Molecular and Cellular Biology of Viruses.* New York: Garland Science, 2019.

Lyons, Sherrie L. *From Cells to Organisms: Re-Envisioning Cell Theory.* Toronto: University of Toronto Press, 2020.

Marquardt, Martha. *Paul Ehrlich.* New York: Schuman, 1951.

Maxwell, Robert A., and Shohreh B. Eckhardt. *Drug Discovery: A Casebook and Analysis.* New York: Springer Science+Business Media, 1990.

McCulloch, Ernest A. *The Ontario Cancer Institute: Successes and Reverses at Sherbourne Street.* Montreal: McGill-Queen's University Press, 2003.

McMahon, Lynne, and Averill Curdy, eds. *The Longman Anthology of Poetry.* New York: Pearson/Longman, 2006.

Mickle, Shelley Fraser. *Borrowing Life: How Scientists, Surgeons, and a War Hero Made the First Successful Organ Transplant.* Watertown, MA: Imagine, 2020.

Milo, Ron, and Rob Philips. *Cell Biology by the Numbers.* New York: Taylor & Francis, 2016.

Monod, Jacques. *Chance and Necessity: An Essay on the Natural Philosophy of Modern Biology.* New York: Alfred A. Knopf, 1971.

Morris, Thomas. *The Matter of the Heart: A History of the Heart in Eleven Operations.* London: Bodley Head, 2017.

Mukherjee, Siddhartha. *The Emperor of All Maladies: A Biography of Cancer.* New York: Scribner, 2011.

————. *The Gene: An Intimate History.* New York: Scribner, 2016.

Needham, Joseph. *History of Embryology.* Cambridge, UK: University of Cambridge Press, 1934.

Neel, James V., and William J. Schull, eds. *The Children of Atomic Bomb Survivors: A Genetic Study.* Washington, DC: National Academy Press, 1991.

Newton, Isaac. *The Principia: Mathematical Principles of Natural Philosophy.* Translated by I. Bernard Cohen and Anne Whitman. Oakland: University of California Press, 1999.

Nuland, Sherwin B. *Doctors: The Biography of Medicine.* New York: Random House, 2011.

Nurse, Paul. *What Is Life? Understand Biology in Five Steps.* London: David Fickling Books, 2020.

O'Malley, C. D. *Andreas Vesalius of Brussels, 1514–1564.* Berkeley: University of California Press, 1964.

O'Malley, Charles, and J. B. Saunders, eds. *The Illustrations from the Works of Andreas Vesalius of Brussels.* New York: Dover, 2013.

Ogawa, Yōko. *The Memory Police.* Translated by Stephen Snyder. New York: Pantheon Books, 2019.

Otis, Laura. *Müller's Lab.* Oxford, UK: Oxford University Press, 2007.

Oughterson, Ashley W., and Shields Warren. *Medical Effects of the Atomic Bomb in Japan.* New York: McGraw-Hill, 1956.

Ozick, Cynthia. *Metaphor & Memory.* New York: Random House, 1991.

Perin, Emerson C., et al., eds. *Stem Cell and Gene Therapy for Cardiovascular Disease.* Amsterdam: Elsevier, 2016.

Pelayo, Rosana, ed. *Advances in Hematopoietic Stem Cell Research.* London: Intech Open, 2012.

Pepys, Samuel. *The Diary of Samuel Pepys.* Edited by Henry B. Wheatley. Translated by Mynors Bright. London: George Bell and Sons, 1893. Available at Project Gutenberg, https://www.gutenberg.org /files/4200/4200-h/4200-h.htm.

Pfennig, David W., ed. *Phenotypic Plasticity and Evolution: Causes, Consequences, Controversies.* Boca Raton, FL: CRC Press, 2021.

Playfair, John, and Gregory Bancroft. *Infection and Immunity.* Oxford, UK: Oxford University Press, 2013.

Ponder, B. A. J., and M. J. Waring. *The Genetics of Cancer.* Amsterdam: Springer Science+Business Media, 1995.

Porter, Roy, ed. *The Cambridge History of Medicine.* Cambridge, UK: Cambridge University Press, 2006.

———. *Greatest Benefit to Mankind. A Medical History of Humanity from Antiquity to the Present.* London: HarperCollins, 1999.

Power, D'Arcy. *William Harvey: Masters of Medicine.* London: T. Fisher Unwin, 1897.

Prakash, S., ed. *Artificial Cells, Cell Engineering and Therapy.* Boca Raton, FL: CRC Press, 2007.

Rasko, John, and Carl Power. *Flesh Made New: The Unnatural History and Broken Promise of Stem Cells.* California: ABC Books, 2021.

Raza, Azra. *The First Cell: And the Human Costs of Pursuing Cancer to the Last.* New York: Basic Books, 2019.

Reaven, Gerald, and Ami Laws, eds. *Insulin Resistance: The Metabolic Syndrome X.* Totowa, NJ: Humana Press, 1999.

Redi, Francesco. *Experiments on the Generation of Insects.* Translated by Mab Bigelow. Chicago: Open Court, 1909.

Rees, Anthony R. *The Antibody Molecule: From Antitoxins to Therapeutic Antibodies.* Oxford, UK: Oxford University Press, 2015.

Reynolds, Andrew S. *The Third Lens: Metaphor and the Creation of Modern Cell Biology.* Chicago: University of Chicago Press, 2018.

Ridley, Matt. *Genome: The Autobiography of a Species in 23 Chapters.* London: HarperCollins, 2017.

Robbin, Irving. *Giants of Medicine.* New York: Grosset & Dunlap, 1962.

Robbins, Louise E. *Louis Pasteur: And the Hidden World of Microbes.* New York: Oxford University Press, 2001.

Rogers, Kara, ed. *Blood: Physiology and*

Circulation. New York: Britannica Educational, 2011.

Rose, Hilary, and Steven Rose. *Genes, Cells and Brains: The Promethean Promise of the New Biology.* London: Verso, 2014.

Roth, Philip. *Everyman.* London: Penguin Random House, 2016.

Rudisill, Valerie Byrne. *Born with a Bomb: Suddenly Blind from Leber's Hereditary Optic Neuropathy.* Edited by Margie Sabol and Leslie Byrne. Bloomington, IN: AuthorHouse, 2012.

Rushdie, Salman. *Midnight's Children.* Toronto: Alfred A. Knopf, 2010.

Ryan, Kay. *The Best of It: New and Selected Poems.* New York: Grove Press, 2010.

Sandburg, Carl. *Chicago Poems.* New York: Henry Holt, 1916.

Sandel, Michael J. *The Case Against Perfection: Ethics in the Age of Genetic Engineering.* Cambridge, MA: Harvard University Press, 2007.

Schneider, David. *The Invention of Surgery.* New York: Pegasus Books, 2020.

Schwann, Theodor. *Microscopical Researches into the Accordance in the Structure and Growth of Animals and Plants.* Translated by Henry Smith. London: Sydenham Society, 1847.

Sell, Stewart, and Ralph Reisfeld, eds. *Monoclonal Antibodies in Cancer.* Clifton, NJ: Humana Press, 1985.

Semmelweis, Ignaz. *The Etiology, Concept,*

and Prophylaxis of Childbed Fever. Edited and translated by K. Codell Carter. Madison: University of Wisconsin Press, 1983.

Shah, Sonia. *Pandemic: Tracking Contagions, from Cholera to Coronaviruses and Beyond.* New York: Sarah Crichton Books, 2016.

Shapin, Steven. *The Scientific Revolution.* Chicago: University of Chicago Press, 2018.

———. *A Social History of Truth: Civility and Science in the Seventeenth Century.* Chicago: University of Chicago Press, 2011.

Shorter, Edward. *Partnership for Excellence: Medicine at the University of Toronto and Academic Hospitals.* Toronto: University of Toronto Press, 2013.

Simmons, John Galbraith. *Doctors & Discoveries: Lives That Created Today's Medicine.* Boston: Houghton Mifflin, 2002.

———. *The Scientific 100: A Ranking of the Most Influential Scientists, Past and Present.* New York: Kensington, 2000.

Skloot, Rebecca. *The Immortal Life of Henrietta Lacks.* London: Macmillan, 2010.

Snow, John. *On the Mode of Communication of Cholera.* London: John Churchill, 1849.

Solomon, Andrew. *Far from the Tree: Parents, Children and the Search for Identity.* New York: Scribner, 2013.

———. *The Noonday Demon: An Atlas of Depression.* New York: Scribner, 2001.

Sornberger, Joe. *Dreams and Due Diligence: Till and McCulloch's Stem Cell Discovery and Legacy.* Toronto: University of Toronto Press, 2011.

Spiegelhalter, David, and Anthony Masters. *Covid by Numbers: Making Sense of the Pandemic with Data.* London: Penguin Books, 2022.

Stephens, Trent, and Rock Brynner. *Dark Remedy: The Impact of Thalidomide and Its Revival as a Vital Medicine.* New York: Basic Books, 2009.

Stevens, Wallace. *Selected Poems: A New Collection.* Edited by John N. Serio. New York: Alfred A. Knopf, 2009.

Styron, William. *Darkness Visible: A Memoir of Madness.* New York: Open Road, 2010.

Swanson, Larry W., et al. *The Beautiful Brain: The Drawings of Santiago Ramón y Cajal.* New York: Abrams, 2017.

Tesarik, Jan, ed. *40 Years After In Vitro Fertilisation: State of the Art and New Challenges.* Newcastle, UK: Cambridge Scholars, 2019.

Thomas, Lewis. *A Long Line of Cells: Collected Essays.* New York: Book of the Month Club, 1990.

———. *The Medusa and the Snail: More Notes of a Biology Watcher.* New York: Penguin Books, 1995.

Vallery-Radot, René. *The Life of Pasteur.* Vol. 1. Translated by R. L. Devonshire. New York: Doubleday, Page, 1920.

Van den Tweel, Jan G., ed. *Pioneers in Pathology.* New York: Springer, 2017.

Vesalius, Andreas. *The Fabric of the Human Body.* 7 Vols. Vol. 1. Book I. *The Bones and Cartilages.* Translated by William Frank Richardson and John Burd Carman. San Francisco: Norman, 1998.

Virchow, Rudolf. *Cellular Pathology as Based upon Physiological and Pathological Histology: Twenty Lectures Delivered in the Pathological Institute of Berlin During the Months of February, March, and April, 1858.* Translated by Frank Chance. London: John Churchill, 1860.

———. *Disease, Life and Man: Selected Essays.* Translated by Lelland J. Rather. Stanford, CA: Stanford University Press, 1938.

Wadman, Meredith. *The Vaccine Race: How Scientists Used Human Cells to Combat Killer Viruses.* London: Black Swan, 2017.

Wapner, Jessica. *The Philadelphia Chromosome: A Genetic Mystery, a Lethal Cancer, and the Improbable Invention of Life-Saving Treatment.* New York: The Experiment, 2014.

Wassenaar, Trudy M. *Bacteria: The Benign, the Bad, and the Beautiful.* Hoboken, NJ: Wiley-Blackwell, 2012.

Watson, James D., Andrew Berry, and Kevin Davies. *DNA: The Secret of Life.* London: Arrow Books, 2017.

Watson, Ronald Ross, and Sherma Zibadi, eds. *Lifestyle in Heart Health and Disease.* London: Elsevier, 2018.

Wellmann, Janina. *The Form of Becoming: Embryology and the Epistemology of Rhythm, 1760–1830.* Translated by Kate Sturge. New York: Zone Books, 2017.

Whitman, Walt. *Leaves of Grass: Comprising All the Poems Written by Walt Whitman.* New York: Modern Library, 1892.

Wiestler, Otmar D., Bernhard Haendler, and D. Mumberg, eds. *Cancer Stem Cells: Novel Concepts and Prospects for Tumor Therapy.* New York: Springer, 2007.

Wilson, Edmund. *The Cell in Development and Inheritance.* New York: Macmillan, 1897.

Wilson, Edward O. *Letters to a Young Scientist.* New York: Liveright, 2013.

Wolpert, Lewis. *How We Live and Why We Die: The Secret Lives of Cells.* London: Faber and Faber, 2009.

Wurtzel, Elizabeth. *Prozac Nation.* New York: Houghton Mifflin, 1994.

Yong, Ed. *I Contain Multitudes: The Microbes Within Us and a Grander View of Life.* London: Bodley Head, 2016.

Yount, Lisa. *Antoni van Leeuwenhoek: Genius Discoverer of Microscopic Life.* Berkeley, CA: Enslow, 2015.

Zernicka-Goetz, Magdalena, and Roger Highfield. *The Dance of Life: Symmetry, Cells and How We Become Human.* London: Penguin Books, 2020.

Zhe-Sheng Chen, et al., eds. *Targeted Cancer Therapies, from Small Molecules to Antibodies.*

Lausanne, Switz.: Frontiers Media, 2020.

Zimmer, Carl. *Life's Edge: The Search for What It Means to Be Alive.* New York: Penguin Random House, 2021.

——. *A Planet of Viruses.* Chicago: University of Chicago Press, 2015.

Žižek, Slavoj. *Pandemic! COVID-19 Shakes the World.* London: Polity Books, 2020.

IMAGE CREDITS

Interior

Page 3: Walther Flemming, "Contributions to the Knowledge of the Cell and Its Vital Processes," *Journal of Cell Biology* 25, no. 1 (April 1, 1965): 3–69, https://doi.org/10.1083/jcb.25.1.3.

Page 70: Courtesy of Dr. Lesley Robertson, Delft School of Microbiology at Delft University of Technology

Page 73: *Micrographia: or some physiological descriptions of minute bodies made by magnifying glasses. With observations and inquiries thereupon,* by R. Hooke. Wellcome Collection. Attribution 4.0 International (CC BY 4.0)

Page 75: *Micrographia: or some physiological descriptions of minute bodies made by magnifying glasses. With observations and inquiries thereupon,* by R. Hooke. Wellcome Collection Attribution 4.0 International (CC BY 4.0)

Page 107: *Archiv für Pathologische Anatomie und Physiologie,* 1847, first issue. Wikimedia Commons, CC BY 1.0

Page 120: The anthrax bacillus: ten examples, as seen through a microscope. Colour photograph, ca. 1948, after A. Assmann, ca. 1876, after R. Koch and F. Cohn, ca. 1876. Wellcome Collection. Attribution 4.0 International (CC BY 4.0)

Page 129: On the mode of communication of cholera, by John Snow. Wellcome Collection. Public Domain Mark

Page 171: Courtesy of the author

Page 174: Courtesy of the author

Page 195: Adapted from Walther Flemming, CC0. Anderson et al. eLife 2019;8:e46962. DOI: https://doi.org/10.7554/eLife.46962

Page 257: "Experimental Evolution of Multicellularity," William C. Ratcliff, R. Ford Denison, Mark Borrello, Michael Travisano, *Proceedings of the National Academy of Sciences* 109, no. 5 (January 2012): 1595–1600; DOI: 10.1073/pnas.1115323109. Courtesy of Michael Travisano, PhD

Page 273: Courtesy of the author

Page 315: Source: Julius Bizzozero, "Ueber

einen neuen Formbestandtheil des Blutes und dessen Rolle bei der Thrombose und der Blutgerinnung," *Archiv für pathologische Anatomie und Physiologie und für klinische Medicin* 90, no. 2 (1882): 261–332.

Page 356: Proceedings of the Royal Society of London. Wellcome Collection. Attribution 4.0 International (CC BY 4.0)

Page 357: Courtesy of the author

Page 495: *Exercitatio anatomica de motu cordis et sanguinis in animalibus,* by Guilielmi Harvei. Wellcome Collection. Public Domain Mark

Page 519: Courtesy of Instituto Cajal del Consejo Superior de Investigaciones Cientificas, Madrid, © 2022 CSIC

Page 547: Example of Deep Brain Stimulation Lead Location and Patient-Specific Volume of Tissue Activated (VTA) Used for Tractography Maps from K. S. Choi, P. Rivia-Posse, R. E. Gross et al., "Mapping the 'Depression Switch' During Intraoperative Testing of Subcallosal Cingulate Deep Brain Stimulation," *JAMA Neurology* 72, no. 11 (2015):1252–60. Courtesy of Dr. Ki Sueng Choi

Page 644: Courtesy of the author

Pages 49, 147, 285, 457, 479, 583: Kiki Smith

Insert

1. Emily Whitehead Foundation

2. National Library of Medicine

3. Rijksmuseum http://hdl.handle.net/10934/RM0001.COLLECT.46995

4. Courtesy of The Rockefeller Archive Center

5. National Archives (111-SC-192575-S)

6. Portrait of Paul Ehrlich and Sahachiro Hata. Wellcome Collection. Attribution 4.0 International (CC BY 4.0)

7. Courtesy of Instituto Cajal del Consejo Superior de Investigaciones Científicas, Madrid, © 2022 CSIC

8. The Thomas Fisher Rare Book Library, University of Toronto

ABOUT THE AUTHOR

Siddhartha Mukherjee is the author of *The Gene: An Intimate History,* a #1 *New York Times* bestseller; *The Emperor of All Maladies: A Biography of Cancer,* winner of the 2011 Pulitzer Prize in general nonfiction; and *The Laws of Medicine.* He is the editor of *The Best American Science and Nature Writing 2013.* Mukherjee is an associate professor of medicine at Columbia University and a cancer physician and researcher. A Rhodes scholar, he graduated from Stanford University, University of Oxford, and Harvard Medical School. He has published articles in many journals and periodicals, including *Nature,* the *New York Times Magazine,* and *The New Yorker.* He lives in New York with his wife and daughters. Visit his website at SiddharthaMukherjee.com.

The employees of Thorndike Press hope you have enjoyed this Large Print book. All our Thorndike, Wheeler, and Kennebec Large Print titles are designed for easy reading, and all our books are made to last. Other Thorndike Press Large Print books are available at your library, through selected bookstores, or directly from us.

For information about titles, please call:

(800) 223-1244

or visit our Web site at:

http://gale.cengage.com/thorndike

To share your comments, please write:

Publisher
Thorndike Press
10 Water St., Suite 310
Waterville, ME 04901